# Studies in Systems, Decision and Control

Volume 294

**Series Editor**

Janusz Kacprzyk, Systems Research Institute, Polish Academy of Sciences, Warsaw, Poland

The series "Studies in Systems, Decision and Control" (SSDC) covers both new developments and advances, as well as the state of the art, in the various areas of broadly perceived systems, decision making and control–quickly, up to date and with a high quality. The intent is to cover the theory, applications, and perspectives on the state of the art and future developments relevant to systems, decision making, control, complex processes and related areas, as embedded in the fields of engineering, computer science, physics, economics, social and life sciences, as well as the paradigms and methodologies behind them. The series contains monographs, textbooks, lecture notes and edited volumes in systems, decision making and control spanning the areas of Cyber-Physical Systems, Autonomous Systems, Sensor Networks, Control Systems, Energy Systems, Automotive Systems, Biological Systems, Vehicular Networking and Connected Vehicles, Aerospace Systems, Automation, Manufacturing, Smart Grids, Nonlinear Systems, Power Systems, Robotics, Social Systems, Economic Systems and other. Of particular value to both the contributors and the readership are the short publication timeframe and the world-wide distribution and exposure which enable both a wide and rapid dissemination of research output.

** Indexing: The books of this series are submitted to ISI, SCOPUS, DBLP, Ulrichs, MathSciNet, Current Mathematical Publications, Mathematical Reviews, Zentralblatt Math: MetaPress and Springerlink.

More information about this series at http://www.springer.com/series/13304

Rolf Dornberger
Editor

# New Trends in Business Information Systems and Technology

## Digital Innovation and Digital Business Transformation

Springer

*Editor*
Rolf Dornberger
School of Business
Institute for Information Systems, FHNW
University of Applied Sciences and Arts
Northwestern Switzerland
Basel, Switzerland

ISSN 2198-4182 ISSN 2198-4190 (electronic)
Studies in Systems, Decision and Control
ISBN 978-3-030-48331-9 ISBN 978-3-030-48332-6 (eBook)
https://doi.org/10.1007/978-3-030-48332-6



This Springer imprint is published by the registered company Springer Nature Switzerland AG
The registered company address is: Gewerbestrasse 11, 6330 Cham, Switzerland

# Preface

What is digital transformation? This question is often asked because no one knows what digital transformation and its impact on our lives are and will be in the future. Therefore, it is a highly demanding challenge to describe in detail what constitutes digital transformation.

The term *digital transformation* is often referred to as *digitalization* (not to be confused with *digitization*, the process of converting analog data into a digital counterpart). In this book, we use the word digitalization to describe the evolution process, in which digital technologies (meaning electronic, information, communication, etc. technologies) enable changes in business and society. These changes can range from continuous and smooth to erratic and disruptive.

On the one hand, new digital technologies are the drivers for changes in classical processes (e.g., text-processing software has made the old mechanical typewriters obsolete). On the other hand, a need for business or organizational processes is looking for the best available digital technology solution to overcome old processes (e.g., a suitable voice-over-IP protocol is in the process of replacing the traditional telephone transmission). In this book, in the first case, we call it *digital innovation*, in the second case, *digital business transformation*. In general, however, both tendencies are often interconnected and not so clearly separated. In order to provide a selection of the diversity of what digital transformation means for our daily lives, this book presents many different digitalization cases—some more focused on digital innovation and some on digital business transformation.

This book was written by professors and researchers who work, research, and teach daily in the areas of (Business) Information Systems, (Business) Information Technology, Computer Science, Business Administration and Management at the School of Business of the University of Applied Sciences and Arts Northwestern Switzerland FHNW. This university of applied sciences focuses on applied research and development while retaining a high level of practical relevance. Thus, the writing style of the chapters of this book follows the philosophy of reconciling the degree of depth and rigor in research with a meaningful translation into relevance in practice.

While the Institute for Information Systems of the School of Business originally focused on *bridging the gap between business and IT*, thus aligning business and IT in the context of organizations, its topics are now expanding extensively to cover a wider range of digitalization: The impact of new ICT-related technologies and IT-supported methods in business and society is investigated. In addition to the classic domains of business information technology, our areas of interest are, for example, agile process management in a digital economy, cloud computing and related digital transformations, cybersecurity and resilience, digital supply chain management, artificial intelligence including computational intelligence (i.e., nature-inspired AI) and symbolic/sub-symbolic AI, Internet-of-Things, human–machine interaction, (social) robotics, management of complex systems, humans in technical systems and organizations, etc. Our passion for the different facets of digitalization eventually led to the idea of writing this second book (after the Springer book "Business Information Systems and Technology 4.0" in 2018).

As the editor of the book, I would like to thank our employer, the University of Applied Sciences and Arts Northwestern Switzerland FHNW and particularly the School of Business for supporting the writing of the book by granting additional hours to the authors. My immense and heartfelt thanks also go to all our authors, who have made excellent contributions to this book with their view of digitalization. My special thanks go to Prof. Dr. Thomas Hanne, Prof. Dr. Uwe Leimstoll, and Prof. Dr. Michael von Kutzschenbach, who have organized a thorough and independent scientific peer review of all the chapters according to internationally recognized high quality standards. Additionally, I would like to thank Vivienne Jia Zhong for coordinating the process of elaborating the contributions of 42 authors, compiled in 21 chapters, efficiently and competently. I would also like to thank Christine Lorgé and Natalie Jonkers for their diligence, competence, and commitment to language and readability. Finally, I would like to thank all our families for their continued patience and great support in allowing us to write this book on weekends and during the night.

Basel, Switzerland Prof. Dr. Rolf Dornberger
January 2020

# Contents

# Digital Innovation and Digital Business Transformation in the Age of Digital Change

**Rolf Dornberger and Dino Schwaferts**

**Abstract** This chapter introduces the central theme of the book and the following chapters. In our book, we present 20 selected examples of digitalization in the age of digital change. The chapters are divided into two parts, called "Digital Innovation" and "Digital Business Transformation". On the one hand, digital innovation includes cases where a new technology stimulates or enables new business opportunities. On the other hand, digital business transformation collects examples of digitalization in which business or management concepts exploit certain technological solutions for their practical implementation. To share this mind-set with the reader, quantum-computing technology will be taken as an example of digital innovation. This will be followed by reflections on digital business transformation, which seeks and uses digital technologies. Finally, the next chapters will be presented accordingly.

**Keywords** Digital innovation · Digital business transformation · Digitalization · Quantum computing · Digital eco-system

## 1 Introduction

It is widely predicted that digitalization will change the world. Already in 2018, the book "Business Information Systems and Technology 4.0: New Trends in the Age of Digital Change" [1] has presented selected cases which show how quickly the world changes and transforms. That book has provided explanations of key principles such as digitization, digitalization, and digital transformation. Therefore, the model in Fig. 1 "Digitalization: Yesterday, today and tomorrow" [2] was introduced. This

R. Dornberger (✉)
School of Business, Institute for Information Systems, FHNW University of Applied Sciences and Arts Northwestern, Peter Merian-Strasse 86, 4002 Basel, Switzerland
e-mail: rolf.dornberger@fhnw.ch

D. Schwaferts
School of Business, Institute for Information Systems, FHNW University of Applied Sciences and Arts Northwestern, Riggenbachstrasse 16, 4600 Olten, Switzerland
e-mail: dino.schwaferts@fhnw.ch

 
R. Dornberger (ed.), *New Trends in Business Information Systems and Technology*, Studies in Systems, Decision and Control 294,
https://doi.org/10.1007/978-3-030-48332-6_1

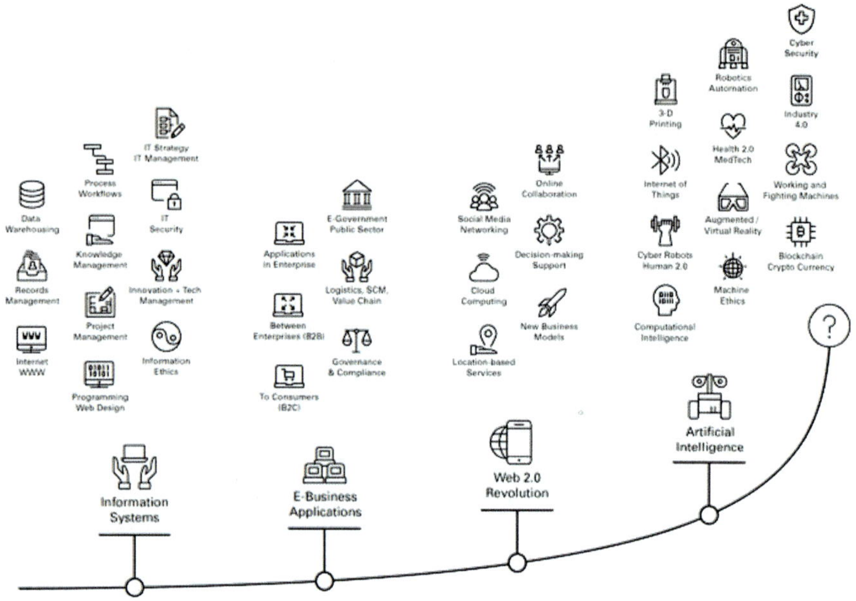


**Fig. 1** Digitalization: yesterday, today and tomorrow [2]

model discusses the development of the information and communication technology over four stages: The sequence begins with (1) Information Systems in the 1980s and 1990s and (2) E-Business Applications in the late 1990s and early 2000s. In addition, the rise of web applications and smartphones from the mid to the end of the first decade of the 21st century, launched the (3) Web 2.0 Revolution. The sequence concludes with the renaissance of (4) Artificial Intelligence from the mid to the end of the second decade of the 21st century.

As illustrated in the model (see Fig. 1), the development of new digital technologies and applications seems to be growing exponentially (e.g., the number of digital technologies steadily increases; every year the number of scientific publications increases worldwide, etc. [3]). This implies that a vast amount of potentially world-changing innovations is constantly being generated. As examples, the authors stated in [1] that in addition to improvements in smarter software, new hardware devices are being created, such as robots and 3D printers, equipping the computers with a kind of a body and reshaping the supply chains. In parallel, the cyber-world is evolving, increasingly encompassing technologies such as virtual and augmented reality (VR/AR), which provide the interface to a digital or cyber world. Many aspects of the world are now becoming "cyber". For example, cryptocurrencies are evolving and expanding established forms of money; and enhancing humans with artificial robotic body parts is no longer a distant dream.

This book goes a step further to provide new examples of how digital technologies and/or business opportunities created by digitalization cause innovation and transformation. It therefore builds on the model with the four stages discussed in [2], but also proposes the two aspects "Digital Innovation" and "Digital Business Transformation", which cross-sectionally cover the four stages. Here, digital innovation means the use of new or existing digital technologies (from another context) to support certain aspects of management or business. However, digital business transformation is—according to Gartner [4]—"the process of exploiting digital technologies and supporting capabilities to create a robust new digital business model."

To present and organize the subsequent chapters of this book, this introductory chapter briefly mentions the two aspects "digital innovation" and "digital business transformation". The first aspect concerns a new technology (such as quantum computing) that is the starting point and enabler for new applications and business solutions. The second aspect concerns business/management ideas or concepts (such as digital eco-systems) that need digital technologies to support and implement them. As a guide for the readership of this book, all the ensuing chapters are briefly outlined and focus either on the first aspect "digital innovation" or the second "digital business transformation".

## 2 The Technology of Quantum Computing as Digital Innovation

### *2.1 Identifying Promising Technologies*

In order to predict whether and how a particular new digital technology can become a promising technology in the sense of enabling one of the most important digital innovations of the future, the important technologies to come must be continuously identified. The question often is: What will come next? In the model [2] in Fig. 1, many readers may wonder what the question mark on the rising curve means. Albert Einstein answered this question very elegantly by saying: "I never think of the future—it comes soon enough."[1] According to this quote, the future seems to be approaching even faster today than in earlier times.

A well-known source and reference for future technologies and trends is the Gartner hype cycle for emerging technologies [5]. It proposes and discusses future technologies and trends in four phases:

1. Being a hype (peak of inflated expectations)
2. Producing a disappointment (trough of disillusionment)
3. Getting a realistic estimation (slope of enlightenment)
4. Leading to common application (plateau of productivity).

[1] https://quoteinvestigator.com/2013/07/23/future-soon/.

Once a new technology or trend appears on the hype cycle, it takes up a particular position on the cycle according to the estimation of its phase and in comparison to other technologies. Additionally, the predicted time until it becomes a common application is noted.

## 2.2 *Quantum Computing as an Example for Digital Innovation*

One technology that has been on the rising slope of the Gartner hype cycle for years is "quantum computing", which has appeared 11 times since 2000. In fact, Gartner predicted that quantum computing would reach its plateau of productivity in a maximum of five years or that over the next five to ten years it would be in general use. Thus, one can assume that the digital technology of quantum computing is an answer to the question mark in Fig. 1 and will become one of the next digital innovations. Why, however, will quantum computing become so important?

Quantum computing [6] is a novel ground-breaking mechanism for computing/processing units, leading to a new computer generation with an immensely higher computing speed. Its principles are based on the phenomena of quantum physics that can be used within a new type of central processing units inside a computer. While the theory is well developed, its realization in new hardware and software to take advantage of its full potential remains difficult. Currently, there is a handful of highly advanced research labs, which possess more or less well-running prototypes of quantum computers. However, there is still a chance that the realization of application-ready quantum-computing devices will not be possible in this century.

Assuming that ready-to-use quantum computing will be realized in the next years, one can say that it will change the world fundamentally. There will be a jump from today's "nice computing power" to "unbelievable, ultimate computing power", in which case the only limitations will be the computing memory and the available algorithms (software) to compute (almost) everything, known as Quantum Supremacy [7]. Some reflections below show why quantum computing is on the way to becoming such an important digital innovation—perhaps the most disruptive digital innovation of this century [8].

Bitcoin and similar cryptocurrencies will lose their motivation [9]. Theoretically, with quantum computing, all bitcoins will be instantly mined, because the computationally expensive computations for adding new blocks to the blockchain system will be carried out instantly. The miner (a specific role in the blockchain system) who gets first access to quantum computing will mine all bitcoins (perhaps introducing stupid transactions to generate new blocks with senseless back and forth transactions). Certainly, there are mechanisms, such as time delays that are integrated into the mining procedure; but today's standard ideas of generating particular cryptocurrencies (through computationally expensive blockchain computations) will have to be completely rethought.

In cryptography, quantum computing will break through almost any current encryption in no time [10]. Certainly, a new discipline will come into play called "quantum cryptography" that adapts and replaces outdated encryption algorithms. Nevertheless, in every encryption and decryption process, quantum computers will always be superior to traditional computing devices, even to today's supercomputers. For example, as cars exchange encrypted commands with their remote keys, it will allow hackers to use quantum computers to unlock not only one car, but also all other cars in the vicinity within a fraction of a millisecond.

DeepMind Technologies is a subsidiary of Alphabet Inc., the mother company of Google. The research of DeepMind focuses on different aspects of extending and applying Artificial Intelligence (AI). A showcase is the gameplay of the ancient Asian board game "Go". The projects AlphaGo, AlphaGo Zero, and Alpha Zero developed AI frameworks [11] that are (almost) always able to beat human world champions and other software in Go [12] and in various other kinds of single- and multi-player games [13]. DeepMind's AI approaches are superior to most other (published) approaches, including other software solutions, as well as to human gameplay strategies in games with (almost) infinite possible moves. The algorithms are special heuristics. Some of them are trained with huge human-generated datasets; some of them use self-training by generating own dataset by applying the game rules. These powerful heuristics search limited parts of the entire search space, where promising next moves seem favorable. However, a quantum computer may be able to search the entire search space (not only limited parts) and discover unknown sequences of moves that could even beat the best powerful heuristics. In other words, the ultimate quantum-computing power has great potential to further enhance today's best AI and decision-making algorithms—but also to make some AI approaches obsolete by instantly scanning the entire search space.

Another major application field in which quantum computing will have a tremendous impact, is modeling and simulation to be known as Quantum Simulation [14]. For example, the knowledge of how the universe, the solar system, and earth evolved over billions of years is still incomplete. Many details are still insufficiently understood. Today, different physical astronomical models and approaches are available to describe the development of the universe. Unfortunately, these models are still partially incompatible, and even supercomputers take weeks to compute results in well-selected sub-cases. Here, quantum computing will be able to compute different models with increasing granularity in almost no time. Then it will be possible to easily compare the results of the different models, improve the models, and gain an understanding about which models are best suited for different cases and how they fit together. In addition to astronomy, which takes place on the large scales (i.e. space and time), quantum computing will also help to understand the effects on the small scales, where quantum physics explores them.

Another discipline in which increased computing speed will have an extraordinary effect is engineering, which mostly uses search and optimization algorithms to computationally solve engineering problems [15]. In the last 40 years, numerical computations have already replaced various hardware tests and field experiments. Today, e.g., the most efficient engines, cars, and airplanes are developed and further optimized on the computer; prototype testing is reduced to a minimum. Furthermore, with quantum computing, it will be possible to simultaneously compute and compare millions of design variants. Each product is designed and optimized on the computer until optimal properties are reached. Then, because of myriads of computational optimization loops, all products of various companies will be quite similar. For example, washing machines, smartphones, and airplanes will be almost identical independently of the producing company. On the other hand, the differentiation will lie in the personalization of engineered products [16]. Users will get an optimal, personalized product that is tailored to their requirements, e.g., a personalized car (shape and size, engine, color, interior, etc.). Hundreds of personal choice features, including an optimized personal product design, can be integrated during the online configuration. Hence, quantum computing will create a new "personalized product business" [17].

Extending the idea of being able to design whatever is technically possible, we will also be able to research and develop new ways to preserve nature [18]. Ideally, we will be able to model, simulate, and optimize countless variants of environmental technologies and sustainable, renewable energy solutions. We will be able to find ways to reduce carbon dioxide emissions and other pollution markers. In addition, we will be able to better understand society as a living organism, and, in turn, enhance the living conditions of billions of people, merely by discovering new technologies and new ways for these solutions to influence the quality of life.

Certainly, all these examples of the benefits of quantum computing are very optimistic. However, as with any technology, quantum computing can also be used wrongfully, which we will not discuss here.

In summary, when digital technologies find their way to common applications, they are adopted as digital innovations. In this chapter, the technology of quantum computing was chosen as an example to explain how a new technology might stimulate many new applications and business opportunities, which is consequently called digital innovation. Later in the book in the Digital Innovation part, several chapters are collected to present specific technologies or technological approaches, which stimulate new applications, businesses, processes, or even thinking.

## 3 Leveraging Digital Business Transformation Through Digital Technologies

The section above describes the concept of digital innovation (used in this book to structure the upcoming chapters) using the example of quantum computing (i.e. how a digital technology might stimulate new applications). This section focuses on the possibility of realizing additional opportunities in business, management, learning, etc. by seeking and then exploiting new specific digital technologies. This is related to the already mentioned definition of digital business transformation [4] as "the process of exploiting digital technologies and supporting capabilities to create a robust new digital business model." However, the wording "business model" is used in a broader context, namely as "business, management or process opportunities", which can be realized, supported, and/or improved by seeking and exploiting suitable digital technologies. In the next section, some reflections are made about rethinking managerial approaches for business transformation in the age of digital change.

### *3.1 Rethinking Managerial Approaches for Business Transformation in the Age of Digital Change*

When reflecting on the most important achievements of the industrial era around the 19th century, steam power, electrification, and the automobile could come to mind first. In parallel with the emergence of these new technologies, new management challenges also arose. According to the economic historian Pollard [19], current management approaches and topics have their origin in the emergence of large-scale organizations during the industrial era, where large companies were explicitly confronted with challenges on another extended scale. E.g. human resource management for mass employment, leadership and management of a big entity, huge financial issues, procurement and sales on a broad level, etc. Following Pollard, the origin of today's management approaches is strongly connected to the emergence of new technologies—in addition to various societal developments, ecological challenges and so forth.

Today, a new type of technology, namely the digital technologies, such as the Internet of Things (IoT), AI, Big Data, Cloud Computing, robotics, etc. have an impact on life and work [20] (e.g., all the IoT technology to be applied in Industry 4.0). Such a transition to a digital society and economy leads to different expectations towards management by both customers and employees. The question could be which new managerial approaches are emerging for business transformation in the age of digital change—to be reflected on in the next section and in the upcoming book chapters under the umbrella term "Examples for an increasing need for Digital Business Transformation".

### *3.2 About Examples in Digital Business Transformation*

The following are some reflections on the particular state of companies and new possibilities offered by digital business transformation (i.e. the exploitation of digital technologies and supporting capabilities assisting company needs). The question is how digital technologies influence companies in the age of digital change.

Searching for management concepts and explanations of why companies in the industrial age are as they are, one might still refer to Coase [21], where he already asked in the 1930s, why and under what circumstances what kind of company would emerge. He concluded that a meaningful enterprise size and a special state of the company are the result of the balance between disadvantages from transaction costs and disadvantages from overhead costs, imperfection in resource allocation, and inflexibility. In the 70s and 80s of the last century, Porter [22] defined the state of a company by the concept of the value chain, where the value added performed in an organization determines the size and the condition of the company. However, around the year 2000, Porter added the influence of digitalization to his model, where the value added is now divided among different partners in a networked structure, which he calls "the system of systems". Examples in business are the automotive industry, which has long reduced its vertical integration, similar to early examples of virtual organizations, which have been building on this concept for more than 20 years [23].

In the age of digital change, the question arises: what exactly is the benefit of having the opportunity to exploit digital technologies to change the style of management and the state of companies? Can structures and processes be better defined to contribute more to the value creation? A particularly great opportunity that digital technologies offer today, is the simple division of tasks, the value chain, the management, the IT and so forth of every entity of the company (e.g., on the level of a project, a department or even the entire company). This means that many tasks—in terms of total value added—could now be divided and elaborated in different entities in parallel, fully supported by digital technologies. These entities could even lie outside the company, following the trend of focusing on one's own unique abilities. Each entity makes its own contribution to value creation by being connected and working together via digital technologies. Here, Porter's system of systems [22] with the flexible composition of services in the value chain and their orchestration or digital eco-systems [24] provide answers on how digitalization makes this possible [25].

Although these short reflections are only taken as an example in digital business transformation, it can be assumed that digital technologies have an important impact on management at various levels. Digital business transformation is strongly related, for example, to managerial approaches, the style of management, state of companies as well as to particular management topics (e.g. HR management or learning management) which are quickly changing nowadays. Therefore, it is indispensable to continuously investigate various upcoming management, business, societal demands, and further challenges in the search for digital technologies and solutions that support them. Selected topics are provided later in this book in the Digital Business Transformation part.

## 4 Organization of This Book

As explained above, this book consists of two parts, which represent the above-mentioned topic fields, namely Digital Innovation and Digital Business Transformation, and which, in turn, compile the more technology-based topics and the more business/application-relevant topics. In the following, a very short description and a brief summary of the book chapters is given.

(a) Examples of emerging Digital Innovation (according to chapter order later in the book)

- Various digitization technologies have been researched for several years and optical character recognition (known as OCR) is nothing new. Nevertheless, an update with the latest machine-learning algorithms proves a new technological leap. "A Survey of State of the Art Methods Employed in the Offline Signature Verification Process" reviews the domain of offline signature verification and presents a comprehensive overview of methods typically employed in the general process of offline signature verification.
- When it comes to innovative idea generation, Design Thinking is mentioned as one of the most widespread and effective methods. However, in addition to all attempts to improve the design thinking processes, a sensible technological implementation and related scenarios, as discussed in "Agile Visualization in Design Thinking", promise a further innovative approach.
- In general, enriching classical algorithms with machine-learning approaches further improves their potential. An example is given in the text mining case "Text Mining Innovation for Business", reflecting on the business innovation enabled by developing text-mining solutions in response to the business needs of companies.
- The sensible use of embedded digital technologies can lead to new applications and generate digital innovations as shown in "Using Mobile Sensing on Smartphones for the Management of Daily Life Tasks", which explores the ability and the potential of the mobile-sensing technology to support daily life task management.
- The use of an advanced software solution in university education provides an advanced tutoring system. "A Dialog-Based Tutoring System for Project-Based Learning in Information Systems Education" presents and discusses the design of an intelligent dialog-based tutoring system that is designed to support students during group projects, to maintain their motivation and give subtle hints for self-directed discovery.
- However, it is often difficult to teach technologically advanced topics, for example object-oriented programming. In order to identify and evaluate the students' learning difficulties, a case of developing an assessment test is presented. "A Human Aptitude Test for Object-Oriented Programming in the Context of AI and Machine Learning" discusses the object-oriented programming paradigm and the potential future developments once AI and machine learning start to steadily increase their influence on the overall design of software systems.
- Methods that belong to AI tend to be among the more challenging subjects for students. A case of evaluating different measures to improve a related university course over ten years is discussed in order to propose best practices. "Adapting the Teaching of Computational Intelligence Techniques for Improving the Learning Outcomes" shows the continuous efforts to integrate nature-inspired AI, and in particular, computational intelligence algorithms into today's university teaching to increase the learning success of students.

- Some technologies force research to discuss them from the philosophical side. Maybe our view of technology is too strongly influenced by a too narrow perception of the world: We will see what we allow to be seen. "Automatic Programming of Cellular Automata and Artificial Neural Networks Guided by Philosophy" discusses an approach called "allagmatic method" that automatically programs and executes models with as little limitations as possible while maintaining human interpretability.
- Some technologies themselves are "hardcore", like robots. They are a hype; they will change the world and businesses. "Facial Recognition and Pathfinding on the Humanoid Robot Pepper as a Starting Point for Social Interaction" examines the facial recognition and navigation capabilities of the robot Pepper and relates them to a better human-robot interaction.

(b) Examples for an increasing need to Digital Business Transformation (according to the chapter sequence later in the book)

- However, robots are not only a new hardware device with extended software capabilities. Organizations might have special process needs which best can be suited by involving and integrating robots. "Social Robots in Organizational Contexts: The Role of Culture and Future Research Needs" describes the organizational contexts currently relevant for social robots addressing the respective cultural challenges.
- Management must innovate and adapt in order to leverage the benefits of digitalization for a digitalized sustainability society. Here, "Digital Transformation for Sustainability: A Necessary Technical and Mental Revolution" outlines an integrated framework linking different levels of digitalness with necessary changes in managerial practice for organizational inquiry.
- Keeping humans-in-the-loop provides many advantages. For example, "Visualization of Patterns for Hybrid Learning and Reasoning with Human Involvement" discusses the traditional roles of humans in hybrid systems and their influence on machine learning and/or knowledge engineering activities.
- Learning can be improved by sensibly using selected technologies. Here, "Modeling the Instructional Design of a Language Training for Professional Purposes, Using Augmented Reality" presents a multi-layered instructional model for language training for professional adult learners.
- The processes in e-commerce are continuously improved by steadily integrating and using new technologies. "Frictionless Commerce and Seamless Payment" discusses a potential solution for improving the user experience and conversion optimization in e-commerce and provides existing practical examples.
- E-commerce has not necessarily led to disintermediation as predicted in the 1990s. "Direct to Consumer (D2C) E-Commerce: Goals and Strategies of Brand Manufacturers" suggests three strategies for brand manufacturers to mitigate conflicts with traditional sales partners while maintaining their presence on the Internet.

- Marketing is another management topic, where new technologies are highly appreciated to strengthen customer retention. "The Digital Marketing Toolkit: A Literature Review for the Identification of Digital Marketing Channels and Platforms" presents a study to support SMEs in their digital business transformation.
- Supply chain management also quickly adsorbs new technologies to improve its processes. "How the Internet of Things Drives Innovation for the Logistics of the Future" discusses an innovative value chain framework relating technology advancement to business innovation in logistics.
- Co-creation in networks is highly supported by digital technologies. As an example, "Recommendations for Conducting Service-Dominant Logic Research" addresses a research gap providing contextual guidance for applying service-dominant logic in research and a framework for innovation in business.
- Many recent business models rely on data as the most important asset, where data privacy is always an issue. "Blockchain Technologies towards Data Privacy—Hyperledger Sawtooth as Unit of Analysis" presents a study that provides insights for selecting a suitable blockchain configuration that would comply with regulatory data privacy requirements.
- "Leadership in the Age of Artificial Intelligence—Exploring Links and Implications in Internationally Operating Insurance Companies" describes a case that demonstrates the need (e.g., with the introduction of AI) for additional management approaches.

## 5 Conclusions

Everywhere, examples and good practices in the field of digitalization appear again and again, which aim to promise a change of the world, stimulated partly by new technologies, partly by business/management/societal needs, often by both. A promising way to identify possible next disruptive leaps in today's age of digital change is to continuously and profoundly explore a broad variety of examples. A small selection is shown in this book.

This chapter is intended to introduce the central theme of the book for the arrangement of the upcoming chapters grouped to two parts. Therefore, the aspect of Digital Innovation is introduced using the example of the upcoming digital technology of quantum computing that will most likely change the world. Then, the aspect of Digital Business Transformation is proposed and reflected as a change of management/business to exploit digital technologies. Finally, these insights are briefly mapped to the following book chapters, which—as will become obvious during the course of the book—show various new trends and challenges in the age of digital change from different perspectives. Maybe one of them will lead to the next disruptive leap in digitalization.

## References


1. Dornberger, R.: Business Information Systems and Technology 4.0: New Trends in the Age of Digital Change. Springer (2018)
2. Dornberger, R., Inglese, T., Korkut, S., Zhong, V.J.: Digitalization: yesterday, today and tomorrow. In: Dornberger, R. (ed.) Business Information Systems and Technology 4.0: New Trends in the Age of Digital Change, pp. 1–11. Springer International Publishing, Cham (2018)
3. Bornmann, L., Mutz, R.: Growth rates of modern science: a bibliometric analysis based on the number of publications and cited references. J. Assoc. Inf. Sci. Technol. **66**, 2215–2222 (2015). https://doi.org/10.1002/asi.23329
4. Digital Business Transformation. https://www.gartner.com/en/information-technology/glossary/digital-business-transformation
5. Fenn, J., Blosch, M.: Understanding Gartner's Hype Cycles. https://www.gartner.com/en/documents/3887767/understanding-gartner-s-hype-cycles
6. Farouk, A., Tarawneh, O., Elhoseny, M., Batle, J., Naseri, M., Hassanien, A.E., Abedl-Aty, M.: Quantum Computing and Cryptography: An Overview. In: Hassanien, A.E., Elhoseny, M., Kacprzyk, J. (eds.) Quantum Computing: An Environment for Intelligent Large Scale Real Application, pp. 63–100. Springer International Publishing, Cham (2018)
7. Harrow, A.W., Montanaro, A.: Quantum computational supremacy. Nature **549**, 203–209 (2017). https://doi.org/10.1038/nature23458
8. Ferrie, C.: Why are scientists so excited about a recently claimed quantum computing milestone? http://theconversation.com/why-are-scientists-so-excited-about-a-recently-claimed-quantum-computing-milestone-124082
9. Hayes, A.S.: Cryptocurrency value formation: An empirical study leading to a cost of production model for valuing bitcoin. Telemat. Inform. **34**, 1308–1321 (2017). https://doi.org/10.1016/j.tele.2016.05.005
10. Hallgren, S., Vollmer, U.: Quantum computing. In: Bernstein, D.J., Buchmann, J., Dahmen, E. (eds.) Post-quantum cryptography, pp. 15–34. Springer, Berlin Heidelberg, Berlin, Heidelberg (2009)
11. Silver, D., Huang, A., Maddison, C.J., Guez, A., Sifre, L., van den Driessche, G., Schrittwieser, J., Antonoglou, I., Panneershelvam, V., Lanctot, M., Dieleman, S., Grewe, D., Nham, J., Kalchbrenner, N., Sutskever, I., Lillicrap, T., Leach, M., Kavukcuoglu, K., Graepel, T., Hassabis, D.: Mastering the game of Go with deep neural networks and tree search. Nature **529**, 484–489 (2016). https://doi.org/10.1038/nature16961
12. Silver, D., Schrittwieser, J., Simonyan, K., Antonoglou, I., Huang, A., Guez, A., Hubert, T., Baker, L., Lai, M., Bolton, A., Chen, Y., Lillicrap, T., Hui, F., Sifre, L., van den Driessche, G., Graepel, T., Hassabis, D.: Mastering the game of Go without human knowledge. Nature **550**, 354–359 (2017). https://doi.org/10.1038/nature24270
13. Lanctot, M., Lockhart, E., Lespiau, J.-B., Zambaldi, V., Upadhyay, S., Pérolat, J., Srinivasan, S., Timbers, F., Tuyls, K., Omidshafiei, S., Hennes, D., Morrill, D., Muller, P., Ewalds, T., Faulkner, R., Kramár, J., De Vylder, B., Saeta, B., Bradbury, J., Ding, D., Borgeaud, S., Lai, M., Schrittwieser, J., Anthony, T., Hughes, E., Danihelka, I., Ryan-Davis, J.: OpenSpiel: A Framework for Reinforcement Learning in Games. ArXiv190809453 Cs. (2019)
14. Georgescu, I.M., Ashhab, S., Nori, F.: Quantum simulation. Rev. Mod. Phys. **86**, 153–185 (2014). https://doi.org/10.1103/RevModPhys.86.153
15. Williams, C.P.: Quantum search algorithms in science and engineering. Comput. Sci. Eng. **3**, 44–51 (2001). https://doi.org/10.1109/5992.909001
16. Berry, C., Wang, H., Hu, S.J.: Product architecting for personalization. J. Manuf. Syst. **32**, 404–411 (2013). https://doi.org/10.1016/j.jmsy.2013.04.012
17. Moon, J., Chadee, D., Tikoo, S.: Culture, product type, and price influences on consumer purchase intention to buy personalized products online. J. Bus. Res. **61**, 31–39 (2008). https://doi.org/10.1016/j.jbusres.2006.05.012

18. Mohseni, M., Read, P., Neven, H., Boixo, S., Denchev, V., Babbush, R., Fowler, A., Smelyanskiy, V., Martinis, J.: Commercialize quantum technologies in five years. Nature **543**, 171–174 (2017). https://doi.org/10.1038/543171a
19. Pollard, S.: The Genesis of Modern Management: a Study of the Industrial Revolution in Great Britain. Harvard University Press (1965)
20. Lasi, H., Fettke, P., Kemper, H.-G., Feld, T., Hoffmann, M.: Industry 4.0. Bus. Inf. Syst. Eng. **6**, 239–242 (2014). https://doi.org/10.1007/s12599-014-0334-4
21. Coase, R.H.: The nature of the firm. Economica **4**, 386–405 (1937). https://doi.org/10.1111/j.1468-0335.1937.tb00002.x
22. Porter, M.E., Heppelmann, J.E.: How smart, connected products are transforming competition. Harv. Bus. Rev. **92**, 64–88 (2014)
23. Porter, M.E.: Location, competition, and economic development: local clus-ters in a global economy. Econ. Dev. Q. **14**(1), 15–34 (2000). https://doi.org/10.1177/089124240001400105
24. Hol, A., Ginige, A., Lawson, R.: System level analysis of how businesses adjust to changing environment in the digital eco-system. In: 2007 Inaugural IEEE-IES Digital EcoSystems and Technologies Conference, pp. 153–158 (2007)
25. Grieder, H., Schwaferts, D.: Digital ecosystem—how companies can achieve a sustainable competitiveness in the future and the digital age. In: Verkuil, A.H., Hinkelmann, K., Aeschbacher, M. (eds.) Digitalisierung und andere Innovationsformen im Management. Edition gesowip, Basel (2019)

# Digital Innovation

# A Survey of State of the Art Methods Employed in the Offline Signature Verification Process

**Michael Stauffer, Paul Maergner, Andreas Fischer, and Kaspar Riesen**

**Abstract** Handwritten signatures are of eminent importance in many business and legal activities around the world. That is, signatures have been used as authentication and verification measure for several centuries. However, the high relevance of signatures is accompanied with a certain risk of misuse. To mitigate this risk, automatic signature verification was proposed. Given a questioned signature, signature verification systems aim to distinguish between genuine and forged signatures. In the last decades, a large number of different signature verification frameworks have been proposed. Basically, these frameworks can be divided into online and offline approaches. In the case of online signature verification, temporal information about the writing process is available, while offline signature verification is limited to spatial information only. Hence, offline signature verification is generally regarded as the more challenging task. The present chapter reviews the field of offline signature verification and presents a comprehensive overview of methods typically employed in the general process of offline signature verification.

**Keywords** Offline signature verification · Image preprocessing · Handwriting representation · Signature classification

M. Stauffer · K. Riesen (✉)
University of Applied Sciences and Arts Northwestern Switzerland, Institute for Information Systems, Riggenbachstrasse 16, 4600 Olten, Switzerland
e-mail: kaspar.riesen@fhnw.ch

M. Stauffer
e-mail: michael.stauffer@fhnw.ch

P. Maergner · A. Fischer
Department of Informatics, University of Fribourg, Boulevard de Pérolles 90, 1700 Fribourg, Switzerland
e-mail: paul.maergner@unifr.ch

A. Fischer
e-mail: andreas.fischer@unifr.ch

A. Fischer
University of Applied Sciences and Arts Western Switzerland, Boulevard de Pérolles 80, 1700 Fribourg, Switzerland





R. Dornberger (ed.), *New Trends in Business Information Systems and Technology*,
Studies in Systems, Decision and Control 294,
https://doi.org/10.1007/978-3-030-48332-6_2

# 1 Introduction

For several hundred years, *handwritten signatures* have been used as biometric authentication and verification measure in a broad range of business and legal transactions around the world. In fact, the first use of handwritten signatures can be traced back to the fourth century, where a signature was used to protect the *Talmud* (i.e. a central text in *Rabbinic Judaism*) against possible modifications. Later on, signatures (i.e. so-called *subscripto*) were used to authenticate documents in the Roman Empire during the period of Valentinian III [14]. More recently, several national and international digitization efforts for electronic signatures have been proposed. Yet, handwritten signatures remain very popular [8, 19, 20]. This is mostly due to the ubiquitous applicability as well as the overall high availability and acceptance of handwritten signatures when compared with electronic signatures.

Due to the high commercial and legal relevance, handwritten signatures also expose a certain risk of misuse. To mitigate this risk, *signature verification* can be employed. That is, a *questioned* signature is compared with a set of *reference* signatures in order to distinguish between *genuine* and *forged* signatures [20]. Traditionally, this task has been conducted by human experts as part of *graphology*, i.e. the study of handwriting. However, due to increasing number of documents, automatic signature verification systems have been proposed already some decades ago [28]. Actually, manual signature verification has been continuously replaced since then. Today, automatic signature verification is still regarded as a challenging pattern recognition task, and thus, remains a very active research field [8].

In general, automatic signature verification approaches can be divided into *online* (also termed *dynamic*) and *offline* (also termed *static*) approaches [19]. In the case of online signature verification, signatures are acquired by means of an electronic input device, such as a digital pen, a digital tablet, or via input on a touch screen (e.g. smartphone or handheld device). Hence, dynamic temporal information about the signing process (e.g. acceleration, speed, pressure, and pen angle such as altitude and azimuth) can be recorded during the handwriting process. In contrast, offline acquired signatures are digitized by means of scanning. As a result, the verification task is solely based on the $(x, y)$-positions of the handwriting (i.e. strokes), and thus, offline signature verification is generally considered the more challenging case. Note that the present chapter focuses only on offline signature verification.

Signature verification systems can also be distinguished with respect to the representation formalism that is actually used, viz. *statistical* (i.e. vectorial) and *structural* representations. The vast majority of signature verification approaches make use of statistical representations [8, 19]. That is, feature vectors or sequences of feature vectors are extracted from handwritten signature images. In early approaches, features are, for example, based on global handwriting characteristics like contour [6, 16], outline [10], projection profiles [30], or slant direction [2, 13]. However, more generic and local feature descriptors (i.e. descriptors that are not limited to handwriting images) were also used such as *Histogram of Oriented Gradients (HoG)* [41] and *Local Binary Patterns (LBP)* [12, 41]. In recent years, features have increasingly been extracted by means of a learned statistical model, viz. *Deep Learning* approaches like *Convolutional Neural Networks (CNNs)* [7, 18, 26]. Regardless of the actual type of feature, different statistical classifiers and/or matching schemes have been employed to distinguish genuine from forged signatures. Known approaches are for example *Support Vector Machines (SVMs)* [10, 12, 41], *Dynamic Time Warping (DTW)* [6, 30], and *Hidden Markov Models (HMMs)* [10, 13].

In contrast to statistical representation formalisms, structural (i.e. string-, tree-, or graph-based) approaches allow representing the inherent topological characteristics of a handwritten signature in a very natural and comprehensive way [27, 33]. Early structural approaches are, for example, based on the representation of stroke primitives [33], as well as critical contour points [4]. In a recent paper, graphs are used to represent characteristic points (so-called *keypoints*) as well as their structural relationships [27]. Hence, both the structure and the size of the graph is directly adapted to the size and complexity of the underlying signature. Moreover, these graphs are able to represent relationships that might exist between substructures of the handwriting.

However, the power and flexibility of graphs is accompanied by a general increase of the computational complexity of many mathematical procedures. The computation of a similarity or dissimilarity of graphs, for instance, is of much higher complexity than computing a vector (dis)similarity. In order to address this challenge, a number of fast matching algorithms have been proposed in the last decade that allow comparing also larger graphs and/or larger amounts of graphs within reasonable time (e.g. [15, 32]).

The goal of the present chapter is twofold. First, the chapter aims to provide a comprehensive literature review of traditional and recent research methods in the field of offline signature verification. Second, the chapter provides a profound analysis of the end-to-end process of signature verification. In particular, the chapter discusses the problems and solutions of data acquisition, preprocessing, feature extraction, and signature classification. Hence, this chapter serves as a literature survey as well as a possible starting point for researchers to build their own process pipe for signature verification.

The remainder of this chapter is organized as follows. Section 2 presents a generic signature verification process. Subsequently, the most important processing steps are individually reviewed in the following sections. First, different publicly available offline signature datasets are compared in Sect. 2.1, while common image preprocessing methods are examined in Sect. 2.2. Next, Sect. 2.3 compares different statistical and structural representations that are eventually used for classification (reviewed in Sect. 2.4). Finally, Sect. 3 presents conclusions and possible trends in the field of signature verification.

## 2 Signature Verification Process

Most signature verification systems are based on four subsequent processing steps, as shown in Fig. 1 [8, 20]. First, handwritten signatures are digitized by means of scanning (in Sect. 2.1 this particular step is discussed in detail). In the second step, signatures are segmented and preprocessed to reduce variations (discussed and reviewed in Sect. 2.2). Next, preprocessed signature images are used to extract statistical (i.e. vectorial) or structural (i.e. string, tree, or graph) representations (discussed and reviewed in Sect. 2.3). Finally, a questioned signature $q$ of claimed user $u$ is compared with all reference signatures $r \in R_u$ by means of a specific dissimilarity measure (discussed and reviewed in Sect. 2.4).

Roughly speaking, if the minimal dissimilarity to all references is below a certain threshold, the unseen signature $q$ is regarded as genuine, otherwise $q$ is regarded as forged. In general, the larger the reference set $R$, the better the accuracy of the verification process. However, the acquisition of arbitrarily large reference sets is often not possible in practice, and thus, limited to a few reference signatures per user. In the following sections, each process step is reviewed separately.

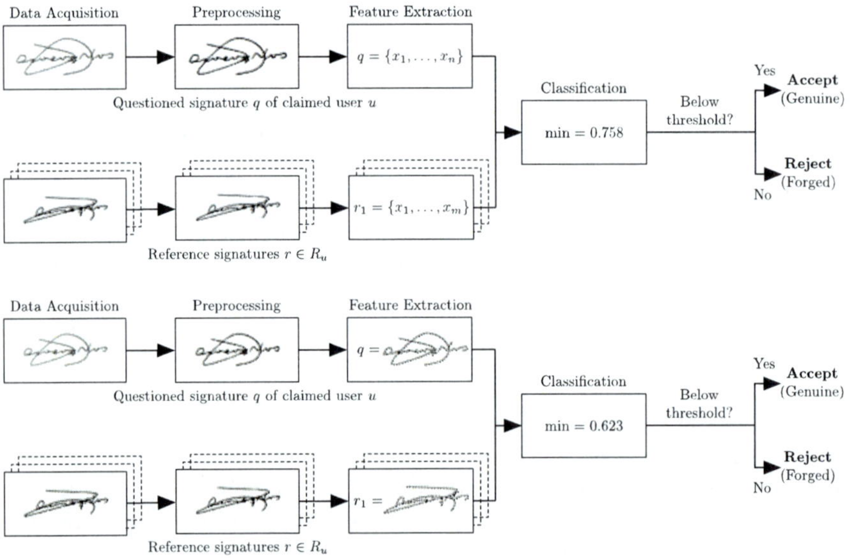


**Fig. 1** Statistical (**a**) and structural (**b**) signature verification processes consist of four major steps: First, handwritten signatures are digitized by means of scanning. Second, noise and variations are reduced by means of different preprocessing algorithms. Next, preprocessed and filtered signature images are represented by either a statistical or a structural representation. In either case, a questioned signature $q$ of claimed user $u$ is compared with the corresponding set of reference signatures $r \in R_u$ of user $u$. If the minimal dissimilarity between $q$ and $r \in R_u$ is below a certain threshold, $q$ is regarded as genuine, otherwise $q$ is regarded as forged

## 2.1 Data Acquisition and Available Datasets

During data acquisition, the available signature images are digitized into a machine-processable format by means of scanning. In most cases, digitized signatures are based on greyscale images, commonly defined as [37]

$$\{S(x, y)\}_{0 \le x \le X, 0 \le y \le Y},$$

where $S(x, y)$ denotes the greyscale level at the $(x, y)$-position of the image and $X$, $Y$ refer to the maximum values of the $x$- and $y$-axis of the image. Typically $S(x, y)$ lies between 0 (= black) and 255 (= white).

In the last decades, a number of different private and public datasets for offline signature has been proposed, see [8] for an exhaustive overview. These datasets can be characterized with respect to their writing style (e.g. Arabic, Bengali, Chinese, Persian, Western, etc.) as well as the number of genuine and forged signatures per user. Table 1 summarizes four widely used and publicly available datasets. In particular, the table shows summaries of three datasets with real world signatures (i.e. MCYT-75 [13, 29], UTSig [34], and CEDAR [21]) as well as one synthetic dataset (i.e. GPDSsynthetic-offline [11]) for two different writing styles (i.e. Western and Persian). Note that GPDSsynthetic is replacing the widely used GPDS-960 dataset [40] that is no longer available due to new data protection regulations.

## 2.2 Preprocessing of Signature Images

Once handwritten signatures are digitally available, preprocessing basically aims to reduce variations caused by the image itself (e.g. noisy background, skew scanning, segmentation from the background) as well as variations caused by inter- and intrapersonal variations in the handwriting (e.g. scaling and slant of handwriting). In most cases, preprocessing is based on several subsequent steps. First, signatures are extracted from the background (e.g. bank checks). Next, a certain filtering and binarization is employed before different normalization techniques are typically applied.

**Table 1** The writing style, the number of users, the number of genuine and forged signatures per user, and the resolution of the images in dpi of widely used datasets for offline signature verification

| Name | Style | Users | Genuine | Forgeries | dpi |
|---|---|---|---|---|---|
| MCYT-75 | Western | 75 | 15 | 15 | 600 |
| GPDSsynthetic-offline | Western | 4000 | 24 | 30 | 600 |
| UTSig | Persian | 115 | 27 | 42 | 600 |
| CEDAR | Western | 55 | 24 | 25 | 300 |

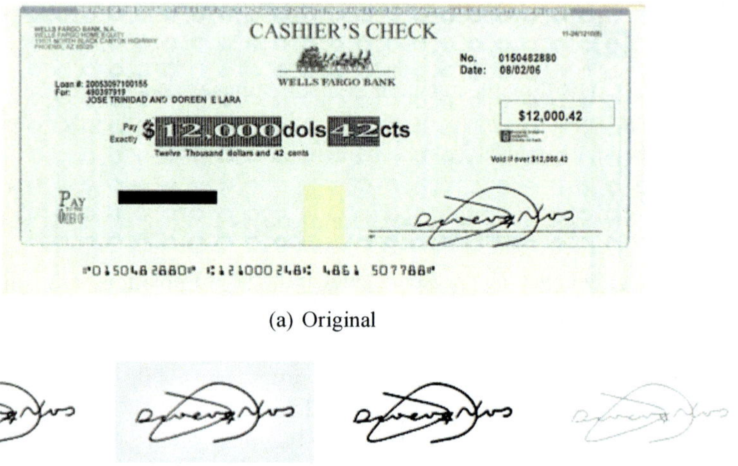


(a) Original

(b) Segmentation (c) Filtering (d) Binarization (e) Skeletonization

**Fig. 2** Common image preprocessing steps for GPDSsynthetic-Offline [11]: **a** bank check with signature,[4] **b** segmented signature image, **c** filtering by means of Difference of Gaussian, **d** binarization by means of thresholding, **e** skeletonization by means of morphological operators

Note, however, that the process sequence is not given *a priori*. The following paragraphs give an overview of some common methods and techniques that are widely accepted as standard preprocessing steps. These steps can be divided into four main phases (see also Fig. 2):

1. Segmentation
2. Filtering
3. Binarization
4. Skeletonization

For an exhaustive overview on preprocessing techniques, see [8, 20].

1. Segmentation

As illustrated in Fig. 2a, the signatures are in many cases part of some larger document (e.g. bank checks or forms), and thus, need to be extracted first. Especially bank checks are regarded as a challenging segmentation problem due to colored background images, logos and preprinted lines. To address this issue, different segmentation approaches have been proposed in the literature [2, 12, 13, 33]. In [9], for example, the segmentation is based on subtracting a blank check from the signed artifact. More generic approaches, are based on *Speeded Up Robust Features (SURF)* features in combination with an Euclidean distance metric [1]. Recently, different Deep Learning techniques based on CNNs have been proposed for this task [22, 34]. In general, the importance of signature segmentation is decreasing due to two reasons: First, the increasing use of blank signature fields and second, the declining importance of bank checks.

2. Filtering

In many cases, scanned handwritten signatures are affected by noisy background, i.e. small pixel fragments that do not belong to the actual signature. The reasons for that can be filthy scanners and papers, smearing as well as non-smooth writing surfaces, to mention just a few. To reduce noise, different filter methods are often employed such as *low pass* filter [4, 12, 24], *median* filter [3], *morphological* filter [2, 6, 13, 16] as well as *Difference of Gaussian* filter [27].

3. Binarization

To clearly distinguish between foreground and background pixels in an image, scanned signature artifacts are often binarized [6, 27]. That is, each pixel of a greyscale image is either assigned black (i.e. ink of signature) or white (i.e. background). In many cases, binarization leads to better classification results as well as overall lower processing times. A number of different binarization approaches have been proposed for this particular task such as *Niblack's* algorithm [4], *Otsu's* method [12, 16–18, 23, 24] as well as *thresholding* [2, 4, 13, 27, 35].

4. Skeletonization

Further image preprocessing approaches are mainly aimed at the reduction of intrapersonal variations in the handwriting. In [2, 3, 13], for example, the handwriting scaling is corrected, while skew and slant are corrected in [3, 23, 24, 30]. Intrapersonal variations can also be reduced by means of skeletonization that is also known as thinning [3, 4, 23, 24, 27].

Note that a number of Deep Learning approaches for signature verification have been proposed recently [8]. In these cases, the neural network is often restricted to a fixed size input, and thus, a certain resizing is necessary [7, 18]. Moreover, the size of training data is often rather small in the case of signature verification which might be a problem in the paradigm of Deep Learning. Hence, data augmentation is often employed in this case to artificially increase the number of available training data [17, 18].

## *2.3 Feature Extraction*

Based on segmented and preprocessed signature images, different types of statistical and structural representations can be extracted. The following two subsections give a brief overview of both representation types.

### 2.3.1 Statistical Representations of Signatures

Statistical representations are extracted from handwritten signature images either on a *global* or *local* scale [37]. In the former case (also known as *holistic approach*), features are extracted on complete signature images, and thus, represented by a single (typically high-dimensional) feature vector $\mathbf{x} \in \mathbb{R}^n$. In the case of local representations, subparts of the handwritten signature images are independently represented by local feature vectors. For example, features are extracted by means of a sliding window that moves from left to right over the whole signature image. Then, at each window position a set of different features is extracted. As a result, a linear sequence of $m$ local feature vectors is acquired for every signature image, i.e. $\mathbf{x}_1, \dots, \mathbf{x}_m$ with $\mathbf{x}_i \in \mathbb{R}^n$.

In early approaches, features are often used to represent different global handwriting characteristics [37]. For instance, features that are based on the direction of the contour, as shown in Fig. 3a [6, 16]. Similarly, the handwriting outline has also been used as representation formalism in [10]. Moreover, the direction of slant, i.e. the inclination of the handwriting, has been found to be useful [2, 13]. Last but not least, projection profiles, i.e. the number of foreground pixels in the horizontal or vertical direction, have been employed as global features in [30].

In contrast, features can also be extracted from a local perspective on the signature [37]. For example, different morphological features have been proposed in [23]. In case of *Gradient, Structural and Concavity (GSC)* features, three types of features are extracted by means of a grid-wise segmentation and concatenated to form a feature vector [35]. Global features (e.g. the signature height, projection profiles, slant angle, etc.) can also be combined with different local features (e.g. grid and texture features), as shown in [3].

Further local features are often based on more generic texture descriptors, i.e. features that can be used to represent arbitrary patterns [37]. In [41], for example, HoG have been employed, as illustrated in Fig. 3b. That is, the direction of gradients is first extracted at different local cell positions and then concatenated to form a single histogram. Likewise, LBP are used to describe circular structures of an image at different neighborhood levels [12, 41], while so-called *surroundedness* features are extracted in [24].

Finally, it is worth mentioning that a number of Deep Learning approaches for feature extraction have been proposed in the last years [18, 26, 36]. That is, rather than extracting handcrafted features by means of a certain predefined procedure, this type of feature representation is learned in an unsupervised manner by means of a CNN. In particular, different CNN architectures have been proposed for this task such as *spatial pyramid pooling* [17], *triplet network* [26] as well as *siamese network* [7].

### 2.3.2 Structural Representations of Signatures

Structural representations are based on powerful data structures such as strings, trees, or graphs [4, 27, 33, 37]. In general, a graph is defined as a four-tuple $g = (V, E, \mu, \nu)$ where $V$ and $E$ are finite sets of nodes and edges, $\mu : V \rightarrow L_V$ and $\nu : E \rightarrow L_E$ are labeling functions for nodes and edges, respectively [31]. In general, the graph size is not fixed and thus graphs are—in contrast to global fea-

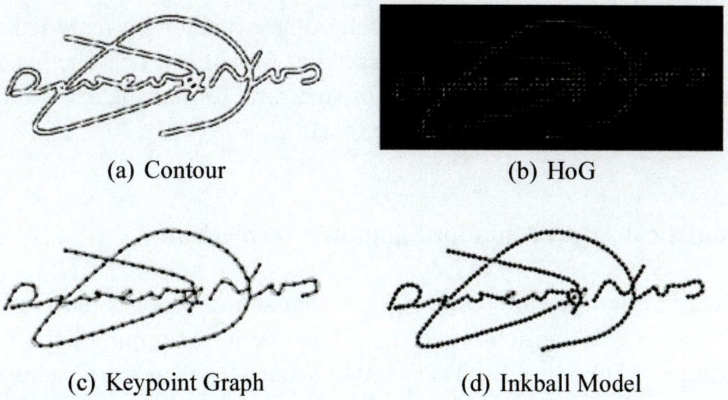


**Fig. 3** Visualization of contour features (see Fig. 3a), HoG features (see Fig. 3b), keypoint graph (see Fig. 3c), and inkball model (see Fig. 3d) of one signature from GPDSsynthetic-Offline [11]

ture vectors—flexible enough to adapt their size to the size and complexity of the underlying handwritten signature image. Moreover by means of edges, graphs are capable of representing binary relationships that might exist in different subparts of the underlying signature.

A vast amount of signature verification approaches still makes use of statistical rather than structural approaches [8, 19, 37]. This is rather surprising as the inherent representational properties of structural formalisms would be well-suited to represent handwriting.

An early structural signature verification approach is based on the representation of signatures by means of stroke primitives [33]. Another approach is based on the detection of characteristics points—so-called keypoints—on contour images [4]. Following this line of research, a recent approach is based on the detection of keypoints in skeletonized signature images, as illustrated in Fig. 3c [27, 37]. In this particular scenario, nodes are used to represent keypoints (e.g. start, end, and intersection points of the handwritten strokes), while edges are used to represent strokes between these keypoints. Other graph-like representations, that have become more popular in recent years are, for example, *inkball* models, as illustrated in Fig. 3d [25]. Similar to keypoint graphs, nodes in inkball models are based on characteristic points of the skeleton (i.e. junction and end points). Note, however, that inkball models are in contrast with keypoint graphs based on rooted trees rather than general graphs.

## 2.4 *Classification of Signatures*

Finally (and regardless the actual representation formalism used), the signature verification is based on a classification task. That is, a questioned signature $q$ of claimed user $u$ is compared with all reference signatures $r \in R_u$ of registered user $u$ (see

Fig. 1). However, the actual method that determines whether $q$ is regarded as genuine or forged depends on the chosen representation formalism, i.e. whether signatures are represented by means of a statistical or structural formalism. The following two subsections give an overview of both approaches.

### 2.4.1 Statistical Approaches for Signature Verification

In case of sequences of feature vectors (i.e. the signature is represented by $\mathbf{x}_1, \ldots, \mathbf{x}_m$), the classification task is often based on a dynamic programming approaches such as, for example, DTW [6, 30]. That is, two sequences of feature vectors (i.e. one sequence for $q$ and one sequence for $r$) are optimally aligned along one common time axis. The minimal sum of alignment costs can then be used as dissimilarity measure between $q$ and $r$. In many cases, however, signatures are represented by holistic and often high-dimensional feature vectors (i.e. the signature is represented by $\mathbf{x} \in \mathbb{R}^n$). Consequently, feature vectors are compared by means of different distance measures like for example Euclidean distance [7, 10, 26], $\chi^2$ distance [12, 16], or Mahalanobis distance [2, 13].

In contrast to these so-called *learning-free* methods, a number of *learning-based* approaches have been proposed as well [12, 13, 17]. That is, a statistical model is trained *a priori* on a (relatively large) training set of reference signatures. Different statistical models have been employed for this task for example SVMs [10, 12, 17, 18, 23, 24, 41], HMMs [10, 13] as well as Neural Networks [3, 23, 24].

Generally, learning-based approaches lead to lower error rates when compared with learning-free methods. However, this advantage is accompanied by a loss of flexibility and generalizability, which is due to the need to learn a statistical model on some training data. In fact, the accuracy of many learning-based approaches is crucially depending on the size and quality of the labelled training data. However, in case of signature verification the acquisition of training data is often expensive and/or otherwise limited.

### 2.4.2 Structural Approaches for Signature Verification

Structural representations, in particular graph-based representations, offer some inherent representational advantages as discussed before. However, this advantage is accompanied with an increased complexity with respect to basic dissimilarity measures. In fact, the exact computation of a graph dissimilarity measure on general graphs is known to be $\mathcal{NP}$-complete, and thus several fast approximations for this task have been proposed in the last decades [31]. In an early work, stroke primitives are first merged locally and then compared by means of a global dissimilarity measure [33]. Moreover, keypoints of contour images are matched by means of a point matching algorithm in [4]. However, the actual structural relationships of the handwriting are not considered during these first structural matching procedures.

More recent structural signature verification approaches make use of fast approximations for *Graph Edit Distance (GED)* [15, 32]. GED measures the minimum amount of distortion (i.e. insertion, deletion, substitution of both nodes and edges) needed to transform one graph into another. In the case of tree-based inkball models, the classification is based on an energy function that measure the amount of deformation, similar to GED [25].

Recently, different approaches have been proposed that allow the combination of statistical and structural classification approaches [26, 38, 39]. In [38], for example, subgraphs are first matched by means of graph matching and then optimally aligned by means of DTW. Another approach is based on an *ensemble* method that allows the combination of a Deep Learning approach (i.e. triplet network) with a fast GED approximation [26]. Finally, *prototype* graph embedding allows to embed graphs into a vectorial representation [39]. This approach combines the representational power of graphs and the mathematically well-founded vector space for dissimilarity computation.

## 3 Conclusion and Outlook

Handwritten signatures have been an important verification measure for many business and legal transactions for several centuries [37]. The popularity of handwritten signatures remains high, even though several international initiatives for electronic signature have been announced. That is, handwritten signatures offer some inherent advantages such as ubiquitous applicability as well as the overall high acceptance and availability. However, handwritten signatures expose also a certain risk of misuse. For this reason, automatic signature verification has been proposed [8, 19, 20]. Most signature verification frameworks are based on four processing steps. First, signatures are digitized by scanning. Second, scanned handwritten signature images are typically preprocessed in order to reduce variations caused by both the artifact (e.g. noisy background, skew scanning) and the handwriting (e.g. scaling of the handwriting). Next, preprocessed signature images are represented either by a statistical (i.e. vectorial) or structural (e.g. graph-based) formalism. Finally, a questioned signature is compared with a set of reference signatures in order to distinguish between genuine and forged signature.

The present chapter provides a comprehensive overview of the state of the art in signature verification based on offline data. This chapter reviews in particular traditional and very recent methods used in the various steps in a generic process of offline signature verification. This analysis can be used either as a literature review or as a starting point for developing an own signature verification engine.

In general, structural approaches offer some inherent representational advantages when compared with statistical formalisms. That is, graphs are able to adapt both their size and structure to the underlying pattern. Moreover, graphs are able to represent a binary relationship that could exist between subparts of the handwriting. However, most signature verification approaches still make use of a statistical rather than structural representation. This is mainly due to the fact that algorithms and methods that need graphs as input have a substantially greater complexity than their statistical counterparts. Yet, in recent years, several powerful and efficient methods for graph processing have been proposed [31]. These methods enable the use of graphs in the field of signature verification. In the future, great potential for further structural signature verification approaches can be expected, especially with the rise of Deep Learning methods for graphs, the so-called Geometric Deep Learning [5].

**Acknowledgements** This work has been supported by the Swiss National Science Foundation project 200021_162852.

## References

1. Ahmed, S., Malik, M.I., Liwicki, M., Dengel, A.: Signature segmentation from document images. In: 2012 International Conference on Frontiers in Handwriting Recognition, pp. 425–429 (2012)
2. Alonso-Fernandez, F., Fairhurst, M., Fierrez, J., Ortega-Garcia, J.: Automatic measures for predicting performance in off-line signature. In: IEEE International Conference on Image Processing, pp. I–369–I–372. IEEE (2007)
3. Baltzakis, H., Papamarkos, N.: A new signature verification technique based on a two-stage neural network classifier. Eng. Appl. Artif. Intell. **14**(1), 95–103 (2001)
4. Bansal, A., Nemmikanti, P., Kumar, P.: Offline signature verification using critical region matching. In: International Conference on Future Generation Communication and Networking Symposia, pp. 115–120. IEEE (2008)
5. Bronstein, M.M., Bruna, J., Lecun, Y., Szlam, A., Vandergheynst, P.: Geometric deep learning: going beyond euclidean data. Signal Process. Mag. **34**(4), 18–42 (2017)
6. Deng, P.S., Liao, H.Y.M., Ho, C.W., Tyan, H.R.: Wavelet-based off-line handwritten signature verification. Comput. Vis. Image Underst. **76**(3), 173–190 (1999)
7. Dey, S., Dutta, A., Toledo, J.I., Ghosh, S.K., Llados, J., Pal, U.: SigNet: convolutional siamese network for writer independent offline signature verification (2017)
8. Diaz, M., Ferrer, M.A., Impedovo, D., Malik, M.I., Pirlo, G., Plamondon, R.: A perspective analysis of handwritten signature technology. ACM Comput. Surv. **51**(6), 1–39 (2019)
9. Djeziri, S., Nouboud, F., Plamondon, R.: Extraction of items from checks. Proc. Fourth Int. Conf. Doc. Anal. Recognit. **2**, 10–13 (1997)
10. Ferrer, M.A., Alonso, J., Travieso, C.: Offline geometric parameters for automatic signature verification using fixed-point arithmetic. IEEE Trans. Pattern Anal. Mach. Intell. **27**(6), 993–997 (2005)
11. Ferrer, M.A., Diaz-Cabrera, M., Morales, A.: Static signature synthesis: a neuromotor inspired approach for biometrics. Trans. Pattern Anal. Mach. Intell. **37**(3), 667–680 (2015)
12. Ferrer, M.A., Vargas, J.F., Morales, A., Ordonez, A.: Robustness of offline signature verification based on gray level features. IEEE Trans. Inf. Forensics Secur. **7**(3), 966–977 (2012)
13. Fierrez-Aguilar, J., Alonso-Hermira, N., Moreno-Marquez, G., Ortega-Garcia, J.: An off-line signature verification system based on fusion of local and global information. In: International Workshop on Biometric Authentication, pp. 295–306. Springer, Berlin (2004)
14. Fillingham, D.: A Comparison of Digital and Handwritten Signatures. Technical report, Massachusetts Institute of Technology (1997)
15. Fischer, A., Suen, C.Y., Frinken, V., Riesen, K., Bunke, H.: Approximation of graph edit distance based on Hausdorff matching. Pattern Recognit. **48**(2), 331–343 (2015)

16. Gilperez, A., Alonso-Fernandez, F., Pecharroman, S., Fierrez, J., Ortega-Garcia, J.: Off-line signature verification using contour features. In: International Conference on Frontiers in Handwriting Recognition. Concordia University (2008)
17. Hafemann, L.G., Oliveira, L.S., Sabourin, R.: Fixed-sized representation learning from offline handwritten signatures of different sizes. Int. J. Doc. Anal. Recognit. **21**(3), 219–232 (2018)
18. Hafemann, L.G., Sabourin, R., Oliveira, L.S.: Learning features for offline handwritten signature verification using deep convolutional neural networks. Pattern Recognit. **70**, 163–176 (2017)
19. Hafemann, L.G., Sabourin, R., Oliveira, L.S.: Offline handwritten signature verification—literature review. In: International Conference on Image Processing Theory, Tools and Applications, pp. 1–8. IEEE (2017)
20. Impedovo, D., Pirlo, G.: Automatic signature verification: the state of the art. Appl. Rev., IEEE Trans. Syst., Man Cybern. Part C (2008)
21. Kalera, M.K., Sargur, S., Xu, A.: Offline signature verification and identification using distance statistics. Int. J. Pattern Recognit. Artif. Intell. **18**(07), 1339–1360 (2004)
22. Kleber Santos Leite Melo, V., Byron Leite Dantas, B.: A Fully convolutional network for signature segmentation from document images. In: International Conference on Frontiers in Handwriting Recognition, pp. 540–545. IEEE (2018)
23. Kumar, R., Kundu, L., Chanda, B., Sharma, J.D.: A writer-independent off-line signature verification system based on signature morphology. In: First International Conference on Intelligent Interactive Technologies and Multimedia, pp. 261–265. ACM Press (2010)
24. Kumar, R., Sharma, J., Chanda, B.: Writer-independent off-line signature verification using surroundedness feature. Pattern Recognit. Lett. **33**(3), 301–308 (2012)
25. Maergner, P., Howe, N., Riesen, K., Ingold, R., Fischer, A.: Offline signature verification via structural methods: graph edit distance and inkball models. In: International Conference on Frontiers in Handwriting Recognition, pp. 163–168. IEEE (2018)
26. Maergner, P., Pondenkandath, V., Alberti, M., Liwicki, M., Riesen, K., Ingold, R., Fischer, A.: Offline signature verification by combining graph edit distance and triplet networks. International Workshop on Structural. Syntactic, and Statistical Pattern Recognition, pp. 470–480. Springer, Berlin (2018)
27. Maergner, P., Riesen, K., Ingold, R., Fischer, A.: A structural approach to offline signature verification using graph edit distance. In: International Conference on Document Analysis and Recognition, pp. 1216–1222. IEEE (2017)
28. Nagel, R., Rosenfeld, A.: Computer detection of freehand forgeries. IEEE Trans. Comput. **C-26**(9), 895–905 (1977)
29. Ortega-Garcia, J., Fierrez-Aguilar, J., Simon, D., Gonzalez, J., Faundez-Zanuy, M., Espinosa, V., Satue, A., Hernaez, I., Igarza, J.J., Vivaracho, C., Escudero, D., Moro, Q.I.: MCYT baseline corpus: a bimodal biometric database. Vis., Image, Signal Process. **150**(6), 395 (2003)
30. Piyush Shanker, A., Rajagopalan, A.: Off-line signature verification using DTW. Pattern Recognit. Lett. **28**(12), 1407–1414 (2007)
31. Riesen, K.: Structural pattern recognition with graph edit distance. In: Advances in Computer Vision and Pattern Recognition. Springer, Berlin (2015)
32. Riesen, K., Bunke, H.: Approximate graph edit distance computation by means of bipartite graph matching. Image Vis. Comput. **27**(7), 950–959 (2009)
33. Sabourin, R., Beaumier, L.: Structural interpretation of handwritten signature images (1994)
34. Sharma, N., Mandal, R., Sharma, R., Pal, U., Blumenstein, M.: Signature and logo detection using deep CNN for document image retrieval. In: International Conference on Frontiers in Handwriting Recognition, pp. 416–422. IEEE (2018)
35. Siyuan Chen, Srihari, S.: A new off-line signature verification method based on graph. In: International Conference on Pattern Recognition, pp. 869–872. IEEE (2006)
36. Soleimani, A., Araabi, B.N., Fouladi, K.: Deep multitask metric learning for offline signature verification. Pattern Recognit. Lett. **80**, 84–90 (2016)
37. Stauffer, M.: From signatures to graphs. Master's thesis, University of Applied Sciences and Arts Northwestern Switzerland (2015)

38. Stauffer, M., Maergner, P., Fischer, A., Ingold, R., Riesen, K.: Off-line signature verification using structural dynamic time warping. In: International Conference on Document Analysis and Recognition (2019)
39. Stauffer, M., Maergner, P., Fischer, A., Riesen, K.: Graph embedding for offline handwritten signature verification. In: International Conference on Biometric Engineering and Applications (2019)
40. Vargas, F., Ferrer, M., Travieso, C., Alonso, J.: Off-line handwritten signature GPDS-960 corpus. In: International Conference on Document Analysis and Recognition, pp. 764–768. IEEE (2007)
41. Yilmaz, M.B., Yanikoglu, B., Tirkaz, C., Kholmatov, A.: Offline signature verification using classifier combination of HOG and LBP features. In: International Joint Conference on Biometrics, pp. 1–7. IEEE (2011)

# Agile Visualization in Design Thinking

**Emanuele Laurenzi, Knut Hinkelmann, Devid Montecchiari, and Mini Goel**

**Abstract** This chapter presents an agile visualization approach that supports one of the most widespread innovation processes: Design Thinking. The approach integrates the pre-defined graphical elements of SAP Scenes to sketch digital scenes for storyboards. Unforeseen scenarios can be created by accommodating new graphical elements and related domain-specific aspects on-the-fly. This fosters problem understanding and ideation, which otherwise would be hindered by the lack of elements. The symbolic artificial intelligence (AI)-based approach ensures the machine-interpretability of the sketched scenes. In turn, the plausibility check of the scenes is automated to help designers creating meaningful storyboards. The plausibility check includes the use of a domain ontology, which is supplied with semantic constraints. The approach is implemented in the prototype AOAME4Scenes, which is used for evaluation.

**Keywords** Symbolic AI · Design thinking · Agile and ontology-aided meta-modeling · AOAME4Scenes · Innovation processes

E. Laurenzi (✉) · K. Hinkelmann · D. Montecchiari · M. Goel
Intelligent Information Systems Research Group, FHNW University of Applied Sciences and Arts Northwestern Switzerland, Riggenbachstrasse 16, Windisch, Switzerland
e-mail: emanuele.laurenzi@fhnw.ch

K. Hinkelmann
e-mail: knut.hinkelmann@fhnw.ch

D. Montecchiari
e-mail: devid.montecchiari@fhnw.ch

M. Goel
e-mail: mini.goel@students.fhnw.ch

E. Laurenzi · K. Hinkelmann
Department of Informatics, University of Pretoria, Lynnwood Rd, Pretoria 0083, South Africa

D. Montecchiari
University of Camerino, Camerino, Italy



R. Dornberger (ed.), *New Trends in Business Information Systems and Technology*, Studies in Systems, Decision and Control 294,
https://doi.org/10.1007/978-3-030-48332-6_3

# 1 Introduction

Visual sketches are an efficient and effective way to communicate ideas [2, 33]. They empower designers to explore the creative design space [36]. Sketching is a traditional and conventional technique used to draw new ideas. Its usage has seen a massive growth after the Renaissance as designed objects have become more complex and more novel [3]. A pioneer on this subject was Leonardo da Vinci. Many of his sketches expressed the ability to communicate with someone else about (1) which artifact should be built, (2) how it should be built, (3) how it should work, as well as (4) the use of thinking and reasoning to support the solution design.

Sketches are not limited to the use as a means of communication but can also assist the cognitive process of conceiving a new solution design. These characteristics make sketching a prominent technique in todays' most widespread innovation processes [45]. In the Google Sprint [14], a whole day is dedicated to the sketch of possible solutions. In Design Thinking [28] sketches are used from an early phase to increase the understanding of the problem, and later to ideate the new solution designs [13, 29].

There are already several approaches in which the sketches are drawn with free hands, e.g. Crazy 8s [23, 43], Collaborative Sketching [8], Sketch Storming [38], CollabSketch [25], SketchStorm [19]. However, freehand sketching can be problematic for the smooth progress of innovation processes. Some people are not capable, hesitant, or even refuse to draw sketches, while others feel uncomfortable when being asked to draw [6]. De Vere et al. [40] associate this discomfort with the lack of wisdom in expressing ideas with sketches. Another problem is the comprehensibility of the drawn sketch. Since ideas are mainly sketched on tangible means (e.g., sticky notes), Hohmann [11] experienced that when any changes are made to sketches, they tend to become messy, and thus incomprehensible.

Such problems motivated the emergence of approaches that use a predefined set of elements such as SAP Scenes [31]. SAP Scenes consists of a collection of predefined graphical elements, the combination of which allows you to sketch storyboards about products or services. Predefined graphical elements help users create visual stories quickly and collaboratively. Therefore, drawing skills are not necessary, and users can focus directly on shaping ideas and scenarios in the form of illustrative storyboards. However, the practice of sketching storyboards with predefined graphical elements has its limitations (see Sect. 2). Such limitations hinder innovation processes, which is also why digital tools are difficult to use in practice.

In this chapter, we aim to show how a new agile and ontology-aided meta- modeling approach can overcome these limitations in order to promote the understanding of problems and ideation in innovation processes. The approach has already been validated in other application domains [15, 17].

The rest of the chapter is structured as follows: Sect. 3 discusses the adopted methodology. Section 4 introduces the state of the art of Design Thinking and visualization tools that allow the creation of sketches. In Sect. 5 , we set the requirements for the proposed artifact for the agile and ontology-aided modeling approach. The approach is then introduced in Sect. 6. The evaluation of the approach is described in Sect. 7. Finally, the conclusion and future work are discussed in Sect. 8.

## 2 Limitations that Hinder the Innovation Processes and Related Work

The first limitation is about the inability of the pre-defined elements to model any underlying reality. SAP Scenes provides "Scenes add-ons" [31] with some additional topic- and industry-specific elements, but they are not sufficient to cover all application scenarios. A real-world example for such limitation is the *Air as a Service* storyboard,[1] which was realized in the context of the EU-funded project DIGI-TRANS [7]. Figure 1 shows a part of the final storyboard. The storyboard supported the transformation of the business model of a German company from the production of ventilation systems to the innovative concept of *Air as a Service*. The new value proposition of the business model becomes a continuous supply of clean air.

In the *Air as a Service* use case, ventilation of large buildings like parking houses were identified as possible application scenarios. However, SAP Scenes did not foresee ventilators or parking lots. The newly needed graphical elements had to be physically created (see Fig. 1).

For physical SAP Scenes, this limitation can be easily overcome by manually drawing new elements. This is different, when the Scenes are modeled with a digital tool. Scene2Model [21] contains the digital representation of SAP Scenes with which users can design digital storyboards. The tool is based on the meta-modeling tool ADOxx [27]. Adding new modeling elements follows the current sequential process of meta-modeling, which reverts the waterfall-like cycle [15] (1) adapt the meta-model; (2) deploy the adapted meta-model in a new tool instantiation; and finally (3) test the new version of the modeling language.

Workshops with innovation experts (see Sect. 5) have confirmed that the sequential engineering process can lead to delays in innovation processes. An agile meta-modeling approach for the quick integration and adaptation of new elements would promote the adoption of digital tools among Design Thinking practitioners.

Another limitation that hinders the practice of Design Thinking is the lack of support from digital solutions in the elaboration of storyboards, especially in the ideation phase. In this phase, a set of storyboards are elaborated and should be available to provide meaningful input to the next phase [1], i.e., the prototype phase. There exist digital tools (e.g. [34, 35]) for online collaboration between participants to work on an idea. However, their use assumes a high level of expertise in innovation processes. In cases where designers (especially the less experienced ones) have a low domain expertise, relevant aspects in the storyboards may be overlooked. A possible

[1]The storyboard was realized in the innovation room of the Herman Hollerith Zentrum (HHZ), University of Reutlingen.

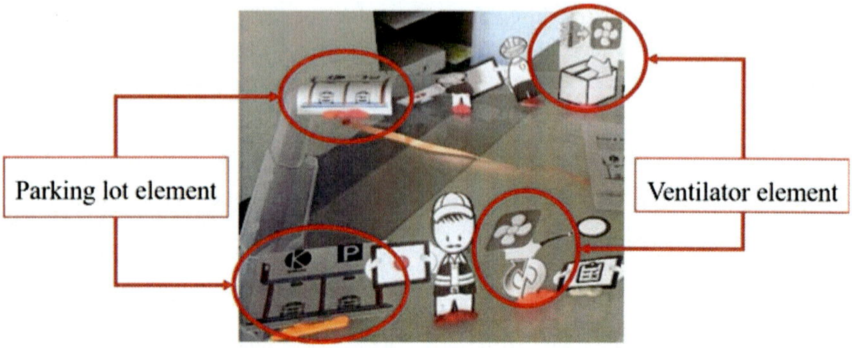


**Fig. 1** Air-as-a-Service storyboard created with an extension of SAP scenes

consequence would be the numerous back and forth loops between the ideation, the prototype and the test phases until a new solution design is refined enough for the Design Thinking process to end [5]. In the worst case, the new solution design will lack realistic foundations that need to be taken into account.

This lack of support can best be addressed by an approach that allows automation of the knowledge stored by SAP Scenes. The objective, in this case, would be to automatically support the creation of plausible and meaningful storyboards. The automation of knowledge can be achieved with the use of symbolic artificial intelligence, i.e., ontologies. The latter have already been used in storyboards of Scene2Model [21, 22] with the purpose of increasing information transparency and clarity of the meaning of storyboards.

In order to address the identified need for agility and automation to support the creation of storyboards, we propose to adopt the agile and ontology-aided meta-modeling approach already conceived by Laurenzi et al. [17]. The approach allows for the integration of new modeling elements and the adaptation of the existing ones on the fly, as well as the seamless alignment of graphical elements to ontologies.

## 3 Methodology

In this work, the chosen methodology is the Design Science Research (DSR) [10, 39], which aims to build an artifact by going through the following phases: awareness of problem, suggestion, development, and evaluation of the artifact and conclusion. In order to increase the awareness of problem, insights were derived from the following three different research activities:

- Research Activity 1 (RA1): Analyzing theories and existing approaches relevant to this research work (reported in Sects. 2 and 4).
- Research Activity 2 (RA2): Modeling a use case by adopting a modeling approach, which makes use of SAP Scenes to create digital storyboards in innovation pro-

cesses (see Sect. 5). During this activity, we confirmed our hypothesis about the need to deal with unforeseen scenarios and derived a list of requirements.
- Research Activity 3 (RA3): Two workshops were held with innovation experts (see Sect. 5) to increase understanding of whether and how the agile and ontology-aided modeling approach can support an innovation process.
- Research Activity 4 (RA4): Next, the conceptualized artifact was implemented in a variation of AOAME [17], which we called AOAME4Scenes. It solves the research problem and satisfies the requirements that were derived in the first three research activities.

The research activities cover the Design Science Research Cycle as proposed by Hevner [10]: RA4 corresponds to the design of the artifact, the rigor cycle (RA1) provides theoretical foundation to the designed artifact and the relevance cycle (RA2 and RA3) ensures the appropriateness of the application context.

## 4 State of the Art

This section first explains what Design Thinking consists of. It then describes existing visualization tools that support innovation processes by sketching.

### *4.1 Design Thinking*

Design thinking is an iterative process that seeks to understand users, challenge assumptions, redefine problems, and create innovative solutions to prototype and test [12]. The innovation process is most helpful to tackle problems that are ill-defined or unknown. During the process, real customers are brought close to design thinkers, who ultimately help solve customers' problems by designing new products or new services for them [29]. According to Schallmo et al. [32], the Design Thinking innovation process is based on four basic principles:

- *Human needs as a starting point* regards humans as the source of innovative ideas to work on.
- *Multi-disciplinarity of teams*: stakeholders from different disciplines work together, starting from the definition of the problem until the solution design.
- *Creative environment* refers to the dedicated creative place to conduct a Design Thinking workshop.
- *Iterative processes*:the iterative nature leads to an improvement of the solutions until a satisfactory stage is reached. It also helps establish a common ground that is comprehensible for all participants from different backgrounds.

As Fig. 2 depicts, the iterative process comprises five phases before a solution is conceived: empathize, define, ideate, prototype and test [18, 20, 28].

**Fig. 2** The design thinking innovation process [5]

The initial phase involves empathizing with the customer or user. Through observations and research the customers' problems are realized, and a concrete issue is determined, which then becomes the basis for the further process. The concrete issue drives the Design Thinking team to concentrate on a single issue. In the next phase, multiple solutions are created and analyzed within a team to find the best possible solution to solve the issue. The solution is shared with the customer to learn their opinion and their level of satisfaction with the idea. The next steps are the construction of the prototype and the testing of its effectiveness with the customers.

Each phase of Design Thinking, albeit for different purposes, may require the visualization of sketches. A common technique for visualization is called storyboarding [4, 42], a sketching method which enables users to visualize a problem or a solution through a storyline. Miron et al. [21]define storyboards as "a quick and valuable means to test the convergence between user needs and the proposed solution". The value comes from making people understand the problem and allowing designers to present their thoughts and imagination to other people. Multiple storyboards can be created to demonstrate different aspects of the problem to different stakeholders involved in the project [23]. In addition, multiple stories can be combined in an application to compare the solutions [31].

## *4.2 Visualization Tools*

Electronic support of sketching refers to the possibility of digitally viewing and using the sketches, saving the scenes electronically for future use, digitally sharing the scene with other team members and working electronically on images. Available digital tools that provide such support can fall into one of the following three categories:

1. *From paper sketches to digital sketches*: with this kind of tool, users first draw the sketches on paper and then bring the image into the system or upload a picture directly to the tool. Mural [24] is an example of such a tool. Once the image is in the tool, it can be used for creating maps and models like scenario maps, empathy maps, business model canvasses, etc.
2. *Sketching on digital screens*: this type of tool allows designers to draw their sketches on a digital screen similar to drawings on paper. FexiSketch [44], for example, enables designers to draw and edit (freehand) sketches on electronic whiteboards using phones, tablets, and desktop versions. The tool also incorporates the import functionality to upload images. Sketchboard [34] and Sketch-

Together [35] offer online whiteboards for teams to work collaboratively and remotely. Designers are empowered to sketch on the digital whiteboard simultaneously. AutoDraw [9] is an online tool that allows a designer not only to draw sketches digitally but also recommend the most similar predefined object to the sketch.

3. *Usage of predefined graphical elements*: this type of tool allows designers to create storyboards with predefined graphical elements. For instance, the desktop tool Scene2Model [22] offers digital graphical elements of SAP Scenes. It additionally empowers users to digitalize a storyboard by importing physical scenes on the tool.

## 5 Requirement Elicitation

The first requirement was derived from recent findings in the literature (see Sect. 2) and is formulated as follows:

- R1: A visualization approach in the context of an innovation process has to enable designers to create digital storyboards with a predefined set of graphical elements. This includes the sub-requirement of making it easier for users who dislike drawing sketches.

Among the advanced digital tools fulfilling this requirement, the Scene2Model is the only one that (1) uses digital predefined elements, (2) enables operations on scenes (e.g. analysis, consistency checking, etc.) and (3) allows semantic enrichment by means of an ontology.

We then used Scene2Model to model a use case and deepen the understanding of the problem in practice. The use case is fictitious and targets a specific application domain—the design of a kitchen for elderly people.

Firstly, we listed all the elements and properties needed to model the desired kitchen. Then, we tried to model these aspects in Scene2Model using SAP Scenes. Every time a desired aspect was missing, the respective table was updated accordingly. The list of missing aspects that could not be modeled was documented, including comments where needed. Additionally, graphical notations for the missing elements were suggested.

It turned out that in this use case eight specific elements were missing: heater, dishwasher, extractor hood, lower cabinet with cooking range, wall cabinet with glass, high cabinet with oven, double door fridge, window as an individual element. Additional domain-specific aspects that could not be modeled were properties such as appliance dimensions, appliance installation height, brand, warranty, and color. Figure 3 shows the "Fridge" element from the kitchen designed in Scene2Model, for which the desired types (i.e., double door and single door fridge) and properties (e.g. dimension) are missing .

For each of the missing elements a graphical notation was suggested and Scene2Model was extended using the ADOxx Development Toolkit [27].

(1) Kitchen designed in Scene2Model

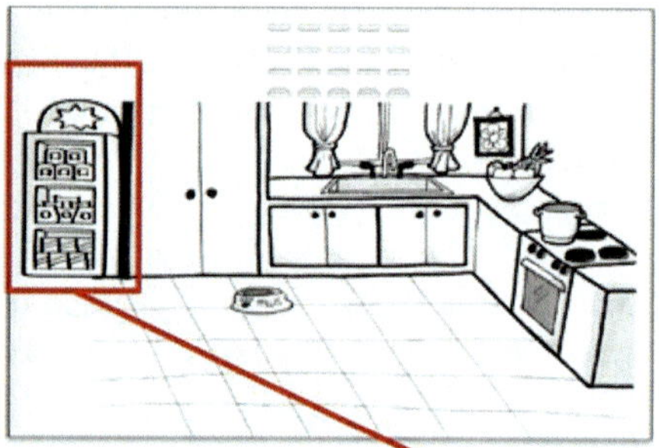

(2) Desired SAP Scenes element and properties and comparison

| | Desired Kitchen | Scene2model |
|---|---|---|
| Data type properties | Furniture Dimensions | - |
| Object type properties | Brand | - |
| | Warranty | |
| | Color | |
| Relations to other class | fridgeHasBrand | sceneobjectHasNote |
| | fridgeHasWarranty | |
| | fridgeHasColor | |
| Subclasses | Double Door Fridge, Single Door Fridge | - |
| Graphical Notation | Double Door Fridge: 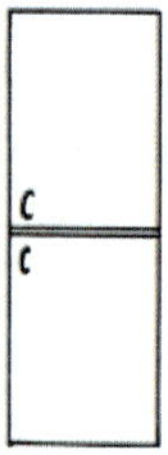 |  |

**Fig. 3** An example of missing elements and properties from the created SAP scene

The experience gained with modeling the use case not only confirmed that the predefined set of elements are not sufficient to target a particular application domain, but also proved the sequential process that is required to incorporate the missing elements. The new elements are first (1) integrated in the meta-modeling tool, then (2) the adapted meta-model is to be deployed in a new instance of the Scene2Model tool, and finally (3) the new set of elements could be used for modeling. This loop had to be repeated when we recognized that the accommodated elements were not properly added.

This experience led us to derive the following requirement:

- R2: A visualization approach in an innovation process has to enable designers to add new domain-specific aspects on the fly to the predefined elements. These aspects include graphical notations and properties such as relations between elements and attributes for elements.

Finally, two workshops were held with four experts from the DIGITRANS project [7] in the innovation room of the Herman Hollerith Zentrum (HHZ),[2] Reutlingen University. As introduced in Sect. 1, DIGITRANS is an EU-funded project and has the purpose to deliver a Design Thinking-based method and a set of tools to support digital transformation processes. This research activity aimed to reach an understanding on whether and how AOAME could be used in an innovation process—in this case business model transformation.

For this, we dedicated the first workshop to the thorough introduction of the agile and ontology-aided meta-modeling approach. As a follow-up, the possible application of AOAME in the digital transformation process was discussed.

In the eyes of the innovation experts, the main benefit of AOAME is the ability to quickly adapt to a modeling language, which includes changes in (1) graphical notations, and (2) in the knowledge retained by the language, i.e., the abstract syntax. This is mainly motivated by the need of integrating domain-specific aspects in each model presented in each phase of the business transformation, e.g. representing a Value Proposition Canvas, a storyboard created with SAP Scenes, a business process in BPMN [26], and an enterprise architecture in ArchiMate [37].

A second workshop dealt with the adoption of AOAME for the digital transformation process. A brainstorming activity was followed by the creation of a storyboard using SAP Scenes. The relevant aspects of AOAME were: the quick adaptation of modeling languages, the integration of modeling elements that belong to different languages, and the automatic ontology creation. According to the innovation experts, the automatic creation of an ontology can be particularly useful at one stage of their innovation process: the Business Model Configuration Simulation. The on the fly-adaptation of a modeling language, conversely, is requested in each phase where the purpose is to visualize a graphical model.

The two workshops raised awareness of the applicability of an agile and ontology-aided meta-modeling approach to support innovation processes. From this, the following two requirements were derived:

- R3: A visualization approach in an innovation process has to be able to accommodate different modeling languages in different innovation phases.
- R4: A visualization approach in an innovation process has to enable the automatic creation of ontologies from graphical models and thus enabling operations on the ontology-based model.

[2] http://www.hhz.de/home/.

## 6 Agile and Ontology-Aided Meta-Modeling for Innovation

The agile and ontology-aided meta-modeling approach seamlessly integrates graphical modeling with an ontology.The approach employs semantic rules, which aid the propagation of language adaptations to the ontology. In turn, the ontology evolves consistently with the changes made on the modeling language and by creating models. Hence, ontology expertise is no longer needed for the creation and maintenance of the ontology or to ensure consistency between models and ontology concepts. The approach is implemented in a modeling tool called AOAME (Agile and Ontology-aided Meta-modeling Environment) [17].AOAME allows the language engineering and modeling activities to interleave as well as adapt a language on the fly [17]. Unlike the current meta-modeling tools, it is no longer necessary to redeploy the modeling language when a change occurs.

AOAME foresees the specifications of modelling languages (notation, abstract syntax and semantics) into three different ontologies: the *Palette Ontology*, *Modeling Language Ontology* and the *Domain Ontology*.

The *Palette Ontology* contains classes, instances, and properties for the graphical representation of modeling constructs (e.g. graphical notation, size, position in the palette). A graphical notation relates to a concept in the abstract syntax. The *Modeling Language Ontology* reflects the abstract syntax, thus containing class taxonomy and

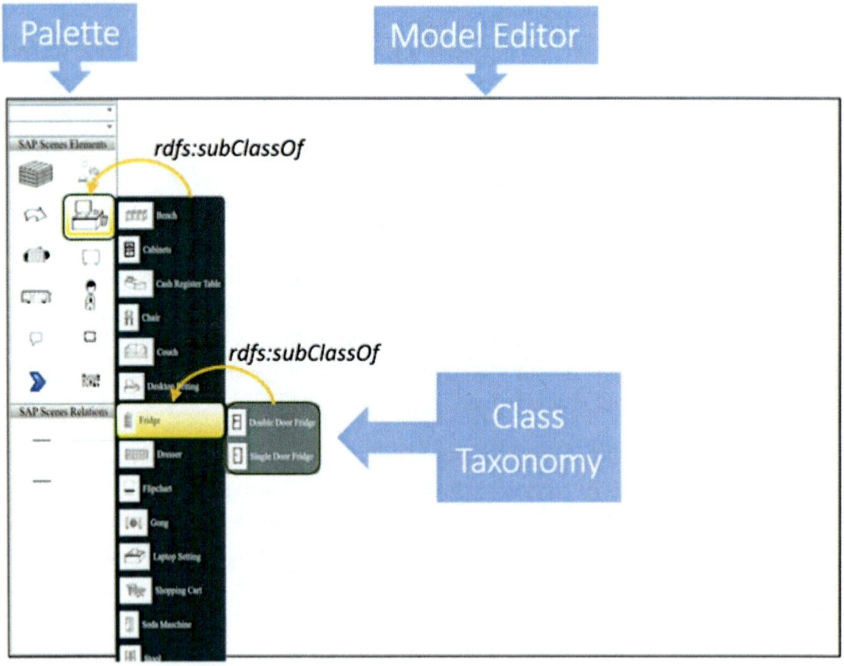


**Fig. 4** Palette and model editor in AOAME4Scenes

relations pertaining to the language view. The *Domain Ontology* reflects the domain of discourse and consists of existing domain knowledge, i.e., classes, relations and constraints. These are used to support the specification of the semantics of classes from the Modeling Language Ontology.

An ontology reflecting SAP Scenes was created. This *SAP Scenes Ontology* together with graphical notations of SAP Scenes were integrated in AOAME resulting in the prototype AOAME4Scenes.

AOAME4Scenes shows all the predefined graphical elements of SAP Scenes from the palette (see left-hand side of Fig. 4). Hence, the storyboards can be created in the model editor (see right-hand side of Fig. 4) by clicking on the graphical elements in the palette. As soon as the graphical element appears in the model editor, the instance is stored in the ontology. The taxonomy of each SAP Scenes Ontology class is displayed graphically by positioning the cursor on the graphical element (see, e.g., the two sub-classes of the "Fridge" element in Fig. 4).

## 7 Evaluation of the Approach

This section describes the evaluation of the agile and ontology-aided meta-modeling approach for innovation processes. For the evaluation, the prototype AOAME4Scenes is used (1) to model a scene (see Sect. 7.1); (2) to show how the ontology—which reflects a scene—can be exploited for the plausibility check of a scene; and (3) to describe how the requirements R1 to R4 are fulfilled.

### 7.1 *Use Case Modeled in AOAME4Scenes*

Figure 5 depicts the actions required to integrate a new element in AOAME4Scenes. The same use case modeled in Scene2Model is considered (see Sect. 5). Two new sub-classes "Double Door Fridge" and "Single Door Fridge" are created for the already existing graphical element "Fridge".

Right-clicking on an existing element allows to select the "Extend" function (see first view of Fig. 5). This action opens a pop-up (second view in Fig. 5), from which the user can enter attributes and relations of the new element, in particular the name, the description and the graphical notations, which can be uploaded.

By clicking on "Create New Element" an instance of new element (called "po:Double Door Fridge" in Fig. 5) is created in the Palette Ontology and a class with the same name but different prefix is created in the Modeling Language Ontology. Then the graphical notation appears in the palette and can immediately be used for modeling (see dashed arrow in the third view in Fig. 5).

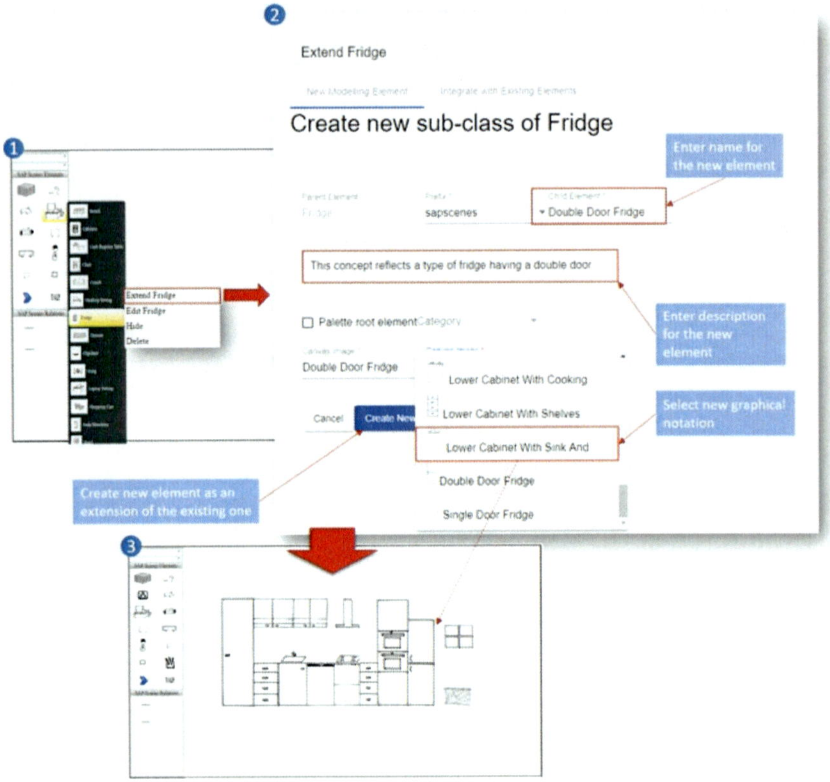


Fig. 5 Extension of SAP Scenes and modeling in AOAME4Scenes

## 7.2 Reasoning for Plausibility Checking of SAP Scenes

An advantage of ontology-based modeling is the machine-interpretability of knowledge. In our case, it is used to automate the plausibility check of the sketched scenes. In the following, the plausibility check is explained through the implementation of a concrete scenario.

Let us consider (1) the ontology-based model of the kitchen scene that is designed in the previous section, and (2) a fragment of a kitchen domain ontology.

Each modeling element is related to classes of the domain ontology. The kitchen domain contains—among others—classes, relations, and constraints regarding electric appliances that are typically used in a kitchen.

We assume that the designers used AOAME4Scenes to create a scene for a smart kitchen as an extension of the scene shown in Fig. 5. Thus, the concept "SmartFridge" exists in the modeling language as a sub-class of "Fridge" and is mapped with a "SmartFridge" concept from the domain ontology. The latter, among others, contains constraints such as "a fridge must have a power supply" and "a SmartFridge must

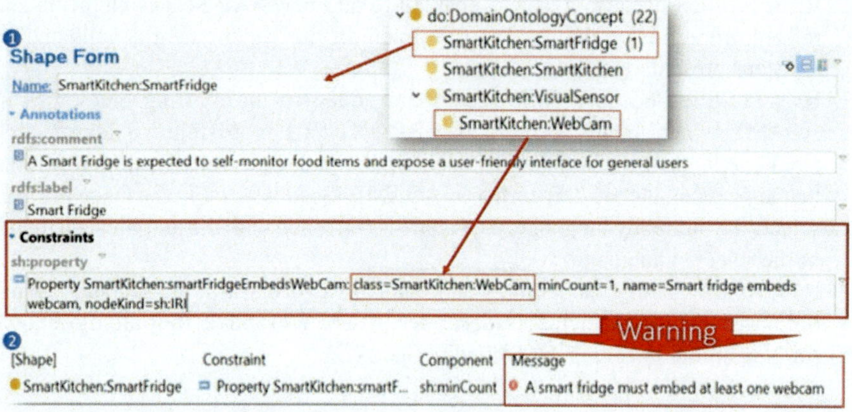


**Fig. 6** Modeling and execution of a domain constraint

embed at least one webcam". The webcam is a sensor that self-monitors food items [30].

Constraints are modeled in the W3C constraint language SHACL (Shape Constraint Language) [41]. Figure 6 shows (1) the constraint property of the visual sensor added on the concept "Smart Fridge"; (2) the execution of the constraint within the kitchen scenario.

The constraint has the name "SmartKitchen:SmartFridge" (i.e., SmartKitchen is the ontology prefix and SmartFridge is the class name in the domain ontology) and contains a property called "SmartKitchen:SmartFridgeEmbedsWebCam".

The constraint is then executed for the plausibility check. The fridge instance in the kitchen scene of Fig. 5 violates the constraint as a webcam is missing. As a consequence, an alert is triggered, which is shown as a message in the second screenshot of Fig. 6. The message reminds the designer to add a webcam in the kitchen scene before starting with the prototype phase. The required element can therefore be added on the fly as a sub-concept of the pre-defined SAP Scenes element "Device" and subsequently added in the smart kitchen scene.

### *7.3 Requirements Evaluation*

In the following, we report the fulfillment of the requirements elicited in this work (see Sect. 5):

1. R1 is fulfilled because AOAME4Scenes enables designers to create digital storyboards with digital SAP Scenes. The fulfillment of this requirement is validated by the use case modeled in Sect. 7.1.

2. R2 is fulfilled because a user is enabled to add new SAP Scenes elements and related properties on the fly. The fulfillment of this requirement is validated by the steps presented in Sect. 7.1.
3. R3 is fulfilled because SAP Scenes are added to the existing ontologies in AOAME, which already accommodate BPMN [17]. It is possible to quickly switch from one language to another by selecting the language from the drop down list that appears on the top-left corner of AOAME4Scenes (see Fig. 5). Once the user selects the modeling language, the palette is refreshed with the graphical elements of the selected language.
4. R4 is fulfilled as (1) the ontology automatically evolves with the adaptations that occur in the graphical elements (described in Sect. 7.1) and (2) the ontology allows for plausibility checking (described in Sect. 7.2).

## 8 Conclusion and Future Work

The presented work showed the relevance of an agile and ontology-aided meta-modeling approach to support the practice of sketching in innovation processes. For this, we focused on the innovation process Design Thinking. Initially we analyzed problems and challenges of innovation processes from literature, workshops, and a use case. From the analysis phase, a set of design requirements was identified that were met to implement the prototype AOAME4Scenes. The latter includes the graphical notations of SAP Scenes and the newly created SAP Scenes Ontology. AOAME4Scenes enables users to integrate new domain-specific aspects (i.e., new elements, relations and attributes) on the fly. Unlike current modeling approaches, unforeseen scenarios can therefore be quickly modeled. Ontologies are automatically created in the background while sketching the scenes. In order to support innovation experts with little experience in a particular domain, constrained domain ontologies are executed over the generated ontology-based scenes for their plausibility check.

The approach was evaluated through the AOAME4Scenes prototype by (1) modeling a domain-specific scene for which the predefined set of graphical elements was not sufficiently expressive; (2) making use of the ontology of the scene for its plausibility check; (3) describing how the designed artifact fulfills the requirements. This evaluation has therefore demonstrated the relevance of the agile and ontology-aided meta-modeling approach in innovation processes in order to foster problem understanding and ideation.

As a future work, we plan to evaluate AOAME4Scenes in concrete innovation projects in order to identify possible improvements. In parallel, research will be launched to investigate the integration of machine learning techniques into the agile and ontology-aided meta-modeling approach [16]. Insights can then be transferred in the context of innovation processes to learn innovation patterns.

## References

1. Beckman, S., Barry, M.: Design and innovation through storytelling. Int. J. Innov. Sci. **1**(4), 151–160 (2009)
2. Bridgeland, D.M., Zahavi, R.: Business Modeling: A Practical Guide to Realizing Business Value. Morgan Kaufmann (2008)
3. Cross, N.: Natural intelligence in design. Des. Stud. **20**(1), 25–39 (1999)
4. Dam, R., Siang, T.: Introduction to the essential ideation techniques which are the heart of design thinking. https://www.interaction-design.org/literature/article/introduction-to-the-essential-ideation-techniques-which-are-the-heart-of-design-thinking. Accessed 01 Dec 2019
5. Dam, R., Siang, T.: Stage 4 in the design thinking process: prototype | interaction design foundation. https://www.interaction-design.org/literature/article/stage-4-in-the-design-thinking-process-prototype/. Accessed 01 Dec 2019
6. Detken, K.: Scenes: a new method and tool to create storyboards. https://experience.sap.com/skillup/scenes-new-method-tool-create-storyboards/. Accessed 01 Dec 2019
7. Digitrans-Consortium: digitrans method for the digital transformation of smes. https://informatik.univie.ac.at/en/research/projects/project/278/. Accessed 01 Dec 2019
8. Fehlau, A.: Collaborative sketching—drawing a bridge between ideation and prototyping. https://experience.sap.com/skillup/collaborative-sketching-drawing-a-bridge-between-ideation-and-prototyping/. Accessed 01 Dec 2019
9. Google Creative Lab: AutoDraw. https://experiments.withgoogle.com/autodraw. Accessed 01 Dec 2019
10. Hevner, A.: A three cycle view of design science research. Scand. J. Inf. Syst. **19** (2007)
11. Hohmann, M.: The final output of the front end innovation process. In: 6th IBA Bachelor Thesis Conference (2015)
12. Interaction design foundation: what is design thinking? https://www.interaction-design.org/literature/topics/design-thinking. Accessed 01 Dec 2019
13. Jonson, B.: Design ideation: the conceptual sketch in the digital age. Des. Stud. **26**(6), 613–624 (2005)
14. Knapp, J., Zeratsky, J., Kowitz, B.: Sprint: How to Solve Big Problems and Test New Ideas in Just Five Days. Simon & Schuster, New York (2016)
15. Laurenzi, E., Hinkelmann, K., Izzo, S., Reimer, U., Merwe, A.V.D.: Towards an agile and ontology-aided modeling environment for DSML adaptation. In: International Conference on Advanced Information Systems Engineering, pp. 222–234. Springer International Publishing, Cham (2018)
16. Laurenzi, E., Hinkelmann, K., Jüngling, S., Montecchiari, D., Pande, C., Martin, A.: Towards an assistive and pattern learning-driven process modeling approach. In: Martin, A., Hinkelmann, K., Gerber, A., Lenat, D., van Harmelen, F., Clark, P. (eds.) AAAI Spring Symposium: Combining Machine Learning with Knowledge Engineering, p. 6 (2019)
17. Laurenzi, E., Hinkelmann, K., van der Merwe, A.: An Agile and Ontology-Aided Modeling Environment. In: Buchmann, R.A., Karagiannis, D., Kirikova, M. (eds.) IFIP Working Conference on The Practice of Enterprise Modeling, pp. 221–237. Springer International Publishing, Cham (2018)
18. Lindberg, T., Meinel, C., Wagner, R.: Design thinking: a fruitful concept for it development? In: Design Thinking, pp. 3–18. Springer, Berlin (2011)
19. Lindley, S., Cao, X., Helmes, J., Morris, R., Meek, S.: Towards a tool for design ideation: insights from use of SketchStorm. In: Proceedings of the 27th International BCS Human Computer Interaction Conference (2013)
20. Lucena, P., Braz, A., Chicoria, A., Tizzei, L.: IBM design thinking software development framework. In: Brazilian Workshop on Agile Methods. Springer, Cham (2016)
21. Miron, E.T., Muck, C., Karagiannis, D.: Transforming haptic storyboards into diagrammatic models: the Scene2Model Tool. In: Proceedings of the 52nd Hawaii International Conference on System Sciences (2019)

22. Miron, E.T., Muck, C., Karagiannis, D., Götzinger, D.: Transforming storyboards into diagrammatic models. In: Chapman, P., Stapleton, G., Moktefi, A., Perez-Kriz, S., Bellucci, F. (eds.) Diagrammatic Representation and Inference, pp. 770–773. Springer International Publishing, Cham (2018)
23. Mohamed, A., Claudia, B.: Designing a digital service to improve the user experience of conference organizers. Master thesis, Chalmers University of Technology Sweden (2017)
24. Mural: Think and collaborate visually. Anywhere, anytime. https://mural.co/. Accessed 01 Dec 2019
25. Nováková, K., Jakubal, V., Achten, H., Matějovská, D.: Collab sketch. eCAADe **2013**(18), 213 (2013)
26. OMG: Business Process Model and Notation (BPMN) Version 2.0. January. Object Management Group OMG, Needham (2011)
27. OMILab: ADOxx Meta-Modelling Platform. https://www.adoxx.org/live/home. Accessed 01 Dec 2019
28. Plattner, H.: An Introduction to Design Thinking Process Guide. The Institute of Design at Stanford, Stanford (2010)
29. Razzouk, R., Shute, V.: What is design thinking and why is it important? Rev. Educ. Res. **82**(3), 330–348 (2012)
30. Sandholm, T., Lee, D., Tegelund, B., Han, S., Shin, B., Kim, B.: Cloudfridge: a testbed for smart fridge interactions (2014). arXiv:1401.0585
31. SAP: Scenes Concept and Building Guide. https://experience.sap.com/designservices/wp-content/uploads/sites/2/2019/07/ScenesConceptAndBuildingGuidepdf-2.pdf. Accessed 01 Dec 2019
32. Schallmo, D., Williams, C.A., Lang, K.: An integrated design thinking approach-literature review, basic principles and roadmap for design thinking. In: ISPIM Innovation Symposium, pp. 1–18. The International Society for Professional Innovation Management (ISPIM) (2018)
33. Sibbet, D.: Visual meetings: how graphics, sticky notes and idea mapping can transform group productivity. Wiley (2010)
34. Sketchboard: Creative team sketching and diagramming on an endless online whiteboard. https://sketchboard.io/. Accessed 01 Dec 2019
35. SketchTogether: Sketchtogether—a real time whiteboard for the web. https://sketchtogether.com/. Accessed 01 Dec 2019
36. Taborda, E., Chandrasegaran, S.K., Kisselburgh, L., Reid, T., Ramani, K.: Enhancing visual thinking in a toy design course using freehand sketching. In: ASME 2012 International Design Engineering Technical Conferences and Computers and Information in Engineering Conference, pp. 267–276. American Society of Mechanical Engineers (2012)
37. The Open Group: ArchiMate 3.0.1 Specification. https://pubs.opengroup.org/architecture/archimate3-doc/. Accessed 01 Dec 2019
38. UXM: A guide to sketch storming—a design game for ideation—UXM. http://www.uxforthemasses.com/sketch-storming/. Accessed 01 Dec 2019
39. Vaishnavi, V., Kuechler, B.: Design science research in information systems. J. MIS Q. **28**(1), 75–105 (2004)
40. de Vere, I., Kapoor, A., Melles, G., et al.: Developing a drawing culture: new directions in engineering education. In: DS 68-8: Proceedings of the 18th International Conference on Engineering Design (ICED 11), Impacting Society through Engineering Design, Vol. 8: Design Education, Lyngby/Copenhagen, Denmark, 15–19 July 2011, pp. 226–235 (2011)
41. W3C: Shapes Constraint Language (SHACL). https://www.w3.org/TR/shacl/. Accessed 01 July 2019
42. Walker, R., Ap Cenydd, L., Pop, S., Miles, H.C., Hughes, C.J., Teahan, W.J., Roberts, J.C.: Storyboarding for visual analytics. Inf. Vis. **14**(1), 27–50 (2015)
43. Wong, N.H.: Making user-focused prototype using design sprint to test, design, and prototype mobile app rapidly. Ph.D. thesis, Tampere University of Applied Sciences (2016)

44. Wüest, D., Seyff, N., Glinz, M.: Flexisketch team: collaborative sketching and notation creation on the fly. In: In: IEEE/ACM 37th IEEE International Conference on Software Engineering, pp. 685–688. University of Zurich (2015)
45. Xin, M.: 3D sketching and collaborative design with napkin sketch. Ph.D. thesis, University of Calgary (2011)

# Text Mining Innovation for Business

**Ela Pustulka and Thomas Hanne**

**Abstract** This chapter reflects on the business innovation supported by developing text mining solutions to meet the business needs communicated by Swiss companies. Two related projects from different industries and with different challenges are discussed in order to identify common procedures and methodologies that can be used. One of the partners, in the gig work sector, offers a platform solution for employee recruitment for temporary work. The work assessment is performed using short reviews for which a method for sentiment assessment based on machine learning has been developed. The other partner, in the financial advice sector, operates an information extraction service for business documents, including insurance policies. This requires automation in the extraction of structured information from pdf-files. The common path to innovation in such projects includes business process modeling and the implementation of novel technological solutions, including text mining techniques.

**Keywords** Digitalization · Text mining · BPMN · Innovation · Anonymization · Gig work · Insurance brokerage

## 1 Introduction

Digital transformation is one of the most important business drivers nowadays [1]. Two concepts come into play in this area: digitalization and digitization. Roughly speaking, digitization is the underlying change in the way we store and use information, where we replace analog/physical storage and data handling with digital information formats and data flows. Digitalization, on the other hand, deals with

E. Pustulka (✉) · T. Hanne
Institute for Information Systems, School of Business, FHNW University of Applied Sciences and Arts Northwestern Switzerland, Riggenbachstrasse 16, 4600 Olten, Switzerland
e-mail: elzbieta.pustulka@fhnw.ch

T. Hanne
e-mail: thomas.hanne@fhnw.ch

 
R. Dornberger (ed.), *New Trends in Business Information Systems and Technology*, Studies in Systems, Decision and Control 294,
https://doi.org/10.1007/978-3-030-48332-6_4

processes that can be completely new and have little to do with previous processes that were the traditional way of doing business on paper or in other physical ways.

Both digitization and digitalization lead to profound changes in the way business and society work. We see both processes as both sources and outcomes of innovation. In digital transformation, the process of change usually takes place organically within companies and originates within the minds of business leaders who shape their business in such a way that they take advantage of the opportunities that arise. Companies then often use external technical expertise to guide the development of new business solutions [2, 3]. In this chapter, we focus on how digital transformation benefits from novel solutions based on text mining, which we develop with our business partners.

The path to innovation is shaped by the business partners who state their needs. Based on our experiences, the following steps in such projects are usually useful and should be carried out in close collaboration with the project partners:

1. creating Business Process Modeling and Notation (BPMN) models [4]
2. obtaining datasets from the business
3. preliminary experiments to assess feasibility, based on the available data
4. designing an anonymization procedure (if needed)
5. obtaining more data
6. prototyping and assessment
7. refinement
8. business adaptation and use.

This approach combines the process with the data that reflect the common perception of text mining and data science defined as a "broader discipline that combines knowledge from information technology and knowledge from management sciences to improve and run operational processes" [5], p. 811.

However, contrary to the assessment made by Aalst and Damiani [5] who expect companies to have many processes with logs that can be mined, we work with small or medium-sized enterprises (SMEs) that have often not documented their processes and do not have such logs, and thus we document the processes manually. We carry out a business process analysis to understand the business requirements and to develop a shared vision of the future with our partners. Visualizing business processes with BPMN is useful because it provides a level of clarity and a communication basis that our partners can relate to. In one case, the analysis led the company to realize that their processes had gaps that were not under their control, which is now being addressed [6].

The second step in our approach is to gain access to datasets for text mining. After initial experiments, we often find that there are not enough data for further analysis, either in terms of volume or data complexity, and we ask for more. Since data privacy needs to be guaranteed [7], we design anonymization solutions that give us clean data that comply with the law. Then, in close collaboration with our partner, we develop prototypes and evaluate them together to see if they meet business needs. Practice shows that a viable solution has to provide high accuracy, with around 90% being sufficient. Finally, the companies take over the prototypes and integrate them into products that offer business value.

Our contribution is the description of the innovation processes and their outcomes, which lead to two business innovations. One is a text mining solution for the gig work sector, which uses text mining for sentiment analysis. The other is ongoing work on the development of an information extraction module for a platform that supports insurance brokerage. In both cases, the development of individual solutions was necessary because there was no suitable standard software available.

In addition to discussing some specific results of the two projects, this chapter points out the partly common procedures and methodologies employed in the two considered cases, which come from different industries and focus on different challenges. In Sect. 2, BPMN is discussed regarding its usefulness in innovation-oriented projects, but also to introduce the reader to the background of the two projects and the respective business domains. In particular, we present the developed BPMN models. Section 3 deals with general aspects of text mining in the considered projects. Three specific application areas of text mining are discussed: the anonymization of documents, sentiment analysis of user comments, and information extraction from unstructured text files. Section 4 discusses our contributions and concludes.

## 2 BPMN Analysis for Applied Research Projects

We use Business Process Modeling and Notation (BPMN) to help us understand the business scenario and requirements [4]. The graphical notation improves the general understanding of the current processes and errors in our judgement can be clearly spotted and corrected. We work in two business areas: evaluation of gig work jobs, see Sect. 2.1 and information extraction serving the needs of financial service advisors (insurance brokers), see Sect. 2.2.

### *2.1 Gig Work Platform Processes*

Greenwood et al. [8], p. 27 define gig-economy platforms as "digital, service-based, on-demand platforms that enable flexible work arrangements". Our business partner in the gig work sector has approximately 150 employees in Europe. The company operates a part-electronic HR recruitment platform and can find employees for temporary work assignments within a few hours of receiving an order from a potential employer. The company uses traditional human-intensive processes to acquire new business (employers) and mostly electronic processes to recruit workers and manage the employment process. We initiated a joint research project focusing initially on improving electronic worker recruitment and retention. Our research looks at the entire business process. We identified some issues based on the priorities specified

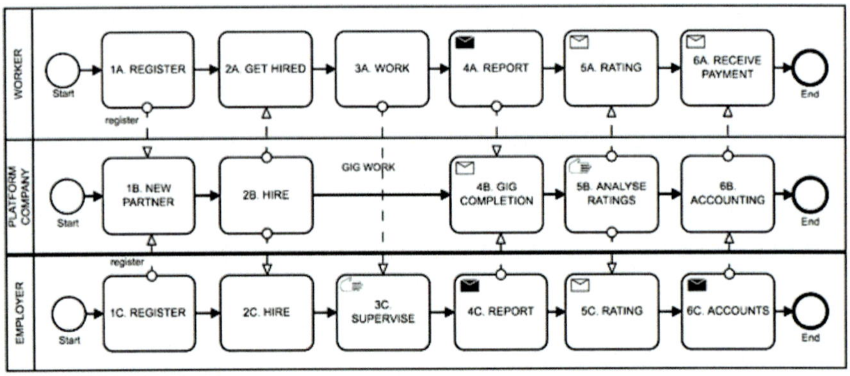


**Fig. 1** Simplified illustration from [6]. Gig work process overview showing the worker lane, the platform company lane, and the employer lane. Processes are numbered 1–6, with letters beside the number, to distinguish between activities performed by the participants. Process 1, REGISTER, is visible in the left part of the figure in all three lanes (1A to 1C). Process 2, HIRE, follows, as 2A–2C. Process 3, GIG WORK, is outside the platform. Process 4, GIG COMPLETION/REPORTING, involves all three lanes and is followed by Process 5, ANALYSE RATINGS (5A to 5C). Finally, Process 6, ACCOUNTING, closes the transaction. All participants are modeled in one pool because they share one marketplace

by the project partner (sentiment analysis and topic analysis of job reviews) and some future work (automated analysis of CVs and assignment of job labels to them).

The company gave us a demo of the platform system and granted access to their Wiki, which included platform documentation [6] in the form of a matrix showing the system functions, both implemented and planned. In a short modeling session we translated those functions into a BPMN model, see Fig. 1. This model was discussed with the business partner, corrected, and frozen. In this case, the process is so closely tied to the underlying platform that it is not changed until a new software version is released, which is rare (from several months to a year).

Figure 1 shows an outline of the platform process which connects employers to employees. The process can logically be divided into interactions with employers and employees, leading to three lanes in the diagram. From left to right, we see process steps that start with the initiation of the business relationship (registration, step 1), followed by hiring (step 2), job execution (work, step 3), gig completion/reporting (step 4), job performance analysis (ratings, step 5) and accounting (step 6). Steps 2 to 6 are repeated for each new temporary assignment for which the prospective employer wishes to recruit.

Text mining will support the Process 5B, Analyze ratings. After a gig, both workers and employers rate each other. They provide a star rating (1–4 stars, with 1 being poor and 4 excellent) and, if the rating is low (1–2 stars), they are required to enter a comment explaining their rating. Comments can also be provided at high rating. Currently, the platform company only communicates the star rating to the worker and to the company (Processes 5A and 5C), together with the confirmed number of hours worked. The job reviews, however, are not used. The business partner wants to communicate them to the parties involved and is in the process of deciding how best to do this, as this needs to be part of the platform system. The project supports this process by prototyping sentiment and topic analysis tools (both based on text mining techniques) and proposing a sentiment dashboard, as detailed in Sect. 3.3.

## 2.2 An Information Extraction Service for Insurance Brokers

The second innovation project discussed in this chapter involves the provision of financial advice services such as decisions regarding insurance contracts. Currently, insurance brokers receive scanned documents from customers and take several minutes per document to manually extract the most relevant information from the document into the software they use to prepare a recommendation regarding insurance-relevant requirements. The scenario is an extension of ideas already in operation (Evia, "expert virtual insurance agent" [9]) and of the trend to automate business processes and replace them with Robotic Process Automation [10]. The business scenario includes three parties: the customer, the broker, and the company providing the extraction service. The envisaged BPMN process is outlined in Fig. 2 (customer interaction not shown). The Extraction Service extracts information from pdfs and produces structured data. This service is being implemented using text mining.

Let us consider insurance policies as documents and focus on those issued in Switzerland. There are up to ten large insurance companies and around 15 types of insurance (vehicle, health, legal, etc.). Each of the insurance types encompasses several possible types of coverage and related data (sum covered, period covered,

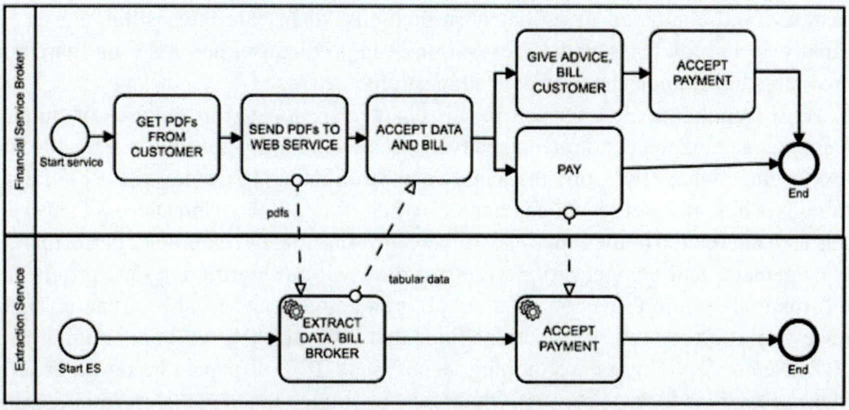


**Fig. 2** An overview of the planned interactions between financial advice brokers and the Extraction Service (ES). The insurance broker receives the paperwork from the customer, sends the documents to the ES and obtains structured information which can be used to prepare a recommendation for the customer. After the billing, the process ends

exclusions, etc.). A policy covers only a selection of clauses and each of these clauses typically has partial costs related to the coverage and the insured sum that are relevant. It is possible to use information extraction technologies based on pdf annotation, such as smartFIX [11], and create an extraction schema for each policy individually. However, the analysis and full implementation take around eight days per policy. Additional complications in the use of current information extraction tools come from the insertion of advertising material and format changes that make the tools unstable. The goal of the project is to design a tool for automated information extraction from pdf documents. The tool will support the business process shown in Fig. 2. Text mining will support information extraction as detailed in Sect. 3.4.

# 3 Text Mining Challenges and Solutions in Applied Research Projects

## *3.1 Tasks and Challenges in Text Mining*

Weiss et al. [12] draw the distinction between data mining and text mining as "numbers vs. text". In text mining, the text is transformed into numbers and then data mining procedures can be applied to a numerical representation. Typical applications of text mining are classification and prediction. In a classification setting, the text is to be divided into "equivalence classes" with shared characteristics. In the prediction setting, statements are to be made about the future, based on the past. Text mining is used in information retrieval [13]. It is also used to cluster and organize document collections, and support browsing [14]. Another application is information extraction, i.e. converting text documents to structured documents that can be searched and compared using database methods, or in data integration, e.g. [15]. Finally, in a predictive scenario, text mining can predict user needs during search or browsing and support the user with suggestions.

As the applications described in this chapter are located in the area of human resource management and financial advice, this section only covers the recent use of text mining in these two domains. Piazza and Strohmeier [16] review the uses of data mining in human resource management. In their opinion, the main success factors in this area are related to the functional dimension (staffing, development, performance management, and compensation), method dimension (algorithms), data provision, information system provision, user support, and ethical and legal awareness. Their review of 100 research contributions finds that 90% of work used data mining and 10% used text mining or web mining. Thus, only 10% of papers reviewed in [16] make provisions for data privacy, which may be implemented via aggregation or depersonalization (anonymization). Our scenarios require anonymization, as detailed in Sect. 3.2.

Text mining is used extensively in the financial sector. Kumar and Ravi [14] describe methods and applications. They find that SVM (support vector machines) are the most popular method, followed by naïve Bayes (NB), k-nearest neighbors (k-NN) and decision trees. Financial applications include foreign exchange prediction, stock market prediction, customer relationship management (CRM), and cybersecurity. In the areas most related to our projects, they only mention the work of Ghani et al. [17] on product attribute extraction. Other work in the area includes [18] describing a system called Midas that uses information extraction, entity resolution, mapping and fusion to integrate public data and support systemic risk analysis in the financial market. Midas can help discover suspicious activities such as co-lending, and risk at the company level, including relationships with other companies, key executives and aggregated financial data. However, systems like Midas are too complex for an SME and they need to be tuned to fit the business scenario. Tuning via expert user interaction is also required in newer information extraction systems [19] and may be a barrier to system adoption.

Text mining scenarios are a challenge to business for several reasons:

- Data may need to be anonymized (legal requirement).
- Large amounts of data are required to build statistical machine learning models that guarantee the required quality of results, and many SMEs may not have enough data to use data mining.
- As text mining is constantly evolving, being aware of the latest developments in research (including new methods or open source software) is a prerequisite.
- Several rounds of prototyping are needed to develop solutions of the required quality, as published methods do not necessarily deliver the best results on company data and need to be extended, tuned, or combined with other published methods.
- Domain specific adjustments are required and result in improved quality. In the projects described in this chapter, this involves the language, specific vocabulary, and other text characteristics (short utterances or even lists). In the sentiment analysis described in Sect. 3.3, there are three languages (German, French, and English) with very short textual comments. In information extraction (see Sect. 3.4) the situation is similar, as we have financial documents in German, French, English, and Italian, including insurance policies, tax statements, bank statements, and ownership certificates.

### *3.2 Data Anonymization*

Text mining requires considerable amounts of data to achieve the quality acceptable for business use. Sensitive personal documents, such as insurance policies or bank statements, are a challenge because they cannot be used directly in text mining due to data protection regulations [7]. In our work (in both projects), we started with a simple approach that was appropriate in the context of sentiment analysis, and developed it further to anonymize financial documents. Providing anonymization software to the business partner was the best way to ensure that we did not handle private data, as the original data remained with its owners.

**Table 1** Pseudo-code of the preprocessing and anonymization algorithm, used with each file, called *pdffile*. Tokenize splits on space. Intersection is the set intersection

```
1  Streets = set (street)
2  Names = set (firstnames)
3  For each pdffile:
4    Process pdffile:
5      generate a csv file containing bounding boxes with coordinates and fontsizes
6    Boxtext = text contained in text boxes from pdffile
7    Tokens = set (tokenize (Boxtext))
8    Streetsinpdf = Tokens.intersection (Streets)
9    Namesinpdf = Tokens.intersection (Names)
10   For term in Tokens:
11     If term in Streetsinpdf: anonymize as STREET
12     If term contains RE "postcode Town": anonymize as TOWN
13     If term contains RE "Swiss car registration number": anonymize as CH00
14     If term in Namesinpdf: find the last name and anonymize as NAME
15     If term is a date and older than 16 years: anonymize as DOB
16     If term contains RE for policy no: anonymize as POLICYNUM
```

In the project with the gig work company, the amount of text that had to be anonymized consisted of only a few thousand short text fragments. The data had random IDs and we only had to remove names, dates and, phone numbers from the text and comments written by the employers and the employees. We used regular expressions (REs) and a dictionary of the thousand most common Swiss names from a public website in order to identify the names in the documents. We also searched for expressions starting with "Mr", "Mrs", "Customer" and their equivalents in three languages. However, this approach did not guarantee high quality, which made us develop a more complex solution designed for financial data.

The anonymization algorithm we developed subsequently relies on the existing information sources available in Switzerland. First, we use an address register provided by the Swiss Post [20], out of which we extract valid street names, here called *street*. Second, we use the register of the most common Swiss given names [21], here called *firstnames*. The algorithm is shown in Table 1. It first loads the *streets* and *first names* into sets (lines 1 and 2). It then traverses the document and uses Pdfminer [22] to extract the bounding boxes (line 5). The text extracted from these is turned into a set of words and set intersection delivers the tokens that are to be anonymized (lines 8 and 9). These tokens are then processed by REs in accordance with customer requirements (lines 11–16).

## 3.3 Sentiment Analysis

The analysis of the gig work platform processes and the discussions with the project partners led to the identification of the need for a sentiment analysis of job reviews written by the workers and their employers. As mentioned above, reviews carry a star rating with 1–2 stars seen as "negative" and 3–4 stars as "positive". However, the star ratings are often not consistent with the textual content of the comments [23], as pointed out by the company. This led to an effort to automatically assign a sentiment value, with classes defined differently: "negative" are the statements containing criticism and "other" are the positive or neutral ones, no matter what the star rating was. "Negative" statements are of interest as they point to problems and can help improve the business and ensure higher customer retention.

A summary of the results [23] is shown in Table 2. In prototyping, manually assigned sentiment labels on 963 job reviews were used, 26% of which were labeled "negative". Machine learning was based on Scikit-learn [24]. The highest accuracy of automated sentiment assignment was 86%, with logistic regression and SVM performing similarly (shown in bold). As the data set is unbalanced, we also show the Matthews correlation coefficient [25].

This initial analysis was discussed with our business partner and led to further development, with additional data annotation and improved quality as shown in Table 3. The resulting accuracy of over 90% is acceptable in the business scenario. Figure 3 shows the learning curve that flattens out as more data is being added to the model.

The next step in the project was to position the use of sentiment analysis prediction in the business. The gig work company operates a help desk and a customer care department that monitor daily activity and look at trends. Similarly, sales management regularly reviews the quality. In the project, a sentiment display was prototyped, part of which is shown in Fig. 4.

**Table 2** Sentiment analysis results as reported in [23] with 963 rows of data. SVM is a support vector classifier, k-NN stands for k-nearest neighbors, Tree is a decision tree, LR is logistic regression and RF is random forest. Accuracy, Matthews' coefficient and the F1 measure are shown for these methods

| | SVM C = 1 | k-NN | Tree | LR | RF |
|---|---|---|---|---|---|
| Accuracy | **0.866** | 0.836 | 0.820 | **0.869** | 0.839 |
| Matthews | **0.634** | 0.555 | 0.515 | **0.646** | 0.551 |
| F1 | **0.913** | 0.893 | 0.881 | **0.914** | 0.897 |

**Table 3** Improved sentiment analysis after further manual sentiment scoring, with 2428 rows of data. Logistic regression performs better than the support vector machine (both in bold). The confusion matrices are shown with the target class containing negative comments in the top left corner

| | SVM C = 1 | k-NN | Tree | LR | RF |
|---|---|---|---|---|---|
| Accuracy | **0.906** | 0.870 | 0.862 | **0.920** | 0.867 |
| Matthews | **0.732** | 0.622 | 0.617 | **0.777** | 0.613 |
| F1 | **0.939** | 0.917 | 0.910 | **0.948** | 0.915 |
| Confusion matrix | **[[428, 154], [74, 1772]]** | [[358, 224], [91, 1755]] | [[402, 180], [154, 1692]] | **[[466, 116], [77, 1769]]** | [[353, 229], [93, 1753]] |

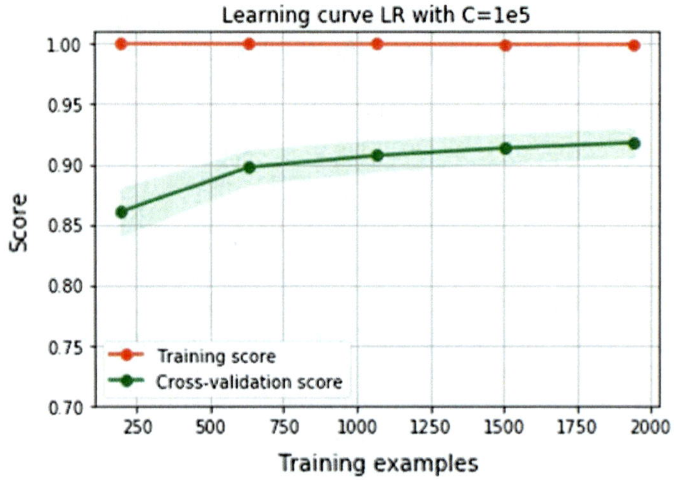


**Fig. 3** The learning curve for logistic regression with the larger dataset shows that adding more data does not improve the results dramatically, but a marginal improvement should still be possible

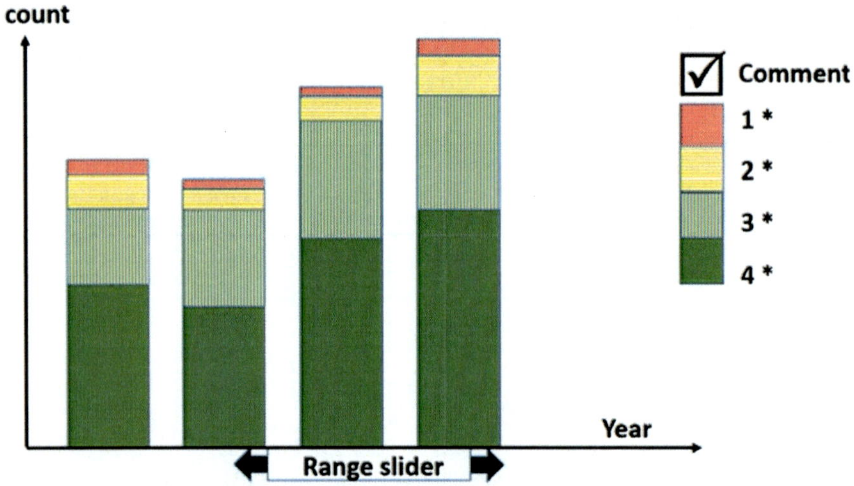


**Fig. 4** A prototype of sentiment visualization showing company sentiment towards workers over time. It is possible to select only the ratings with textual comments (tick box top right), select a time range by using the slider, or view the text details and reclassify the comments (i.e. change negative to other or the other way around, not shown)

Ongoing work is addressing the topics appearing in the sentiment analysis. We are testing the use of biterms [26] to identify topics in multilingual text. The idea is to match the topics that are found automatically with worker and company concerns expressed via other channels such as face to face or helpdesk contact.

### 3.4 *Extraction of Structured Information from Unstructured Text*

Text mining in the financial advice sector (insurance brokerage) supports the development of an Extraction Service (ES) shown in Fig. 2. The service will parse pdf documents from customers and deliver structured data for financial advice brokers. Information extraction is an established research area [15] with recent work including [19]. Our approach builds in particular on the work on document similarity, such as recent CV to job matching done by Xu and Barbosa [27], and builds a domain specific solution consisting of the following elements:

- A knowledge base reflecting the internal company schemas and an external dictionary of the terminology used in the area.
- Document anonymization as shown in Table 1.
- Document sectioning, based on font size and layout.
- Calculating text similarity measures derived from documents and the knowledge base and using them to improve the sectioning and identify features and algorithms that work well in our context.
- Selecting the features to be used in segmentation and matching and using an optimization algorithm to find the best features and algorithms, inspired by results reported by Fua and Hanson [28].
- Refining our solutions to suit the business need and extending them to further document types.

Based on a large set of anonymized documents from our business partner, we are entering the document analysis phase that will identify potential solutions. In addition, a knowledge base capturing company schemas is to be built.

## 4 Conclusions

Our work applies the latest research findings in text mining to business scenarios where innovation is needed. Applied research projects shape a new business reality and help companies capture new markets or tame new technologies. These new developments are part of digital transformation as they change business processes and offload manual data processing to artificial intelligence solutions.

We have performed business process analysis and specified the business requirements. Text mining has already delivered reliable methods for sentiment analysis and document anonymization. We are now working on topic analysis and information extraction and trying out various methods to find the best combinations of text features and data mining algorithms. Our future work will be to tailor the known solutions to the domain of interest and to use machine learning.

For the considered fields of application (sentiment analysis, document anonymization, extraction of structured information) and text mining problems in general, there are many methods available but no standard approach will solve all the problems equally well. Usually, it depends on the specific data which approaches work well enough and which do not. Therefore, a comprehensive evaluation of the methods, the individual preprocessing of data and the fine-tuning of the methods is suggested.

Unfortunately, the limited budget of research projects does not allow for extensive explorations of alternative approaches. In addition, time constraints often suggest developing a solution that is sufficiently efficient, but not necessarily optimal.

**Acknowledgements** Funding was provided by the Swiss Commission for Technology and Innovation CTI (now Innosuisse): 25826.1 PFES-ES and 34604.1 IP-ICT, and by the FHNW.

## References

1. Davidovski, V.: Exponential innovation through digital transformation. In: Proceedings of the 3rd International Conference on Applications in Information Technology, pp. 3–5. ACM, New York, NY, USA (2018). https://doi.org/10.1145/3274856.3274858
2. Ivascu, L., Cirjaliu, B., Draghici, A.: Business model for the university-industry collaboration in open innovation. Procedia Econ. Finan. **39**, 674–678 (2016). https://doi.org/10.1016/S2212-5671(16)30288-X
3. Nambisan, S., Lyytinen, K., Majchrzak, A., Song, M.: Digital innovation management: reinventing innovation management research in a digital world. MIS Q. **41**, (2017). https://doi.org/10.25300/misq/2017/41.1.01
4. Weske, M.: Business Process Management: Concepts, Languages, Architectures. Springer (2012). https://doi.org/10.1007/978-3-642-28616-2_7
5. Van der Aalst, W., Damiani, E.: Processes meet big data: connecting data science with process science. IEEE Trans. Serv. Comput. **8**, 810–819 (2015). https://doi.org/10.1109/tsc.2015.2493732
6. Pustulka-Hunt, E., Telesko, R., Hanne, T.: Gig work business process improvement. In: 2018 6th International Symposium on Computational and Business Intelligence (ISCBI), pp. 10–15 (2018). https://doi.org/10.1109/iscbi.2018.00013
7. Calder, A.: EU GDPR: A Pocket Guide. IT Governance Publishing, Ely, Cambridgeshire, United Kingdom (2018). https://www.jstor.org/stable/j.ctt1cd0mkw
8. Greenwood, B., Burtch, G., Carnahan, S.: Unknowns of the gig-economy. Commun. ACM **60**(7), 27–29 (2017). https://doi.org/10.1145/3097349
9. CNET. Your Next Insurance Agent Will Be a Robot. https://cacm.acm.org/careers/197572-your-next-insurance-agent-will-be-a-robot/fulltext. Accessed April 23, 2019
10. Reich, M., Braasch, T.: Die Revolution der Prozessautomatisierung bei Versicherungsunternehmen: Robotic Process Automation (RPA). In: Handbuch Versicherungsmarketing, pp. 291–305. Springer (2019). https://doi.org/10.1007/978-3-662-57755-4_17
11. smartFix. https://www.insiders-technologies.de/home/products/input-management/general-incoming-mail/smart-fix.html. Accessed 23 April 2019
12. Weiss, S.M., Indurkhya, N., Zhang, T.: Fundamentals of Predictive Text Mining. Springer Publishing Company, Incorporated (2016). https://doi.org/10.1007/978-1-84996-226-1
13. Baeza-Yates, R.A., Ribeiro-Neto, B.: Modern Information Retrieval. Addison-Wesley Longman Publishing Co., Inc, Boston, MA, USA (1999)
14. Kumar, B.S., Ravi, V.: A survey of the applications of text mining in financial domain. Know.-Based Syst. **114**, 128–147 (2016). https://doi.org/10.1016/j.knosys.2016.10.003

15. Baumgartner, R., Flesca, S., Gottlob, G.: Visual web information extraction with lixto. In: Proceedings of the 27th International Conference on Very Large Data Bases, pp. 119–128. Morgan Kaufmann Publishers Inc. (2001)
16. Piazza, F., Strohmeier, S.: Domain-driven data mining in human resource management: a review. In: Proceedings of the 2011 IEEE 11th International Conference on Data Mining Workshops, pp. 458–465. IEEE Computer Society (2011). https://doi.org/10.1109/icdmw.2011.68
17. Ghani, R., Probst, K., Liu, Y., Krema, M., Fano, A.: Text mining for product attribute extraction. SIGKDD Explor. Newsl. **8**, 41–48 (2006). https://doi.org/10.1145/1147234.1147241
18. Burdick, D., Hernández, M.A., Ho, C.T.H., Koutrika, G., Krishnamurthy, R., Popa, L., Stanoi, I., Vaithyanathan, S., Das, S.R.: Extracting, linking and integrating data from public sources: a financial case study. IEEE Data Eng. Bull. **34**, 60–67 (2011). https://doi.org/10.2139/ssrn.2666384
19. Staar, P.W.J., Dolfi, M., Auer, C., Bekas, C.: Corpus conversion service: a machine learning platform to ingest documents at scale. In: Proceedings of the 24th ACM SIGKDD International Conference on Knowledge Discovery and Data Mining, pp. 774–782. ACM (2018). https://doi.org/10.1145/3219819.3219834
20. Die Post. https://www.post.ch/de/geschaeftlich/themen-a-z/adressen-pflegen-und-geodaten-nutzen/adress-und-geodaten. Accessed 11 April 2019
21. Vornamen in der Schweiz. Vornamen der Bevölkerung nach Geschlecht, Schweiz, 2017. BFS-Nummer: su-t-01.04.00.12. Accessed 23 April 2019
22. Pdfminer. Yusuke Shinyama. Pdfminer is a tool for extracting information from PDF documents. https://github.com/euske/pdfminer. Accessed 23 April 2019
23. Pustulka-Hunt, E., Hanne, T., Blumer, E., Frieder, M.: Multilingual sentiment analysis for a swiss gig. In: 2018 6th International Symposium on Computational and Business Intelligence (ISCBI), pp. 94–98 (2018). https://doi.org/10.1109/iscbi.2018.00028
24. Avila, J.: Scikit-learn cookbook: over 80 recipes for machine learning in python with scikit-learn. Packt Publishing, Birmingham, UK (2017). https://doi.org/10.1214/009053604000000067
25. Matthews, B.W.: Comparison of the predicted and observed secondary structure of T4 phage lysozyme. Biochim. et Biophys. Acta (BBA)-Protein Structure **405**, 442–451 (1975)
26. Yan, X., Guo, J., Lan, Y., Cheng, X.: A biterm topic model for short texts. In: Proceedings of the 22nd International Conference on World Wide Web, pp. 1445–1456. ACM (2013). https://doi.org/10.1145/2488388.2488514
27. Xu, P., Barbosa, D.: Matching résumés to job descriptions with stacked models. In: Advances in Artificial Intelligence: 31st Canadian Conference on Artificial Intelligence, Canadian AI 2018, Toronto, ON, Canada, May 8–11, 2018, Proceedings 31, pp. 304–309. Springer (2018). https://doi.org/10.1007/978-3-319-89656-4_31
28. Fua, P., Hanson, A.: An optimization framework for feature extraction. Mach. Vis. Appl. **4**, 59–87 (1991)

# Using Mobile Sensing on Smartphones for the Management of Daily Life Tasks

**Dilip Menon, Safak Korkut, Terry Inglese, and Rolf Dornberger**

**Abstract** Today, all smartphones contain a variety of embedded sensors capable of monitoring and measuring relevant physical qualities and quantities, such as light or noise intensity, rotation and acceleration, magnetic field, humidity, etc. Combining data from these different sensors and deriving new practical information is the way to enhance the capabilities of such sensors, known as sensor fusion or multi-modal sensing. However, the authors hypothesize that the sensing technology that is embedded in smartphones may also support daily life task management. Because one of the biggest challenges in mobile sensing on smartphones is the lack of appropriate unified data analysis models and common software toolkits, the authors have developed a prototype for a mobile sensing architecture, called Sensing Things Done (STD). With this prototype, by applying multimodal sensing and gathering sensor data from performing a specific set of tasks, the authors were able to conduct a feasibility study to investigate the hypothesis set above. Having examined to what extent the task-related activities could be detected automatically by using sensors of a standard smartphone, the authors of this chapter describe the conducted study and provide derived recommendations.

**Keywords** Mobile sensing · Task management · Smartphone sensors · Automatic task detection · Daily life tasks

## 1 Introduction

With the fast development of mobile hardware and software, smartphones are becoming increasingly more powerful and "smarter". Nowadays, smartphones are more than a communication tool for their users. In fact, they are life-dominating multi-functional tools that offer not only a camera and a media player, but also a calendar, e-mail, a fitness tracker and, most importantly, access to the Internet. In

D. Menon · S. Korkut (✉) · T. Inglese · R. Dornberger
School of Business, Institute for Information Systems, FHNW University of Applied Sciences and Arts Northwestern Switzerland, Peter Merian-Strasse 86, 4002 Basel, Switzerland
e-mail: safak.korkut@fhnw.ch



R. Dornberger (ed.), *New Trends in Business Information Systems and Technology*, Studies in Systems, Decision and Control 294,
https://doi.org/10.1007/978-3-030-48332-6_5

addition, smartphones not only respond to their users' wishes and needs, but they also sense and act in a sophisticated way to support users. In fact, the driver that makes data smarter for this application area is called *sensing technology*. However, the challenge of mobile sensing technology is to identify and retrieve the right sensor data and appropriately process and analyze that data using efficient methods [1].

In practice, it remains unclear, to what extent the standard sensing capability of a conventional smartphone would be sufficient and feasible to offer its user additional benefits. At the end of the first decade of the 21st century, an attempt was made to extend smartphones to Personal Digital Assistants [2] to support professionals, e.g., by reminding them of appointments or assisting them with navigation. Now, at the end of the second decade of the 21st century, users' expectations of managing their daily tasks are much more complex. The calendars can be overloaded, appointments have to be shared with several participants, and multiple activities have to be organized and tracked at once. It seems only logical to use a smartphone that can continuously collect data sensing its user's behavior, process this data into activity-related information, and validate the derived information with the tasks to be performed within a demanding task management application.

This chapter is divided into four main parts. Section 2 describes the capabilities of sensor technologies and their potential applications for daily task management. Section 3 discusses a mobile sensing architecture that enables such a sensor-supported task management system. In Sect. 4, the authors introduce the developed Sensing Things Done (STD) prototype and the conducted feasibility study [3] that examined to what extent task-related activities could be detected automatically using standard smartphone sensors. Further, Sect. 5 describes the experiments performed within the said feasibility study contextualizing the analyzed data. Finally, the authors discuss the results and further research applications.

### 1.1 Relevance Within and for Mobile Sensing Systems

The present study, in spite of its limitations, fits well within the new stream of research of the so-defined intelligent positive computing services, such as those proposed by Lee et al. [4], which inevitably implies a holistic perspective, including a wide range of disciplines requiring a cross-disciplinary collaboration among computing, design, human computer interaction, and psychology. In fact, according to Lee et al., one of the main components of mobile sensing systems is the primary agency for behavioral change in human beings [4].

Laport-Lopez et al. [5] differentiate between two categories: People-centric and environmental-centric. They highlight the still open areas in which interdisciplinary mobile sensing systems can be explored: Privacy, formal argumentation, scalability in crowded environments, and the use of devices that are able to detect and recognize emotional information.

In this chapter, the authors share an approach to how mobile sensing systems can be designed and how data can help drive changes in daily behavior.

## 2 Sensor Technologies and Capabilities

Over the last two decades, smartphone-suitable sensors have become smaller and more accurate while using less power [6]. Today, smartphones contain a variety of embedded sensors capable of monitoring and measuring all relevant physical quantities such as light or noise intensity, motion, e.g., rotation and acceleration, magnetic fields, humidity, etc. Combining data from different sensors and deriving new information is the way to enhance the capabilities of sensors—this is known as *sensor fusion* or *multimodal sensing* [7]. Furthermore, smartphones use open interfaces to external web services to further process the aggregated data. For example, the combination of *location sensor* data (e.g., GPS data) with *location services* (e.g., map services) allows tracking and providing location-based information, such as recently visited places. By combining different *motion sensors*, also partially extended with external services, smartphones can gather information about the user's physical activities, such as counting steps or calculating the speed of transportation (also where GPS is not available). *Environment sensors,* such as ambient light sensors, air pressure or humidity sensors, capture information from the user's environment. Further, the most common smartphone sensors and their usage are briefly categorized with respect to positioning, movement, environment, and other aspects.

*Location and position sensors* are the primary source for determining the location of a smartphone. Outdoors, using the systems of GPS, GLONASS, Galileo, or BeiDou, a triangulation with satellites offers a positioning accuracy of a few meters. Based on location-based information from the telecom cell towers and Wi-Fi hotspots, this generally also works indoors, and increases the accuracy of the positioning to centimetres. Connecting a smartphone via other communication means, such as Bluetooth or Near Field Control (NFC) sensors, increases the accuracy of the local positioning even further. Moreover, magnetic field sensors measure the Earth's geomagnetic field determining the orientation of the device relative to the magnetic North [8]. The rotation of the mobile device is computed by using the angular rate of the mobile device's gyroscope on the three physical axes (vertical, lateral, and longitudinal) [8, 9]. Other types of sensors, providing a type of local positioning, are the proximity and distance sensors. To trigger context-based activities, they help detect whether the smartphone is being held near the user's face. For this purpose, either infrared or capacitive proximity, or even light sensors are used.

*Motion sensors* detect physical movements. For example, by measuring the acceleration of the device relative to the earth's gravitational force, the accelerometer can be used to detect its local orientation and—sometimes directly, sometimes with particular post-processing—also the tilt, shake, and rotation or swing actions of the device [8]. In addition, the previously mentioned gyroscope often acts as an additional motion sensor that measures the change (or the rate of change) in translation or rotation in or around the three physical axes.

*Virtual sensors* are no physical hardware sensors, but rather software sensors. They can mimic real sensors by using data from one or more hardware sensors, combining and enriching the sensed data to provide specific new information. For example, the virtual sensor *linear acceleration* uses the information of the accelerometer on all three physical axes and deducts the force of gravity to get pure device acceleration without gravitational effects. Thus, such virtual sensors can estimate significant (extraordinary) motions that serve as tilt detectors and step counters with higher precision compared to the motion sensors alone [8].

*Environment sensors* scan the surrounding environment of the device like the ambient temperature, light, etc. Sometimes smartphones have an embedded pressure sensor, which measures the barometric pressure used to determine the altitude and measure the change in altitude, e.g., to determine the change in the floor level of a person walking up or down the stairs inside a building [10]. Finally, a humidity sensor measures the relative ambient humidity like a hygrometer [11].

*Other sensors* or sensor-like components are the microphone, camera, touchscreen, fingerprint sensor, etc. In fact, microphones can record and identify the sound and noise of the environment, as well as help recognize the voice of the speaker or the (known) conversation partner [12]. The camera not only performs digital image processing (including face recognition) and video recording, but also detects ambient light. Touch sensors perceive the user input (touchpoint) on the screen and identify a fingerprint with a particular type of touch sensor. However, touch sensors are not considered in this research.

To further elaborate on the variety of mobile sensors, the authors present a mobile sensing architecture in the next section.

## 3 Mobile Sensing Architecture

Mobile sensing architecture provides three layers respectively phases, where sensors collect, process, and finally present the data—in this case as activity information to the smartphone users. Lane et al. [6] characterize these three phases as *sense, analyze/learn* and *inform/share/persuade*, which the authors further explain below.

### *3.1 Sensing Phase*

Modern smartphones provide a variety of hardware sensors able to receive raw data of physical quantities. In the sensing phase, the most common activity-tracking smartphone sensors are microphones, GPS, and accelerometers. It is possible to improve the activity detection by combining data from these and other sensors (i.e. sensor fusion). Sensing takes place either at certain events or continuously. However, continuous and/or concurrent sensing has a huge impact on the energy consumption of the smartphone and its battery life [13]. Additionally, the sensing of data may be more difficult to perform due to external conditions, e.g., background noise, and the possible context change, while switching between different apps for instance.

### 3.2 *Analyzing and Learning Phase*

In the analyzing and learning phase, the biggest challenge is that there is no unique approach to meaningful processing of activity detection data. Raw data from the sensors are only worth something if it is possible to turn them into correct information and additional knowledge for the user—here, to detect and manage user tasks. These relations from data to information are often referred to as the DIKW Pyramid [14], which stands for data, information, knowledge, and wisdom (where the generation of wisdom is not considered). It is possible to use approaches such as statistical inference and data mining to learn from data. In the past, various established data mining algorithms have been used (mostly validated with given test scenarios); however, recently, methods belonging to machine learning and other artificial intelligence approaches are increasingly applied to improve the quality of activity detection. Thus, it is possible to speed up the learning phase and find unknown patterns. In addition, with the potential of big data, these methods are becoming prevalent in research and industry [15]. When processing the sensor data in a particular way, it is possible to achieve context modeling. For example, assessing sound patterns recorded by the microphone, the detected noise surrounding the user can help predict the environment (like other people, places, and vehicles). However, such CPU-intensive calculations in real time limit the smartphone's battery life [6].

### 3.3 *Informing/Sharing and Persuading Phase*

The informing/sharing and persuading phase allows activity feedback to generate awareness of the information brought to the users through the mobile sensing application. This information can be either individual statistics or social- and community-based comparisons. Mobile sensing applications often have the potential to change the user's behavior in a positive way. However, as mobile sensing applications generally track and process a huge amount of data (including sensitive information), secure and sensible data-handling is highly relevant.

### 3.4 *Sensing Scale in Mobile Sensing Architecture*

Lane et al. [6] introduced three different types of sensing scales, which can be mapped to the mobile sensing architecture.

1. Applications that focus on *personal sensing*, designed to process data of a single individual, such as tracking particular exercise activities or personal itineraries.
2. *Group sensing* applications collect data from two or more people who share a common goal or interest, e.g., when two people meet in real time.

3. *Community crowd sensing* is the next step to leverage the power of sensor data in general to common knowledge. Combining many people's data into a big data collection can provide a comprehensive overview of certain aspects of knowledge. For example, to see how much the local residents were exposed to pollution, Predic et al. [16] combined the air quality data collected from sensor networks of cities with people's activity data collected by smartphones. Further, Rana et al. [17] used the microphones of all involved smartphones as the sensors to monitor the environment for noise pollution.

# 4 Feasibility Study of Applying Mobile Sensing for Task Management

In addition to the lack of appropriate unified data analysis models and common software toolkits, due to privacy issues, the possibilities of sharing and accessing public data repositories to test and train the software are very limited. To overcome these limitations, the authors created a feasibility study [3] designing a smartphone prototype that derives sensor data from performing particular tasks and processes it through multimodal sensing. The authors explain the study in further detail in the following sections.

## *4.1 Requirements of the Mobile Sensing Prototype for Detecting Tasks*

The developed mobile sensing prototype provides an easy-to-use daily task management system via to-do lists, using smartphone sensors, and a mobile sensing architecture. Thus, the activity information from the multimodal smartphone sensor data is processed in order to recognize and inform the user about the tasks performed (as daily achievements) and assist in managing the remaining tasks.

With this daily task management system, the users receive real-time feedback that is comparable to their perception, thus, the intention is to create awareness and motivate the users, reducing the cognitive load of managing tasks alone [18]. To use mobile sensing for tracking activities relevant for daily task management throughout the day, two factors determine the relevance: (1) *frequency* of activities and (2) *locations* where the persons spend time to perform these activities.

Thus, the identified set of detectable relevant daily activities contains being at work, studying, performing household tasks (e.g., vacuum cleaning), shopping as well as performing physical activities like running, walking, and sleeping.

Further, the authors also analyzed the potential for automatic detection of these activities; and automatic or at least semi-automatic detection (in the sense that post-processing detects the task) is possible by sensing the following aspects:

- Detect the *location*, where only unique tasks can be performed most likely, e.g., work takes place most likely at the office.
- Identify *physical activities* of the user that are unique to a particular task, e.g., running provides a particular sensor data pattern distinguishable from all other activities.
- Sense the *environment,* and leverage the activity detection, e.g., analyzing the microphone data to detect an operating TV and assume that the person is watching TV.

Furthermore, following Bulling et al. [19] who defined particular characteristics of *human activity recognition*, these requirements are also adapted for the development of the prototype:

- *Generalization* means that the activity recognition must work independently from the user, thus, it is important to track the smartphone and the user-performed activities, not the persons themselves.
- *Recognition* states that the sensor data streams are continuous but activity recognition analyses are possible in batches of records (i.e. snapshots, e.g., segmented per minute).
- *Execution* prevents activity recognition analyses from running continuously. As the prototype is for demonstration purposes only, it is also possible to perform the analyses with a time shift or even offline so as not to be limited by the computing power of the smartphone.

As a prerequisite for the prototype, the tracked sensor data does not need to be processed immediately, but only in batches of records and with a time shift. Despite the ubiquity of the modern mobile devices, their computing capacity could be considered a limitation on a larger scale of task detection [5].

Additionally, the feasibility study assumes that, even with standard smartphones and common data analytics methods and tools, the management of daily tasks can be enabled (respectively improved) by additionally using smartphone sensor data. In addition, the aspect of user privacy could become a problem in a real-life application scenario, where a possible solution could be to keep the obtained data on a local system without sending information to the server. However, for this feasibility study, the participants willingly shared their data [5].

## 4.2 Implementation of the Prototype

Based on the above described mobile sensing architecture, the Sensing Things Done (STD) prototype supports processes that use mobile sensing as described by Lane et al. [6]: Two continuously running process cycles are responsible for sensing the sensor data (sense cycle) and analyzing the sensor data for activity detection (analysis cycle). The third process (inform) shows the user the detected activities associated with the intended tasks.

Following the requirements, the first STD prototype version uses different common software components (web services), such as the habit tracker app Habitica API [20], the SensingKit—a multi-platform mobile sensing framework [21], Google Play Service—a proprietary background service for Android devices [22], the Weka Machine Learning software for Java [23], and Android OS. In addition, it is not their origin or their providers that are important, but their functionality. Thus, other services providing the same functionality can be used instead of the mentioned ones.

## *4.3 Architecture of the STD Prototype*

Figure 1 shows the architecture of the STD prototype and how the different components relate to each other, including the data flow. The blue parts are specially encoded for the STD, while the other components are generally available services, which were used with the STD. Here, Habitica Task Management and Google Play Services are the only external services that the STD accesses online. All other components are implemented on the smartphone, where the SensingKit retrieves the raw sensor data.

### 4.3.1 Database Component—The Sensing Things Done STD DB

The prototype uses a single *database*, called STD DB in Fig. 1, to store all persistent data: User configurations regarding tasks and locations as well as the sensor data records retrieved from the smartphone sensors, and the (pre-)analyzed data.

Particular task-specific information needs to be configured when initializing the prototype. Each task that needs to be detected is related to its potential locations. Additionally, each task is described by generic activity patterns, which are related

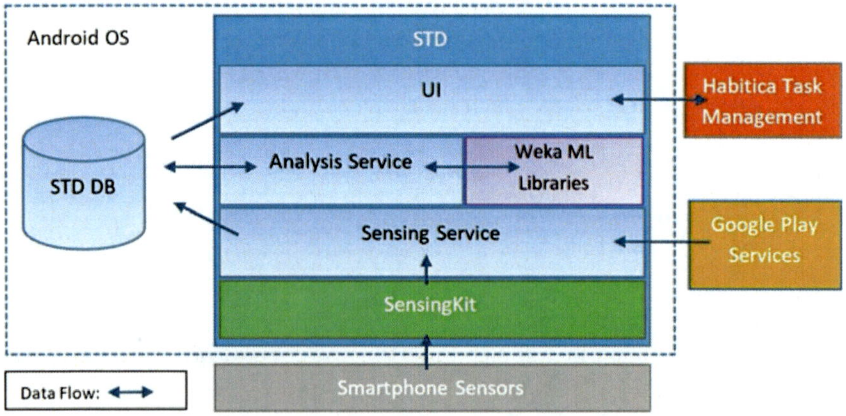


**Fig. 1** Architecture of the Sensing Things Done (STD) prototype: component view

to possible sensor data matching these patterns, e.g., vacuum cleaning involves slow movements within a limited area. Further, the tasks that are prone to particular environmental aspects are described with additional detectable characteristics, such as particularly loud background noise during vacuum cleaning.

### 4.3.2 Components Related to the Sensing Phase

The *sensing service* component captures the sensor data provided by SensingKit, which collects and stores the raw sensor data (see Fig. 1), allowing the analyses of time shifted (offline) data. This Sensing Service also enriches the sensor data with information from external background services, e.g., Google Play Service.

*Geo-fences* are virtual representations of geographic areas in the real world and geolocation services can detect the current location of a smartphone. For example, in the STD (see Fig. 1), the Google Play Service is used to provide the location of the smartphone with high accuracy—by choosing the best combination of location sensing via sensor fusion, e.g., GPS, Wi-Fi, etc. When activated, the sensing application predicts a location with a given radius around the exact coordinates (called geo-fence). For the prototype, an outdoor geo-fence area with a diameter of 50 m was tested as a sufficient upper limit, while with an optimal Wi-Fi connection indoors, an accuracy of 5 m and less was applied. Google Play Service sends a geo-fence notification to the STD, which stores the visited location along with all additional available information (like additionally predicted activities processing the motion sensor data) locally in its database.

Since the requirements of the prototype allow the sensor data from the smartphone sensors (see Fig. 1) to be captured in periodic intervals (by SensingKit) and stored in the database as sets of snapshots (by Sensing Service), the reason why data is not continuously captured and stored is, firstly, because the sensing interval is much higher than the information needed. Secondly, there is no guarantee that the smartphone sensors will be fully synchronous, because they often have different sampling rates, up to firing only once when values have changed (e.g., the light sensor). Thus, data synchronization and consolidation is necessary, for example, by taking these snapshots every five seconds to ensure good granularity that allows just enough snapshots to determine an activity at a specific moment.

### 4.3.3 Components Related to the Analyzing and Learning Phase

In the STD, the authors have developed a so-called Analyzing Service. It retrieves the geo-fence-enriched sensor data records from the database and, before it writes back the processed information, it filters out the data records that seem to be inaccurate. Furthermore, the geo-fencing service delivers ambivalent outcomes, resulting in too large a confidence range to accurately locate the smartphone and relate it to a particular task. If the confidence range is too large, the potential number of false positives increases too much.

Using the Google Play Service in the STD provides a kind of an activity analyzer working in a similar way as the geo-fencing analyzer. Based on the sensed data, the activity analyzer tries to recognize particular activities, such as step counting. This information is also delivered back to the Sensing Service, which then writes this information with all other data into the database.

This way, the Analyzing Service receives all this information (filtered sensor data records, processed location information (geo-fencing), and pre-analyzed activity information) and processes it with the help of a specific machine learning component, here, the Weka Machine Learning suite [24], which, in turn, provides a library of machine learning algorithms. In the STD, this tool is used for the identification of specific activity patterns.

### 4.3.4 Components Related to the Information/Sharing and Persuading Phase

Finally, the prototype performs the task detection on demand, when the user opens the review screen in the user interface (UI). In general, the steps performed in the UI component are:

1. Retrieval of all current open tasks from the task management system. Here, the Habitica Task Management service is connected to the UI component.
2. Evaluation of the match between the characteristics of this task and the sensor data, the retrieved location, and the processed activity pattern (for each task).

As a result, the STD assumes the complexity of the various steps until it provides mobile-sensing-supported task management to the user. Raw sensor data (motion, noise, etc.) with different sampling rates are retrieved, bundled (by the SensingKit), and further enriched with geolocation information as well as partly pre-processed (by Google Play Service) geo-fencing information and activity pre-detection. All data is stored in the database to study the feasibility regardless of the smartphone's computing power. By using machine learning (by Weka) the activity detection is assigned to a task detection. A task management tool (Habitica) controls the task list for completion, while the STD combines all the information and matches the sensed activities to the outstanding tasks.

## 5 Experiments in the Feasibility Study for Mobile Sensing

In order to investigate to what extent task-related activities are detectable automatically by using smartphone sensors, the authors evaluated the STD prototype with mixed evaluation methods of testing and descriptive scenarios. Four test persons used the prototype in given everyday scenarios, which are to be detected as part of their tasks. This feasibility study for task detection by mobile sensing used standard (not high-end) smartphones—an LG Nexus 5 and a Sony Xperia Z5 Compact with

Android 5 and 6—and tested different aspects involving mobile sensing for activity or location detection [3].

## 5.1 Location and Motion Detection Using Standard Sensing Services

### 5.1.1 Location Detection

Many tasks and therefore the underlying activities performed by humans, are primarily detectable by identifying the location of the activity and/or the speed of motion of a particular activity. This can be performed well by using external standard sensing services, such as the Google Play Service.

First, the authors tested the accuracy of the location detection, determining its so-called confidence radius. For this purpose two pre-configured locations were selected using the research scenario described by Stikic et al. [25]. Because of the locations where the experiments took place (generally Wi-Fi-equipped indoor locations in buildings), the confidence radius of the geo-fence locating of smartphones ranged between less than four meters to a maximum of 20 m. Consequently, task-based activities performed in specific indoor locations more than 20 m away from each other could be detected by only using the sensed location. The authors conducted a similar experiment to determine the confidence radius in outdoor scenarios to distinguish particular tasks only by the location of the performed activity.

### 5.1.2 Motion Detection and Step Counting

It is possible to clearly detect some activities because they are explicitly related to specific activity patterns. Such activities are walking and running, where the steps taken might be well identified and pre-processed by external smartphone services, such as step counting via Google Play Service.

In the experiment, the authors tested the capability of counting steps by comparing the values of a hardware step counter (a Fitbit device) and the smartphone step counter using mobile sensing. Different users walked from one location to another forth and back at a 400-m linear distance at a speed of around 5 km/h. Each change of a geo-fence state was captured by noting the time, the current coordinates, and the location detection accuracy using the geo-fence service by Google Play Service. The step count with the Fitbit was 1283 steps, whereas with the smartphone step counter it averaged 1311 steps (+2.2%). Extending the walking distance with some detours (e.g., up and down stairs) the step count was 1689 steps with the Fitbit and 1711 steps with the smartphone (+1.3%).

These tests showed that the external mobile step sensing and counting services including geo-fences were very accurate in the given scenarios. Motion detection and step counting were used to predict and distinguish particular tasks, e.g., walking and running from bike cycling. The confidence radius of localizing the activity was higher inside (with an acceptable range between four and 20 m, due to Wi-Fi sensing) than outside (using only GPS, telecom). Additionally, performing tests at different speeds of movement showed that the confidence radius did not necessarily increase with the speed of the movement.

Furthermore, the test scenario was extended by making the walking patterns more complex: Walking to the train station, waiting for and then taking the train. Once again, the mobile sensing step counter showed similar results compared to the Fitbit tracker (+2.2% is still the maximum deviation).

## 5.2 *Detecting Particular Tasks Using Mobile Sensing*

Task management is facilitated when certain tasks contain unique activities and/or unique sets of properties of these activities (e.g., distinct location and speed of movement). Using location detection to identify a particular location or at least a particular geo-fence location such as outdoor or indoor, many activities can already be clearly assigned to particular tasks.

### 5.2.1 Activities in Outdoor Scenarios

The authors chose four outdoor activities to be detected by mobile sensing, namely walking, running, biking, and resting. These activities were distinguished by their speed of motion: *resting* has a speed of motion of 0 km/h, *walking*—of about 5 km/h (> 0 to 6 km/h), *running*—around 8 km/h (6 to 10 km/h), and *biking*—around 15 km/h (10 to 20 km/h).

In the experiment, the activity recognition by external services successfully detected the four activities in different scenarios identifying exactly which motion activities the humans were performing. Except, the sensed motion data for the resting phase sometimes did not deliver a speed of 0 km/h, but slightly more. To solve this problem, a threshold for *resting* had to be set at about 0.1 km/h.

### 5.2.2 Activities in Indoor Scenarios

In this scenario, the four tested activities were *sitting*, *walking around in buildings*, *vacuum cleaning*, and *sleeping*. Because external activity detection services did not provide a sufficient activity recognition to distinguish these four activities alone, the STD component Activity Analysis was additionally trained to identify these activity patterns using the Weka machine-learning component. Furthermore, in order to distinguish the activities with zero motion speed (i.e. sitting and sleeping), sensor fusion was used to enrich the location detection with ambient light detection; and, in order to distinguish vacuum cleaning from walking and sitting, sensor fusion was used to enrich the motion detection with ambient noise detection.

To identify these indoor activities, several test runs were re-done using different smartphones. Different training and classifier methods belonging to machine learning were tested [3] (not examined in this chapter). The results showed that, with the given setup of using three detection methods in parallel (location, Google activity, and Weka analyzer with Naïve-Bayes method) an acceptable output was reached, giving immediate feedback to the user. However, quite often the trained models failed to recognize the vacuum cleaning.

### 5.2.3 Detecting Activities and Related Tasks in a Day Scenario

The day scenario included these four activities: (1) Getting up in the morning; (2) Commuting by bike and train to work; (3) Buying bread from the configured location after work, and (4) Returning home.

In the performed experiments, these activities (respectively tasks) were detected sufficiently well (i.e. the performed activities more or less corresponded to the tracked ones) with only some minor deviations: The data from the wrist-worn step counter differed by around 15% due to the fact that the mobile was not always carried with the experiment participant. During biking, the STD tracked 5 min, while the user was on a cycle for around 6 min.

## *5.3 Managing Tasks by Sensing Locations and Activities*

In another experiment, the authors created particular tasks in the Habitica task management system and mapped them to activities and/or locations as shown in Table 1, before the user performed the related activities. Additionally, a particular value range was assigned in order to characterize the task as fulfilled.

The task list contained more entries than the user could accomplish in a day. This was intended for crosschecking in case false positives were detected. All users carried their phones in their pockets. The performance or non-performance of a task was detected either correctly or marked as a wrong detection. Some activities (e.g., "Buy milk") were not detected as the supermarket was quite big and the geo-fence was not touching the visited area. Additionally, the first minute of cycling was not detected. The cycling activity was inaccurate, probably due to a very slow speed, so that even the step goal was not reached. The step counter of the Fitbit tracker and the STD showed a huge difference that could not be explained compared to other results.

**Table 1** Managing tasks by sensing locations and activities

| Task list | Assigned motion activity | Assigned location | Value ranges |
|---|---|---|---|
| Buy milk | Rest | Supermarket location A | Performed |
| Buy beer | Rest | Supermarket location B | Performed |
| Cycling | Biking | Outdoors | >60 min |
| Daily walk | Walking | Outdoors | >3000 steps |
| Jogging | Running | Outdoors | >20 min |
| Sleep enough | Rest | Home location C | >420 min |
| Check habitica | Rest | Indoors | Performed |
| Talk to customer | Rest | Office location D | Performed |

## 6 Summary and Conclusions

People have to perform many tasks every day, while they usually carry their smartphones (or other smart devices). The key question in this research was: "Why not use the smartphones to support managing the tasks to be performed?" The smartphones contain various embedded sensors to sense different physical variables such as location, acceleration, ambient light, and noise. Using these sensor data—fused and pre-processed by connected external services and assessed by machine-learning algorithms—various activities performed by the smartphone users can be detected and quantified such as the speed of movement, number of steps, exact location with environmental factors, e.g., ambient light and noise.

The idea behind this research was to use the capabilities of mobile sensing to detect specific human activities and match them with the tasks on a given task list. The authors created the Sensing Things Done (STD) prototype to study to which extent task-related activities are detected by mobile sensing using standard smartphones. The developed prototype proposed a general solution, by designing a system, using smartphone sensors to detect the person's daily activities, and providing the evaluated activity information to a task management system.

It was shown that many different activities and related locations could be detected by the smartphone, such as—in outdoor scenarios—resting, walking, running, and cycling as well as—in indoor scenarios—resting, walking around, vacuum cleaning, and sleeping. These activities were clearly mapped, as far as possible, to tasks such as *daily walk*, *jogging*, but also *vacuum cleaning* (task and activity were the same), *buying milk* or *talking to a customer*.

The described approach worked on the level of a feasibility study, assuming that it was possible to map the activities uniquely to specific tasks or at least to a unique set of activities and additional sensed information (like ambient light or noise). The developed prototype thus made it possible to combine synergies from various existing systems (hardware sensors, external smartphone services, integrated machine-learning components etc.).

Testing the feasibility, the three designed and conducted experiments showed the quality of sensed data and processed activity and location detection, as well as the general ability to identify a variety of human activities just while they are carrying their smartphones.

Hence, the feasibility study showed the following positive aspects to confirm the underlying principles by using the STD prototype:

1. Good accuracy in tracking general (mostly physical) activities,
2. use of geo-fences allowing easy configuration of location-based tasks,
3. accomplishing detected tasks by easily providing the users with a sense of accomplishment, and
4. the modularity of the prototype combining external smartphone services with integrated machine learning components (reuse in future research).

However, the weaknesses of the feasibility study and the STD were:

1. Users carried smartphones differently, which affected the detection of particular activities, such as sleeping, and they could not effectively be detected if the smartphone was placed in different places or even outside the bedroom.
2. Particular activities, such as vacuum cleaning, needed extensive training data (with different persons and different vacuum cleaners) to be distinguished from resting or walking around.
3. It was still necessary to manually assign activities to tasks before an automatic detection would be possible, and is known as participatory sensing [6].

Finally, the feasibility study showed that the principles (implemented in the STD prototype) could be used for tracking locations and activities and thus detecting accomplished tasks. The area of daily task management, combined with mobile sensing, worked and showed great potential. The benefit of mobile sensing for daily activity recognition has been confirmed. In addition, using smartphones (or other smart devices) for activity detection has great potential as well and due to the rapid evolution of their "smartness" (sensor quality, computing power, interconnectivity, etc.) even more so in the future. Furthermore, the principles and the modularity of the prototype could be extended and adapted for future research.

## 7 Future Research

This research was primarily inspired by the fact that mobile sensing is not sufficiently researched and applied in the area of personal productivity management. There is a need to address the phenomenon of task overload by automatically assisting people with task management. While the authors have evaluated the prototype with some test users on given short-term scenarios, the effects on user acceptance, user experience and long-term benefits have to be tested and statistically evaluated in long-term studies with larger test groups. This can provide further insights into improving task management from a user's perspective.

With respect to mobile sensing, the evaluation of the STD showed that activity recognition for certain activities could be improved by researching new combinations of machine-learning methods and tools to more efficiently learn particular activity patterns. The STD is only a research prototype. With the recent rise of the popularity of wearable tech, the porting of STD into a wearable system could lead to new possibilities potentially diminishing the differences among the activity patterns of different users and their perceived locations.

A restraining factor for doing such intensive analyses on the smartphone might still be the energy consumption, thus using cloud computing for data analytics might be a better option. Great potential lies in leveraging the data collected by the use of external systems. By using this data, one can possibly add more context and accurate information about the activities of the person. Nevertheless, a final word must be devoted to the privacy of users and how to resolve the question of how to ensure the security of the analyzed data and that only authorized users can access them.

## References

1. Baesens, B.: Analytics in a Big Data World: The Essential Guide to Data Science and Its Applications (Wiley and SAS Business Series). Wiley, Hoboken, New Jersey (2014)
2. Time.com: First smartphone IBM Simon. http://time.com/3137005/first-smartphone-ibm-simon/
3. Menon, D.: Analysis of Smartphone-Ambient Data to Manage Daily Life Tasks. Unpublished Master Thesis, Olten (2016)
4. Lee, U., Han, K., Cho, H., Chung, K.-M., Hong, H., Lee, S.-J., Noh, Y., Park, S., Carroll, J.M.: Intelligent positive computing with mobile, wearable, and IoT devices: literature review and research directions. Ad Hoc Netw. **83**, 8–24 (2019). https://doi.org/10.1016/J.ADHOC.2018.08.021
5. Laport-López, F., Serrano, E., Bajo, J., Campbell, A.T.: A review of mobile sensing systems, applications, and opportunities. Knowl. Inf. Syst. 1–30 (2019). https://doi.org/10.1007/s10115-019-01346-1
6. Lane, N.D., Miluzzo, E., Lu, H., Peebles, D., Choudhury, T., Campbell, A.T.: A survey of mobile phone sensing. IEEE Commun. Mag. **48**(9), 140–150 (2010)
7. Ayub, S., Bahraminasab, A., Honary, B.: A sensor fusion method for smart phone orientation estimation. In: 13th Annual Post Graduate Symposium on the Convergence of Telecommunications, Networking and Broadcasting. Liverpool (2012)
8. Google: Android Sensor Overview. http://developer.android.com/guide/topics/sensors/sensors_overview.html
9. Allan, A.: Basic Sensors in IOS: Programming the Accelerometer, Gyroscope, and More. O'Reilly Media Inc, Sebastopol, California (2011)
10. Liu, G., Iwai, M., Tobe, Y., Matekenya, D., Hossain, K.M.A., Ito, M., Sezaki, K.: Beyond horizontal location context: measuring elevation using smartphone's barometer. In: Proceedings of the 2014 ACM International Joint Conference on Pervasive and Ubiquitous Computing: Adjunct Publication, pp. 459–468 (2014)

11. Khan, W.Z., Xiang, Y., Aalsalem, M.Y., Arshad, Q.: Mobile phone sensing systems: a survey. IEEE Commun. Surv. Tutor. **15**(1), 402–427 (2013)
12. Wang, R., Chen, F., Chen, Z., Li, T., Harari, G., Tignor, S., Zhou, X., Ben-Zeev, D., Campbell, A.T.: StudentLife: assessing mental health, academic performance and behavioral trends of college students using smartphones. In: Proceedings of the 2014 ACM International Joint Conference on Pervasive and Ubiquitous Computing, pp. 3–14 (2014)
13. Priyantha, B., Lymberopoulos, D., Liu, J.: Enabling energy efficient continuous sensing on mobile phones with littlerock. In: Proceedings of the 9th ACM/IEEE International Conference on Information Processing in Sensor Networks, pp. 420–421 (2010)
14. Rowley, J.: The wisdom hierarchy: representations of the DIKW hierarchy. J. Inf. Sci. **33**(2), 163–180 (2007)
15. Chen, H., Chiang, R.H.L., Storey, V.C.: Business intelligence and analytics: from big data to big impact. MIS Q. **36**(4), 1165–1188 (2012)
16. Predić, B., Yan, Z., Eberle, J., Stojanovic, D., Aberer, K.: ExposureSense: integrating daily activities with air quality using mobile participatory sensing. In: 2013 IEEE international conference on pervasive computing and communications workshops (PERCOM workshops), pp. 303–305 (2013)
17. Rana, R.K., Chou, C.T., Kanhere, S.S., Bulusu, N., Hu, W.: Ear-phone: an end-to-end participatory urban noise mapping system. In: Proceedings of the 9th ACM/IEEE International Conference on Information Processing in Sensor Networks, pp. 105–116 (2010)
18. Allen, D.: Getting Things Done: The Art of Stress-Free Productivity. Penguin, New York (2015)
19. Bulling, A., Blanke, U., Schiele, B.: A tutorial on human activity recognition using body-worn inertial sensors. ACM Comput. Surv. CSUR. 46(3), Article No. 33 (2014)
20. Habitica V3 API Documentation. https://habitica.com/apidoc/
21. SensingKit. A multi-platform mobile sensing framework. https://www.sensingkit.org/
22. Google LLC: Google Play services—Apps on Google Play. https://play.google.com/store/apps/details?id=com.google.android.gms&hl=en_US
23. Weka 3—data mining with open source machine learning software in Java. https://www.cs.waikato.ac.nz/ml/weka/
24. Witten, I.H., Frank, E., Hall, M.A., Pal, C.J.: Data Mining: Practical Machine Learning Tools and Techniques. Morgan Kaufmann (2016)
25. Stikic, M., Larlus, D., Ebert, S., Schiele, B.: Weakly supervised recognition of daily life activities with wearable sensors. IEEE Trans. Pattern Anal. Mach. Intell. **33**, 2521–2537 (2011)

# A Dialog-Based Tutoring System for Project-Based Learning in Information Systems Education

**Hans Friedrich Witschel, Prajakta Diwanji, and Knut Hinkelmann**

**Abstract** This chapter discusses the design of a dialog-based intelligent tutoring system for the domain of Business Information Systems education. The system is designed to help students work on group projects, maintain their motivation, and provide subtle hints for self-directed discovery. We analyze the domain of Business Information Systems—which we find to be "ill-defined" in the sense that e.g. multiple conflicting solutions may exist and be acceptable for a given task. Based on an extensive collection of requirements derived from previous work, we propose a solution that helps both groups find solutions and individuals reflect on these solutions. This combination ensures that not only the group's result is valid, but also that all group members reach the defined learning goals. We show how the complexity of the domain can be captured in a rather simple way via constraint-based engineering and how machine learning can help map student utterances to these constraints. We demonstrate the intended working principles of the system with some example dialogs and some first thoughts about backend implementation principles.

**Keywords** Intelligent tutoring systems · Constraint-based domain modeling · Conversational interfaces

---

H. F. Witschel (✉)
School of Business, Institute for Information Systems, FHNW University of Applied Sciences and Arts Northwestern Switzerland, 4600 Olten, Switzerland
e-mail: hansfriedrich.witschel@fhnw.ch

P. Diwanji
University of Camerino, Camerino, Italy
e-mail: prajakta.diwanji@unicam.it

K. Hinkelmann
School of Business, M.Sc. Business Information Systems, FHNW University of Applied Sciences and Arts Northwestern Switzerland, 4600 Olten, Switzerland
e-mail: knut.hinkelmann@fhnw.ch



R. Dornberger (ed.), *New Trends in Business Information Systems and Technology*, Studies in Systems, Decision and Control 294,
https://doi.org/10.1007/978-3-030-48332-6_6

# 1 Introduction

(Accompanied) self-study plays an important role in the development of autonomy and self-learning competence of students—especially if one assumes that only the active and self-determined "construction" of knowledge (instead of passive consumption) leads to sustainable learning success [21].

In this chapter, we aim to contribute to improved personalized learning experiences of students in the domain of Business Information Systems. In this domain, as in many others, a very promising approach to teaching consists of project-based learning [17]—where students work in groups on complex, real-world problems ("cases") over a longer period of time.

While a project-based learning approach holds great promise for effective learning, there are several concerns. Firstly, students' motivation needs to be maintained, also in the face of difficulties and over the entire duration of a project. It is known that a good tutor/facilitator can achieve this [11], but with larger numbers of students, the tutor becomes a scarce resource.

Here, the promise of Intelligent Tutoring Systems (ITS, [19]) comes into play—the promise of delivering individualized and adaptive coaching to students via intelligent software. There is a rise of interest in ITS that are based on dialogs in natural language [20] due to the natural learning experience that they can help create.

Therefore, our aim is to outline a dialog-based ITS solution for the domain of Business Information Systems (BIS). The primary goal of this solution is to foster intrinsic motivation which has been shown to be a key ingredient to learning success [22].

Hence, the remainder of this chapter is structured as follows: in Sect. 2, we describe the background of this work. In Sect. 3, we summarize the requirements that a dialog-based intelligent tutor in the BIS domain must meet, and then we present a system architecture (Sect. 4) that will contribute to this. We finally demonstrate the utility of our dialog-based ITS through some sample dialogs that this system might help realize and use them to highlight the most important didactic and motivational considerations.

# 2 Background

In the following, an introduction to the various background aspects of our work is given: starting from insights into motivational factors, it moves on to cover problem-/project-based learning and finally analyzes important characteristics of the BIS domain.

## 2.1 Learning and Motivation

Previous work in education has clearly shown that motivation plays a key role in learning [27]. Furthermore, *intrinsic motivation* has been shown to have a more sustainable effect than extrinsic motivation: Ryan and Deci [22] state that intrinsic motivation i.e. the internal drive or curiosity brings about better learning outcomes among students than rewards/punishment and external impetus such as grades.

However, intrinsic motivation is often a fragile thing. For instance, in a survey of 140 students of the MSc Business Information Systems of the University of Applied Sciences and Arts Northwestern Switzerland (FHNW), only 35% of participants stated that they "often" or "always" completed the tasks assigned as class preparation [9]. In addition to the most frequently mentioned lack of time, students also mentioned motivation problems and excessive demands with some topics/tasks as reasons for a lack of preparation.

How can intrinsic motivation be increased? Some researchers provide practical tips, including e.g. giving emotional support, empathizing with negative feelings, using a certain type of language and giving positive and constructive feedback [15].

In addition to these affect-oriented behaviors, the design of learning experiences usually follows the imperatives of the well-known *self-determination theory (SDT)* [23, 26]. SDT suggests that there are three central psychological needs that should be satisfied to nurture intrinsic motivation: autonomy—the sense of freedom in steering one's own learning process, competence—the sense of mastery of concepts, and relatedness—the sense of shared purpose within a group. The following sections will discuss how certain didactic settings can help to satisfy these needs.

## 2.2 Problem- and Project-Based Learning

Problem-based learning [11] or project-based learning [17] are examples of teaching strategies that claim to increase intrinsic motivation [11]. In both approaches, students learn content and thinking strategies when solving problems. Project-based learning usually implies a larger scope—using authentic/real-world complex problems—and group work, while problem-based learning occurs individually and on smaller-scaled problems.

Especially project-based learning can be said to meet the requirements that the self-determination theory (SDT, [23]) places at the center of motivation:

- Autonomy—having a sense of freedom: in a project, students decide for themselves which solution strategies to apply and which steps to take next [24].
- Competence—having a sense of mastery: the successful completion of a project gives students clear feedback that they mastered a complex problem
- Relatedness—having a sense of shared purpose: most projects are carried out in groups where learners have to constantly defend their findings against the rest of the group, which in turn helps to build knowledge [11].

The *role of the problem* has been emphasized as an important motivational factor in problem-/project-based learning [8, 11]. In order to foster competence, problems should be "optimally challenging", i.e. not so difficult that they could lead to frustration, but hard enough to generate curiosity and prevent boredom [8]. The notion of optimally challenging tasks can be traced back to Vygotsky's concept of the Zone of Proximal Development [29] that describes the tasks that a learner cannot yet carry out completely alone, but with some (though not too much) help from a more experienced tutor. These are usually the tasks that are perceived as optimally challenging.

Problems that require communication and/or multidisciplinary effort will lead to a stronger feeling of relatedness [11]. Furthermore, good problems are realistic [11] and complex [12].

In problem-/project-based learning, the *role of the teacher* changes to that of a *facilitator* who is essentially "an expert learner, able to model good strategies for learning and thinking" [11]. A popular mode of facilitation relies on the so-called cognitive apprenticeship [4] where the facilitator first demonstrates how to solve a task ("modeling"), then helps ("scaffolding"), and finally merely watches students ("fading").

## 2.3 Dialog-Based Intelligent Tutoring Systems

With larger groups of students, the facilitator or tutor becomes a scarce resource. It would be desirable to provide individual support and assistance for students at all times, but this cannot always be the case.

As a result, interest in dialog-based systems such as educational chatbots is currently growing, promising a personalized and adaptive education at low cost [13]. Dialogs in natural language are considered a promising way in coaching since they represent a natural way of communication and enable the easy integration of additional participants (e.g. fellow students) into a coaching session [13].

Especially in the area of Intelligent Tutoring Systems (ITS), advances in dialog-based tutors are trying to mimic human facilitators [20, 28]. By observing experienced tutors, Olney et al. [20] found out their conversational strategies, including their ways of offering cognitive apprenticeship.

## 2.4 Business Information Systems: An "Ill-Defined Domain"

As pointed out in [10], in some domains the construction of Intelligent Tutoring Systems is not as straightforward as in others. These domains are termed "ill-defined domains".

### 2.4.1 Characteristics of BIS and Ill-Defined Domains

An important characteristic of ill-defined domains is the existence of multiple and controversial solutions and the lack of a clear way to determine whether a solution is acceptable [16]. Although [16] provides examples from law, ethics, music and architecture—and contrasts them with the field of physics as an example of a well-defined domain—multiple solutions are also typical in the field of Business Information Systems (BIS). While there might be certain criteria that an acceptable solution must fulfill in that domain, they are by no means exhaustive and that implies that occasionally solutions might be controversial, i.e. rated as very good by one BIS expert and as average by the next.

In addition to other characteristics, such as the lack of a formal theory, another important property of ill-defined domains concerns typical tasks: Lynch et al. [16] identify design tasks (e.g. music composition) as ill-defined—since their goal is often novelty—as well as "analytical tasks" (e.g. medical diagnosis)—where conclusions often have to be drawn on the basis of incomplete information in changing environments.

Both design and analytical tasks are common in BIS: in fact, project tasks in BIS typically involve the analysis of a case, i.e. a situation in a company—which is usually described in an incomplete fashion—followed by the design of a novel solution that remedies some of the weaknesses that the case analysis has revealed.

In practice, both types of tasks can be guided by certain principles and/or formalisms, which we will take up later to develop tutoring approaches for BIS projects:

- **Modeling**: in BIS, the analysis of a case is often performed by making use of modeling languages. These make it necessary to simplify the observed reality and often involve a graphical notation that can help visualize the result of the analysis. An example of such a modeling language is the entity-relationship (E/R) model [2], which captures entities and relationships in real life to inform the design of databases. The Business Process Model and Notation (BPMN, [3]) also falls into this category—it helps to describe business processes visually. The creation of E/R or BPMN models is a very common task for BIS students. The modeling languages help students to simplify reality and thus relate new situations to known ones; they also impose certain rules, the violation of which is inacceptable.
- **Design**: While models can help to analyze a situation and spot weaknesses (e.g. in a business process), the development of solutions or remedies for these weaknesses is a creative process. Theoretically, solutions can be built (e.g. programmed) from scratch, but in reality, BIS students will apply known solutions at different levels of abstraction—ranging from abstract architecture or design patterns to concrete web services or code snippets. When doing so, they need to be prepared to defend the fit of a chosen solution element to the requirement that it should meet, or the weakness that it should remedy. The quality of such a fit and the corresponding argument is an ill-defined notion where disagreement prevails: we all know that certain (software) solutions support some persons very well, but are not considered useful by others.

A final characteristic of ill-defined domains is that of overlapping sub-problems [16]: problems cannot be decomposed into independent sub-problems. This has implications for collaborative learning: groups that need to solve problems in ill-defined domains cannot hope that each member will simply provide a part of the solution—a good solution requires teamwork because every choice that the group makes for a sub-problem may affect other sub-problems. The group should keep an eye on the big picture. Clearly, especially the "Design" task in BIS has this property: components of information systems are usually arranged in complex architectures with various interdependencies. Changing one component often requires changing other components.

### 2.4.2 Potential Solutions to Tutoring in Ill-Defined Domains

Both of the typical BIS tasks outlined above have in common that they might have multiple and potentially controversial solutions. However, they also have in common that, as a rule, an agreement can be reached on what properties a good solution should *not* exhibit: models should not violate the rules of their modeling language (e.g. BPMN) and certain design choices are considered bad practices because they generally fit poorly to the given problem.

The approach of constraint satisfaction [18] allows to create ITS for such tasks: instead of formulating "the correct solution" and analyzing students' deviations—as is done in so-called "model tracing" [1]—a constraint-based system will accept all solutions that do not violate any constraint. A constraint is usually formulated as a rule in the form "If A holds true, then B should also hold true".

In the BIS domain, this allows to assess design tasks e.g. by saying that if a case contains a problem A, then any of the solution approaches B, C or D should be chosen. It also allows to assist students in modeling, as has been done successfully in [25], an ITS that supports the design of entity relationship models via constraint-based modeling.

In addition to constraint satisfaction, [10] mentions several other possible ways to accommodate ill-defined tasks in the design of ITS:

- Structuring learning around real-life cases
- Supporting metacognition, i.e. helping learners assess their own contributions (e.g. via some heuristics that allow to approximately detect quality deficits, but without a formal or detailed model of a solution).
- Using interactive narratives, i.e. putting learners into stories (e.g. cases) and letting them influence the story's direction. This can be seen as a dialog-based form of simulation.
- Structuring learning around collaboration, i.e. fostering group work and peer learning by e.g. guiding collaboration and/or offering a dialog-based system that acts as a peer, instead of a tutor who needs to be more knowledgeable than the users/learners.

The approach of cognitive apprenticeship for tutoring is seen as a promising approach to providing computer-based tutoring [7], especially when using prompts that invite students to reflect on their ideas [6]—which again links to the above-mentioned metacognition support.

## 3 Summary of Requirements

As mentioned above, the aim of this paper is to propose a "Digital Self-Study Assistant" (DSSA) to accompany students in a new way during their self-study. This shall help create a natural learning environment for students through a communication method and a channel (e.g. WhatsApp) that students are familiar with and, on the other hand, it can be adapted to the individual needs of students.

Table 1 summarizes the insights from literature that we will use to design the DSSA. In the last column, the corresponding design choices are enumerated, each with a number in brackets.

Summarizing these choices and putting them in a better order provides a clearer picture of how the DSSA will behave:

> The DSSA will be available to groups and individuals (3) who would like to receive support in solving a semester assignment (1). The DSSA will be able to support projects in the field of Business Information Systems that involve tasks of modeling and/or solution design (5).
>
> Initially, the DSSA will wait for students to choose an aspect of their assignment that they are currently working on and wish to discuss (2); later, the DSSA might decide to delve deeper into topics that require more attention (8).
>
> For a given aspect of the project (chosen by the students), the DSSA will support the **problem-solving phase** by asking students questions to help them reflect about certain important constraints that their solution should satisfy (6). If the DSSA detects or suspects a weakness in students' answers, it will try to provide constructive feedback (4). This may also include elements of cognitive apprenticeship (9), e.g. by pointing them to videos where modeling is performed or by scaffolding via some hints.
>
> Later, in a **reflection phase**, the DSSA will ask students to reflect on their project outcome (7)—this might well be on an individual level to ensure that students have an equally good understanding of the group's solution.

A general requirement is that the DSSA must at all times be aware of the students' level of knowledge and know their zone of proximal development—which is not an easy task [7]. As pointed out below, this will be the task of the student model within the entire system architecture.

**Table 1** Insights and derived design choices for the DSSA

| Area | Insight | Derived design choice for DSSA |
|---|---|---|
| Motivation | Project-based learning (PBL) is effective and supports motivation [11, 17] | (1) support PBL |
| | Autonomy fosters motivation [22, 24] | (2) leave initiative mostly to students |
| | Relatedness fosters motivation [11, 22] | (3) support group chats as well as individual chats |
| | Competence fosters motivation [22] helps build motivation [15] | (4) give positive and constructive feedback |
| Domain specifics | BIS has analysis (modeling) and design as main task types (see [16]) | (5) prepare dedicated dialog scripts for the two task types |
| | BIS solutions are "ill-defined", but constraints can often be formulated (as done e.g. in [25] | (6) use constraint-based modeling where possible |
| | Ill-defined tasks can be supported using metacognitive/reflective approaches [10] | (7) assist students in assessing their own project work → an incomplete domain model without formal reasoning on constraints might suffice |
| Good tutoring practice | Expert tutors dynamically determine topics [20] | (8) dynamically decide which topics to focus on |
| | Cognitive apprenticeship is effective, especially when using "self-monitoring prompts" [6, 7] | (9) support modeling and scaffolding via reflective approaches |

## 4 System Overview

Which ingredients—models and algorithmic components—does the DSSA require and how should they be orchestrated in order to deliver the type of interaction discussed in the previous section?

Figure 1 shows our proposed system architecture. It is a slight variation of the standard architecture discussed in [19] and is composed of: the domain model which contains the knowledge to be acquired, the student model which holds general and learning-related information about students, and the tutoring model which contains the didactic strategies for the tutoring. In order to provide more insight into how the tutoring model operates, we have divided it into an actual tutor (deciding how to shape the dialog), the interface where the dialog is delivered, and the interpreter that recognizes important concepts, learning progress, or difficulties from the student's answers.

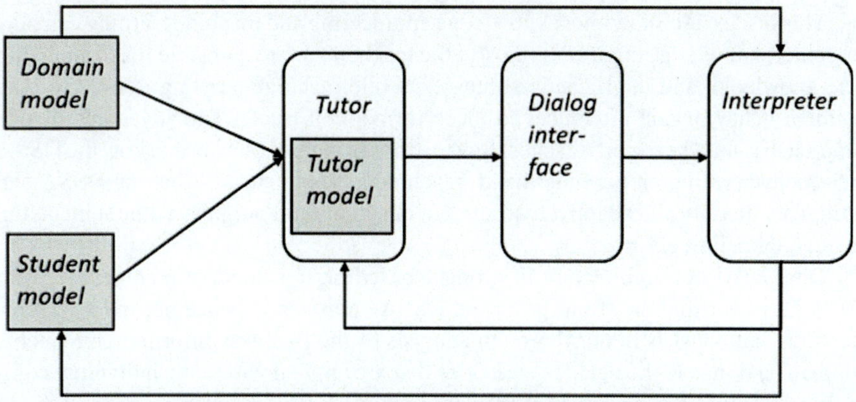


**Fig. 1** Proposed system architecture

The interplay of these components is planned as follows:

- The **domain model** needs to hold representations of important concepts and relationships of a particular sub-domain of BIS, presumably in the form of an ontology. Constraints, which are also part of the domain model, will be expressed with the help of the ontological concepts. We foresee a manual creation of the domain model—since it only needs to contain the most important constraints and the concepts needed to express them, the effort for a particular course should be manageable.
- The **student model** holds general information about the student, as well as summaries of student utterances that the student has performed while interacting with the DSSA (see the description of the Interpreter below).
- The **tutor** (or tutor model) has the task to provide an interaction with the student by generating text in natural language. To do this, it needs (a) a general strategy (or script) for leading the conversation (see design choice (5) in Sect. 3), (b) a (semi-)formal summary of what the student said in his/her previous utterance(s)—as provided by the Interpreter—including constraint violations and c) an estimate of the student's Zone of Proximal Development (ZPD, [29]) as derived from the student model.
- The **Interpreter** creates a (semi-)formal representation of the student's utterance. This includes the identification of ontological concepts and the detection of constraint violations.

As mentioned above, the domain model will be knowledge-engineered while the Interpreter will be designed to mostly use machine learning. Regarding the tutor, the dialog scripts for the two major BIS tasks will be generated by experts, while the estimation of a student's ZPD will again be based on machine learning—Bayesian networks are a popular choice for learning and inferring students' learning progress [5].

This discussion shows how knowledge engineering and machine learning can play together: knowledge engineers (mostly the teachers) are responsible for formalizing the knowledge and the didactic strategies, while machine learning is used to map student behavior and utterances to these formal constructs. The advantage of this approach is that the expert (teacher) stays always in control of what and how the DSSA will deliver while, on the other hand, machine learning ensures that the DSSA can still react flexibly and adaptively to student questions and progress without incurring a huge modeling effort.

Obviously, our architecture is strongly based on the standard architecture from [19]. The contribution of our proposed DSSA approach is hence not mainly architectural, but consists in our thorough analysis of the Business Information Systems domain and thus in the detailed strategies that are applied within the individual components. This is demonstrated in the next section.

## 5 Demonstration of DSSA Working Principles

To better understand how the tutor component from Sect. 4 works, we demonstrate here an example dialog which is based on a partial domain model that will be explained in the following.

The course for which the dialog is designed is called "Business Intelligence". One of its major learning goals is that students should be enabled to analyze questions or information needs of business stakeholders and translate them into a so-called multi-dimensional model [14]—i.e. a modeling task. The modeling task is part of a semester assignment where the information needs must be elicited from a real-life case study.

A sub-goal refers to the students' ability to decide whether elements of the model fit together, or, in technical terms: whether measures and dimensions are true to the chosen grain of the model. Without going deeper into this (see [14] for details), we can specify a domain model for this aspect.

The domain model consists of a constraint, which in turn requires the definition of four core concepts. The concepts are "grain", "fact", "measure" and "dimension". For the moment, the only relationship between these concepts that needs to be modeled is the one between "fact" and "grain", which we can label as an equivalence relation (the grain corresponds directly to what one defines as a fact).

With these concepts, we can formulate a constraint as follows—where the previously defined concepts are highlighted in **bold**:

IF a group chooses **grain** A
AND the group chooses **dimension** B
THEN the **dimension** B must be true to the **grain** A.

An analog constraint can be formulated for measures, but this is not needed for the following dialog. Note that this constraint is only formulated in natural language and we do not, at this point, foresee a formal representation of it.

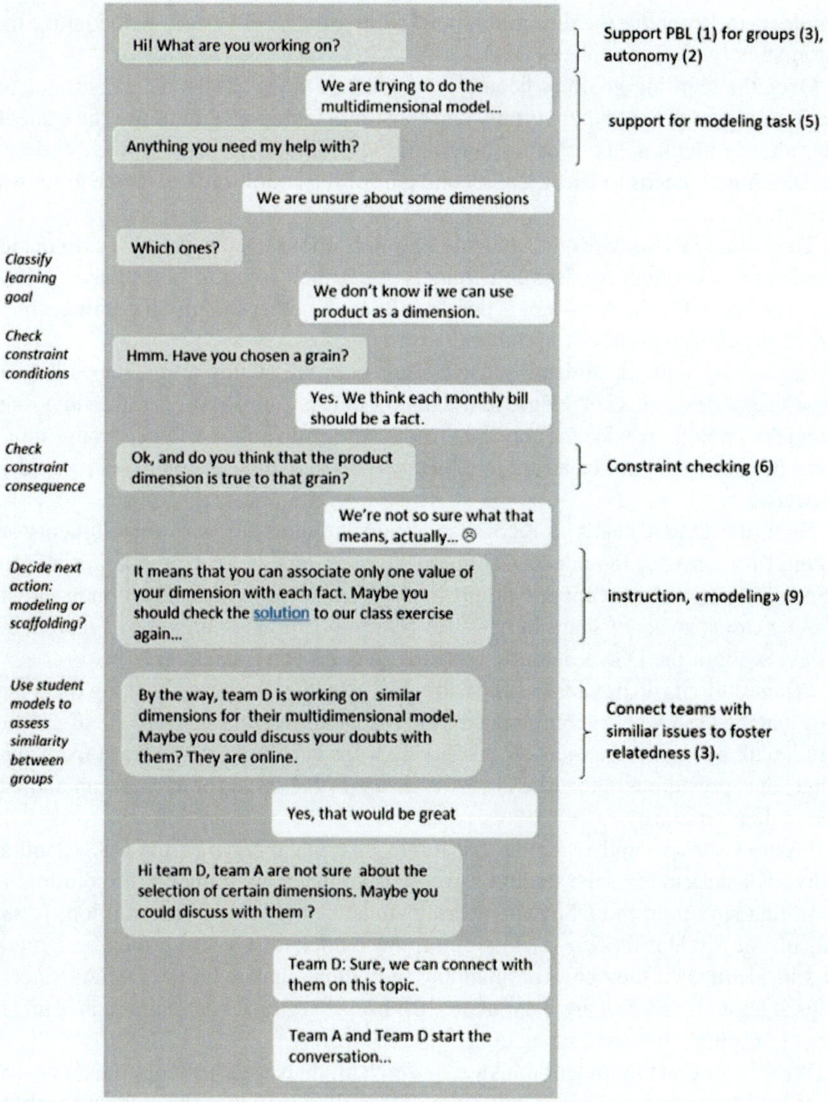


**Fig. 2** Example chat between a group and the DSSA in the problem-solving phase

Figure 2 shows an example dialog between a group of students and the DSSA. The dialog itself is in the middle. On the left side, some explanations were added that point to the internal processing that the DSSA performs, while on the right side, it is highlighted how the design choices from Sect. 3 become visible within the dialog.

We can see the rough dialog script that the DSSA follows: first, it needs to figure out the learning goal that the students are trying to achieve. As the students decide on the topic for the discussion (autonomy), the DSSA needs to recognize this automatically—a classification task.

Once the learning goal has been identified, the DSSA checks the constraints for this learning goal. Since there is only one constraint in this case and since the students have already mentioned one part of its condition (a dimension that they have chosen), the DSSA now needs to check the second part of the condition, i.e. which grain was chosen.

Then, the DSSA invites the students to reflect about the constraint, i.e. to decide whether the consequence of the constraint is true in their solution. Note that the DSSA does not know the answer—i.e. it does not know whether the product dimension is true to the chosen grain of monthly bills (see Fig. 2).

Several variants of students' answers are possible at this point, depending on the students' level of knowledge. In the dialog in Fig. 2, this level is rather low—the group does not know what the concept "truth to the grain" means. Other groups might know this (or might not be aware that they do not) and choose a more self-confident answer.

Here, the DSSA needs to recognize (again via classification) that students are asking for help with the concept as such. This triggers an act of modeling, i.e. the DSSA explains the concept and points to a resource where its application is shown. In other cases, an act of scaffolding—i.e. more subtle hints—might be triggered, but always without the DSSA actually knowing the or a correct answer.

When students do not answer (possibly because they are busy checking the materials that the DSSA has recommended), the DSSA also makes them aware of another group with a similar situation, so that both groups can connect and learn from each other. This presupposes that the DSSA is aware of the issues of each group and can assess their similarity.

Figure 3 shows another dialog that might take place between the DSSA and an individual student after the student's group has finished working on the solution.

In that situation, the DSSA invites the student to reflect on the solution. It will consult the student model—at this point, the models of all group members might be still identical if they have not had any individual chats with the DSSA. Since it finds at least one conversation about the "truth-to-the-grain" constraint in the student model, it brings this issue up again.

We foresee that the student's answer—which might be a rather lengthy essay—can be checked using some simple heuristics. The DSSA may give the student feedback about the result of this check. However, discriminative feedback about the quality of the solution needs to come from a human tutor. What counts here is to trigger the act of reflection on the student's part.

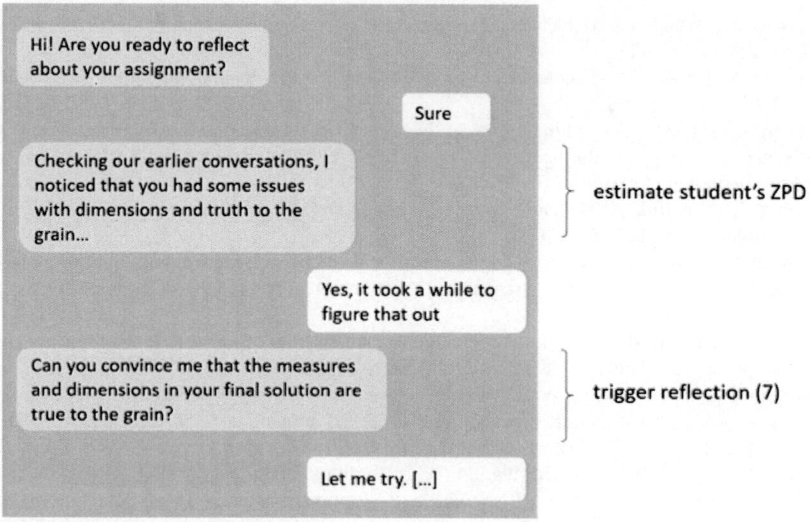


**Fig. 3** Example chat between an individual student and the DSSA in the reflection phase

## 6 Conclusions

This chapter shows how didactical and motivational considerations can lead to the design of a reflective dialog that should support students in project-based learning in the "ill-defined" domain of Business Information Systems (BIS).

The presented concept tries to keep the engineering effort low by focusing on the essentials of learning goal-related constraints. We recommend using machine learning to map student utterances to concepts and constraints. Dialog scripts are derived from the typical tasks of the BIS domain and support processes that were found to be effective in ill-defined domains. We also pay special attention to motivational aspects to keep students engaged throughout an entire project. We have used example dialogs to illustrate how all the previously mentioned aspects can be realized in practice.

We will implement the DSSA in the near future and plan to use it to support a semester assignment in one of the courses at our school. By analyzing the real (anonymized) dialogs that will emerge, we hope to improve its design further.

## References

1. Aleven, V.: Rule-based cognitive modeling for intelligent tutoring systems. In: Nkambou, R., Mizoguchi, R., Bourdeau, J. (eds.) Advances in Intelligent Tutoring Systems, pp. 33–62. Springer Science and Business Media, Berlin, Heidelberg (2010)
2. Chen, P.P.S.: The entity-relationship model—toward a unified view of data. ACM Trans. Database Syst. (TODS) **1**(1), 9–36 (1976)

3. Chinosi, M., Trombetta, A.: BPMN: an introduction to the standard. Comput. Stand. Interfaces **34**(1), 124–134 (2012)
4. Collins, A., Brown, J.S., Holum, A.: Cognitive apprenticeship: making thinking visible. Am. Educ. **15**(3), 6–11 (1991)
5. Conati, C.: Bayesian student modeling. In: Nkambou, R., Mizoguchi, R., Bourdeau, J. (eds.) Advances in intelligent tutoring systems, pp. 281–299. Springer Science and Business Media, Berlin, Heidelberg (2010)
6. Davis, E.A.: Scaffolding students' knowledge integration: prompts for reflection in KIE. Int. J. Sci. Educ. **22**(8), 819–837 (2000)
7. Dennen, V.P.: Cognitive apprenticeship in educational practice: research on scaffolding, modeling, mentoring, and coaching as instructional strategies. Handb. Res. Educ. Commun. Technol. **2**(2004), 813–828 (2004)
8. Di Domenico, S.I., Ryan, R.M.: The emerging neuroscience of intrinsic motivation: a new frontier in self-determination research. Front. Hum. Neurosci. **11**, 145 (2017)
9. Diwanji, P., Hinkelmann, K., Witschel, H.F.: Enhance classroom preparation for flipped classroom using AI and analytics. In: Proceedings of the 20th International Conference on Enterprise Information Systems (ICEIS), pp. 477–483 (2018)
10. Fournier-Viger, P., Nkambou, R., Nguifo, E.M.: Building intelligent tutoring systems for ill-defined domains. In: Nkambou, R., Mizoguchi, R., Bourdeau, J. (eds.) Advances in Intelligent Tutoring Systems, pp. 81–101. Springer Science and Business Media, Berlin, Heidelberg (2010)
11. Hmelo-Silver, C.E.: Problem-based learning: what and how do students learn? Educ. Psychol. Rev. **16**(3), 235–266 (2004)
12. Jonassen, D.H., Hung, W.: All problems are not equal: implications for problem-based learning. Essential Readings in Problem-Based Learning, pp. 7–41 (2015)
13. Kerry, A., Ellis, R., Bull, S.: Conversational agents in E-Learning. In: International Conference on Innovative Techniques and Applications of Artificial Intelligence, pp. 169–182 (2008)
14. Kimball, R., Ross, M.: The Data Warehouse Toolkit: The Complete Guide to Dimensional Modeling. Wiley, New York (2011)
15. Kusurkar, R.A., Croiset, G., Ten Cate, O.T.J.: Twelve tips to stimulate intrinsic motivation in students through autonomy-supportive classroom teaching derived from self-determination theory. Med. Teach. **33**(12), 978–982 (2011)
16. Lynch, C.F., Ashley, K.D., Aleven, V., Pinkwart, N.: Defining ill-defined domains: a literature survey. In: Intelligent Tutoring Systems (ITS 2006): Workshop on Intelligent Tutoring Systems for Ill-Defined Domains (2006)
17. Markham, T.: Project based learning a bridge just far enough. Teach. Libr. **39**(2), 38 (2011)
18. Mitrovic, A.: Modeling domains and students with constraint-based modeling. In: Nkambou, R., Mizoguchi, R., Bourdeau, J. (eds.) Advances in Intelligent Tutoring Systems, pp. 63–80. Springer Science and Business Media, Berlin, Heidelberg (2010)
19. Nkambou, R., Mizoguchi, R., Bourdeau, J. (eds.): Advances in Intelligent Tutoring Systems, vol. 308. Springer Science & Business Media, Berlin, Heidelberg (2010)
20. Olney, A.M., Graesser, A.C., Person, N.K.: Tutorial dialog in natural language. In: Nkambou, R., Mizoguchi, R., Bourdeau, J. (eds.) Advances in Intelligent Tutoring Systems, pp. 181–206. Springer Science and Business Media, Berlin, Heidelberg (2010)
21. Reinders, H., et al.: Towards a classroom pedagogy for learner autonomy: a framework of independent language learning skills. Aust. J. Teach. Educ. (Online) **35**(5), 40 (2010)
22. Ryan, R.M., Deci, E.L.: Intrinsic and extrinsic motivations: classic definitions and new directions. Contemp. Educ. Psychol. **25**(1), 54–67 (2000)
23. Ryan, R.M., Deci, E.L.: Self-determination theory and the facilitation of intrinsic motivation, social development, and well-being. Am. Psychol. **55**(1), 68 (2000)
24. Stefanou, C., Stolk, J.D., Prince, M., Chen, J.C., Lord, S.M.: Self-regulation and autonomy in problem-and project-based learning environments. Act. Learn. High. Educ. **14**(2), 109–122 (2013)
25. Suraweera, P., Mitrovic, A.: An intelligent tutoring system for entity relationship modelling. Int. J. Artif. Intell. Educ. **14**(3, 4), 375–417 (2004)

26. Trenshaw, K.F., Revelo, R.A., Earl, K.A., Herman, G.L.: Using self-determination theory principles to promote engineering students' intrinsic motivation to learn. Int. J. Eng. Educ. **32**(3), 1194–1207 (2016)
27. Usher, A., Kober, N.: What Is Motivation and Why Does It Matter? Center on Education Policy (2012)
28. VanLehn, K., Jordan, P.W., Rosé, C.P., Bhembe, D., Böttner, M., Gaydos, A., Makatchev, M., Pappuswamy, U., Ringenberg, M., Roque, A., et al.: The architecture of Why2-Atlas: a coach for qualitative physics essay writing. In: International Conference on Intelligent Tutoring Systems, pp. 158–167 (2002)
29. Vygotsky, L.S.: Mind in Society: The Development of Higher Psychological Processes. Harvard University Press, Cambridge, Massachusetts (1980)

# A Human Aptitude Test for Object-Oriented Programming in the Context of AI and Machine Learning

**Rainer Telesko and Stephan Jüngling**

**Abstract** Many current IT systems are implemented based on the object-oriented (OO) programming paradigm, which over more than two decades has proved to be one of the most successful mechanisms for code re-use and the most powerful extension mechanisms used in many software components and systems. Combined with a solid understanding of business principles and good communication skills, OO is still considered to be one of the core skills in the design of platforms and systems that drive our current IT landscape. The self-evaluation test, which we developed as an early indicator for prospective Business Information Technology (BIT) students, revealed insights about the skill level of beginners and serves as a starting point to reflect on abstraction skills in the context of the current digitalization and the increase in artificial intelligence (AI) components. The article explains the relevance of OO thinking on different levels of abstraction in the context of the lifecycle of current system architectures and provides an outlook on how these abstraction skills can be re-used when switching from an OO development paradigm into a new area where AI and machine learning will steadily increase their influence on the overall design of software systems.

**Keywords** Aptitude · Test · Object-oriented analysis and -design · Kohonen feature map · OO and AI · Transfer learning · Architectural thinking in AI · Boxology

R. Telesko (✉) · S. Jüngling
School of Business, Institute for Information Systems, FHNW University of Applied Sciences and Arts Northwestern Switzerland, Riggenbachstrasse 16, 4600 Olten, Switzerland
e-mail: rainer.telesko@fhnw.ch

S. Jüngling
e-mail: stephan.juengling@fhnw.ch



R. Dornberger (ed.), *New Trends in Business Information Systems and Technology*, Studies in Systems, Decision and Control 294,
https://doi.org/10.1007/978-3-030-48332-6_7

# 1 Introduction

OO is a very powerful paradigm and revolutionized the way how software is constructed. In principle, it is not only a special way of structuring code compared to the procedural programming paradigm, but also a very powerful way of reusing code that encapsulates data and behavior.

In the OO software paradigm, it is important to recognize that there are only two ways to add functionality: either by inheritance or by delegation.

Inheritance is the principle of re-use that can be used to build up hierarchies of specialization. New functionality is grouped into additional classes that inherit the behavior from all parent classes and could even override some of the existing methods in the case of inadequate behavior or behavior that needs to be adapted.

Delegation is concerned with the interaction and collaboration of classes. That is where OO design patterns come into play as reusable solutions across different problem domains. One of the most prominent examples is the Model-View-Controller or MVC design pattern [1], which is used to synchronize a number of different views that should be consistent.

Given these fundamental design patterns embedded in our current IT systems, students need to learn how business information systems are designed, built, and integrated into an overall system landscape of a company, and must therefore be able to understand and apply these OO related principles.

This chapter is organized as follows. In Sect. 2, the motivation and background for conducting an OO self-test (called OO SET) is explained, Sect. 3 describes the setup, implementation, and results of this test in detail. Section 4 explores the collected data of OO SET using AI techniques. The last chapter is dedicated to the rapid development of AI and Machine Learning and their impact on "traditional" or old-fashioned development paradigms like OO. This closes the circle again, because the knowledge that forms the basis of the OO SET questionnaire should not be static, but must be constantly adapted to optimally prepare future students for the challenges in IT jobs.

# 2 Background and Context

The Bachelor of Science FHNW in Business Information Technology (BSc BIT) is a computer science degree program that focuses specifically on how IT systems are designed and implemented, and how they can be used for specific business purposes. It aims to develop the specific skills required to bridge the gap between business and IT (see Fig. 1). The BIT program aims to promote digitalization with a particular focus on innovative IT, modeling techniques and the usage of communication for an effective business–IT-alignment.

The program was launched initially in 2014 as an English equivalent to the German "BSc Wirtschaftsinformatik". Already successfully established in Basel, the program

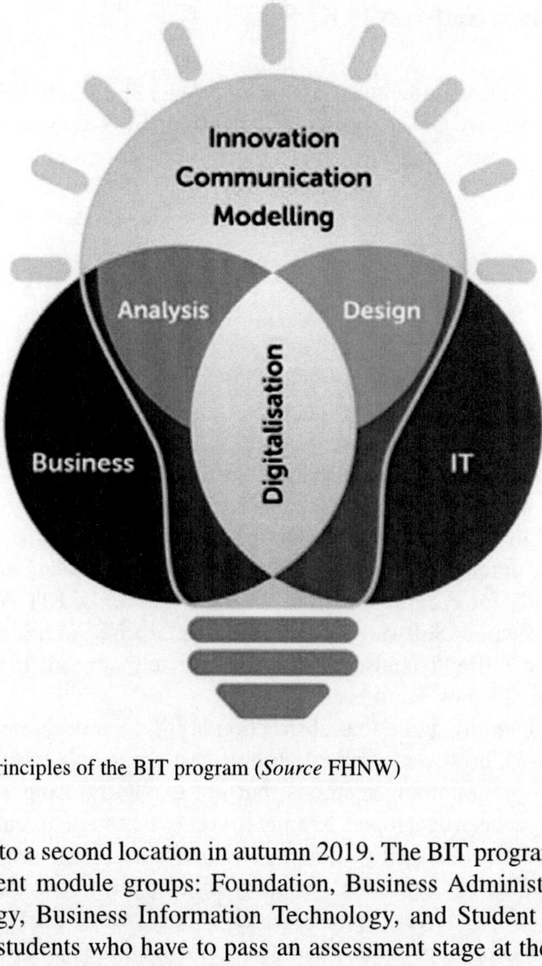


**Fig. 1** Design principles of the BIT program (*Source* FHNW)

was expanded to a second location in autumn 2019. The BIT program is divided into the five different module groups: Foundation, Business Administration, Information Technology, Business Information Technology, and Student Projects. It was observed that students who have to pass an assessment stage at the end of the first semester often fail in the modules Programming, Mathematics, and Introduction to Business Information Technology. This was the starting point to establish a self-evaluation test (SET) that particularly addresses the areas of logical and analytical thinking, understanding algorithms, mathematics, and abstract thinking. This self-test is anonymous and can serve prospective students as a first glimpse of skills that will be needed during the BIT studies.

Although initial results were promising that the self-test could play the role of an indicator to pass the assessment stage, we decided to extend the scope from the assessment stage to the entire course of study, thus covering content from a wider variety of IT modules.

## 3 The Aptitude Self-test "OO SET"

This section deals with an in-depth discussion of the OO SET, a test which particularly addresses skills related to OO analysis and OO design. The section describes all steps of the in-house OO SET project, namely motivation, design and implementation, results, and outlook.

### *3.1 Motivation*

SET was conducted with registered student volunteers as well as applicants interested in finding out whether BIT is the right choice. In order to focus more on the specific skills that prevail not only in the assessment stage but also in the higher semesters of the IT and business information technology modules, we recently focused our research interest in setting up a test (OO SET) that deals particularly with the kind of abstraction skills that dominate the OO-paradigm. OO-analysis and OO-design (OOA, OOD) are among the most important concepts in software engineering and are crucial not only for Programming and Introduction to the BIT courses but also for many others such as Software Engineering, Enterprise Architecture, Database Technology or the entire IT landscape of an enterprise that needs to optimize its IT as part of the digitalization of all business areas.

Researchers generally agree that abstraction ability is a necessary skill for OO programming [2–5]; however, a reliable tool to test a person's level of abstraction ability related to programming methods, but not explicitly using a programming language has not yet been developed. Kramer [6] states that students can be stimulated to improve their abstraction ability through activities such as modeling, mathematical study, and explaining to each other how they accomplish to solve a problem in a specific context. Other sources suggest that the act of creating storyboards, labelled diagrams, and concept maps require the use of abstraction [7]. If these statements are true, these activities would also serve as OO test scenarios.

Traditional OOP testing methods focus heavily on interpreting and generating code, which corresponds to the levels "Apply" (using information in new situations) and "Analyze" (drawing connections among ideas) according to the taxonomy of Bloom [8]. However, Or-Bach and Lavy [5] were able to test for abstraction ability in an OO context without using programming syntax by allowing the students to answer questions in plain English writing. Given our preference of using multiple choice (MC) questions, we have to find a scenario where the understanding of OO concepts can be assessed and related to the appropriate taxonomy levels of Bloom.

## 3.2 *Design and Implementation*

The OO SET was implemented with Google Forms (30 MC questions). As a "test" domain the card game Blackjack was selected because it is widely known and can be relatively easily explained for a short assignment such as the questionnaire. Before answering the questions, the test students had some time to familiarize themselves with the basic rules of the Blackjack game and basic OO knowledge (presented as a tutorial with examples). This aspect is important as OO is not self-explanatory and completely new for a considerable number of participants. In order to get a clearer picture of the students' aptitude and to support a more sophisticated evaluation, every question was assigned to an OO concept category and level according to Bloom. Students with and without pre-knowledge in programming are enrolled in the BIT program. In order to "simulate" such a situation for the OO SET, two test groups—total beginners and BIT students of the first semester with some knowledge based on the current programming module—were considered.

The first part of the questionnaire asks for information about participants, such as age, prior OO knowledge, gender, etc. and covers a basic overview of relevant OO principles. The second part of the questionnaire (see Fig. 2) deals with questions related to core OO concepts. The selection of the suitable OO concepts discussed and tested in the questionnaire was made in close cooperation with BIT lecturers and is based on similar field research [9, 10]. The list of understandings of OO concepts used and tested in the questionnaire includes: classes, objects, classes vs. objects,

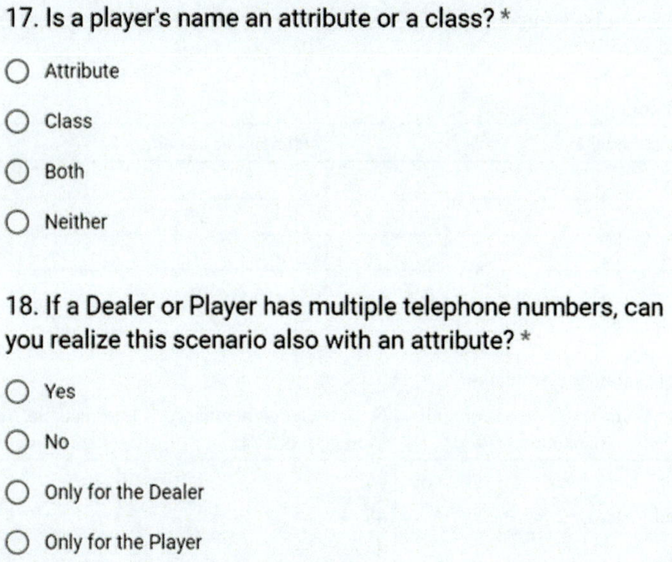

17. Is a player's name an attribute or a class? *

- Attribute
- Class
- Both
- Neither

18. If a Dealer or Player has multiple telephone numbers, can you realize this scenario also with an attribute? *

- Yes
- No
- Only for the Dealer
- Only for the Player

**Fig. 2** Example questions for Questionnaire (part 2)

attributes, classes vs. attributes, methods in classes, parameters of methods, inheritance, multiplicity, encapsulation, and relationships between classes (association, aggregation, composition).

## *3.3 Questionnaire Results*

A total of 24 male and 3 female students participated over a two-week period, with the following distribution of their self-described prior knowledge of OOP: 5 students with no OOP knowledge, 12 with beginner's knowledge, 5 intermediate, and 5 advanced.

After the adjustment, the breakdown of the points per Bloom level (indicative of the level of difficulty) for the questionnaire was as follows (see Table 1). The "Remember" level means the ability to recall facts, the "Understand" level is about the ability to explain ideas or concepts and the "Analyze" level is concerned with breaking the concept into parts and understanding how each part is related to another.

The average questionnaire score across all student groups (i.e. based on their self-indicated level of OO knowledge) was 55%. Intuitively, students that indicated they had prior OO knowledge (i.e. "intermediate" or "advanced" levels) performed better than those with little or no prior OO knowledge, as can be seen in Table 2. In addition, this table illustrates that the students performed best on questions mapped to the "Remember" level of the Bloom taxonomy, and worst on those mapped to the "Analyze" level.

It is important to note that all student groups achieved an average passing grade for the "Remember" questions (67% average across all student groups), showing that

**Table 1** Bloom question levels

| Bloom question level | Total number of points |
|---|---|
| Remember | 23 |
| Understand | 20 |
| Analyze | 12 |
| All levels | 55 |

**Table 2** Questionnaire evaluation

| Bloom question level | Average score (all students) (%) | No knowledge/beginner average score (%) | Intermediate/Advanced average score (%) |
|---|---|---|---|
| Remember | 67 | 65 | 70 |
| Understand | 52 | 48 | 60 |
| Analyze | 37 | 33 | 43 |
| All levels | 55 | 52 | 61 |

it is feasible to test the abstraction ability in an online, MC-based format. This has not been shown in any of the reviewed sources and is an encouraging outcome of our research.

## 3.4 First Validation of OO SET

In order to validate our hypothesis that a solid understanding of the OO paradigm is a valid indicator for students to be able to successfully pass the exam in programming as well as the assessment stage, a subset of OO related questions was selected and compared to the grades in programming. Although the situation of our current students in the assessment stage and the prospective students is different—our students had lectures in OO and programming, the prospective students get only a very rough understanding of OO as part of the OO SET—the correlation of the grades in programming with OO related questions was analyzed.

The OO-related questions were extracted from the exam introduction to BIT, where the understanding of OO is necessary to transfer textual business descriptions into UML models. Entities need to be recognized from textual descriptions and compiled into UML class diagrams or Entity Relationship Diagrams. In both cases, it is important to distinguish entities from attributes, similar to the questions in the OO SET.

In the programming exam, the different questions were dedicated to code reading, data types, arrays, inheritance, OOP and JavaFX. Interestingly, the average number of points for the coding-related categories was around 50% while the percentage for OOP and inheritance was only 32%.

Figure 3 shows the correlation between the total numbers of points from OO questions of the 54 students from the BIT exam, versus the overall points in programming, both normalized to one. The scatterplot substantiates that there is at least an indication of a positive relationship between programming skills and an understanding of OOA and OOD concepts.

## 3.5 Next Steps

The initial results with the OO SET are encouraging and show that such a test is feasible and will help the university optimize its resources while at the same time raising students' awareness of the specifics and challenges of BIT.

Future work will be mainly related to testing by further extending and tuning the question database. In detail, we will investigate which questions should be part of the core set of the OO SET bearing in mind that the time for getting acquainted with OO should be kept short. In order to achieve this goal, we will also analyze the outcomes of the programming module in detail to better understand the weaknesses of beginners when learning a programming language.

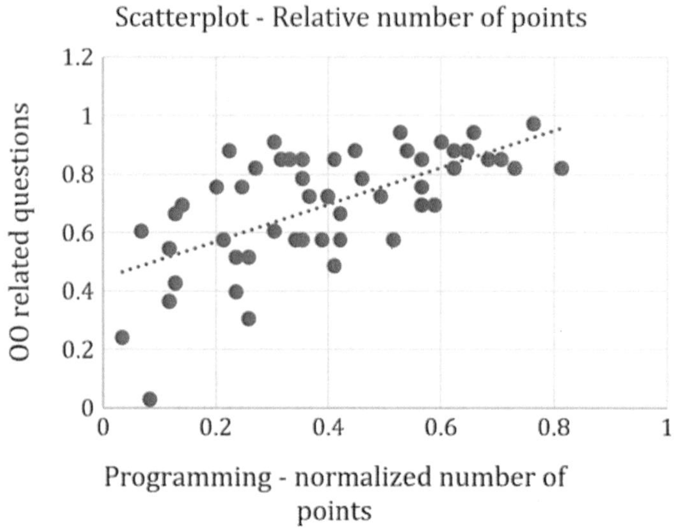


**Fig. 3** Correlation of skills in OO and programming

## 4 Use Cases for AI in the Context of Aptitude Testing

The first round of experiments with OO SET confirmed the feasibility of the approach. The next step is now to investigate how our approach can be extended with AI to optimize core processes of the BIT program, namely admission and teaching. For this purpose, we are investigating two use cases (UC 1 and UC 2). UC 1 deals with the prediction of student performance based on metadata and learning activities. UC 2 is about finding clusters of students with similar performance, especially those who are unable to pass the assessment stage. Information from both UC can be used as "add-in" to optimize resources in teaching and to focus on difficult cases in the admission process.

### *4.1 Motivation*

An interesting approach is the use of AI methods for aptitude tests. Such an AI-enriched aptitude system can "learn" the relationship between important parameters of influence and the aptitude of the student and can therefore be used for new, untrained situations. Basically there exist two AI paradigms, the symbolic and the sub-symbolic approach [11]:

- Symbolic approach (expert systems): For example, knowledge is "coded" with "if-then-else" rules that are executed during runtime when facing a specific situation. The main advantage is that all knowledge is clearly documented and traceable.

However, building such a knowledge base requires well-equipped experts and continuous efforts to manage the rule set.

- Sub-symbolic approach (Artificial Neural Networks—ANN): The main idea here is often referred to as "learning by example", i.e. define a training set and calibrate the network with a learning algorithm, so that both existing examples (e.g. patterns) can be retrieved and also the generalization of new, unknown examples is possible. The basic principle is derived from interconnected neurons in the brain. The main advantage is that even beginners can quickly build such an AI system, based on a representative number of examples with the drawback of getting a "black box" solution that does not provide justification or transparency for the calculated output and often achieves surprising or incorrect results.

For performance prediction and aptitude tests in general, neural networks are one of the preferred approaches in the literature (see examples for UC 1 below), because data sets for training a neural network can be easily generated by analyzing students' results from earlier terms. However, neural networks are not a "silver bullet" for every scenario and the effort to identify decisive input parameters, to obtain a data set of appropriate size, to select and "optimize" the neural network architecture, should not be underestimated.

## 4.2 UC 1: Performance Prediction

Considerable research was done to predict the performance of students using neural networks. Relevant contributions in this context are:

- Plagge [12] investigated the application of different feed-forward neural networks (mostly multilayer perceptrons) to predict the outcomes of first year students based on a sample pool of data from Columbus State University from 2005–2010. The accuracy rate for a student's return to the second year with the best-found learning strategy and network architecture approached 75%. Among the data used for training the network, Plagge experimented with a mix of student meta-data (age, gender, ethnicity, etc.), family-related data (father/mother highest education level, estimated family contribution, etc.) and performance data from the past (high school and fall grade point average, entrance test results, etc.).
- Okubo et al. [13] experimented with input data from learning activities (based on the e-learning system Moodle, the e-portfolio system Mahara and the e-book system BookLooper) in order to predict final grades. Because the data (such as quiz results, slide and course views, etc.) were collected every week, a recurrent neural network was used that was trained with backpropagation through time. Using the log data from 108 students until the sixth course week, the accuracy of prediction was above 90%.
- As opposed to Plagge [12], Wei [14] used only monitoring data during the course (preparation time for class, completion of homework, review time after class,

attendance, assignment score, etc.) for predicting the grade. For the sample of 800 students, the best result for predicting the grade was around 80%.

Bearing in mind the outcome of the work mentioned above, there will be two promising research directions for the application of neural networks with data from the OO SET in the future. One research direction is based on the fact that our aptitude test is already a good predictor for the programming course (see Fig. 3). When using a feed-forward neural network, the input data is composed of the student's meta-data (part 1 of the questionnaire), i.e. age, gender, highest degree of education, country, level of previous knowledge (beginner, intermediate, advanced). The output calculates the performance, i.e. pass/fail of the programming course or the assessments stage as a whole. Training data can be obtained from students who have completed the questionnaire and finished the assessment stage in the meantime. If the accuracy of the prediction needs improvement, one could add further input data related to past grades, e.g. high school grades. A second research direction could aim to replicate the experiment of Okubo et al. [13] with the FHNW e-learning environment to predict the result for the programming course. FHNW is using Moodle as an e-learning system; respective data can be collected by weekly conducted multiple choice-based quizzes, coding questions, and slide views. Moodle offers a PHP-based infrastructure for retrieving the data.

The success of both research directions will depend on the ability to appropriately "code" the input data for use with a neural network and to set up a representative data set for training.

## *4.3 UC 2: Clustering Similar Students*

Another use case of interest is to identify clusters of students, especially those who are unable to pass the assessment stage. Information from such clusters can be used in the admission process as an "early-warning system" by paying attention to factors common to a large group of students, such as previous school education, unsatisfactory analytical skills, language problems, etc.

The Kohonen Feature Map (KFM) is a neural network approach that is a good alternative to conventional clustering techniques, especially if the clusters are not known in advance [11, 15]. A KFM is a two-layered network (see Fig. 4) that is trained using unsupervised learning to produce a low-dimensional (typically two or three dimensions for visualization purposes) representation of the input space. With a KFM it is possible to "learn" the semantic organization of arbitrary concepts represented by input vectors and to visualize similarities between various input data, in our case student data. Outlined below is the UC 2 experiment with a KFM which is planned as the next step in our research.

This ability to represent topological properties of the input space in a map allows to identify clusters of students. As in multilayer perceptrons, we represent the input data as a vector, in our case again the student's meta-data is enriched with detailed

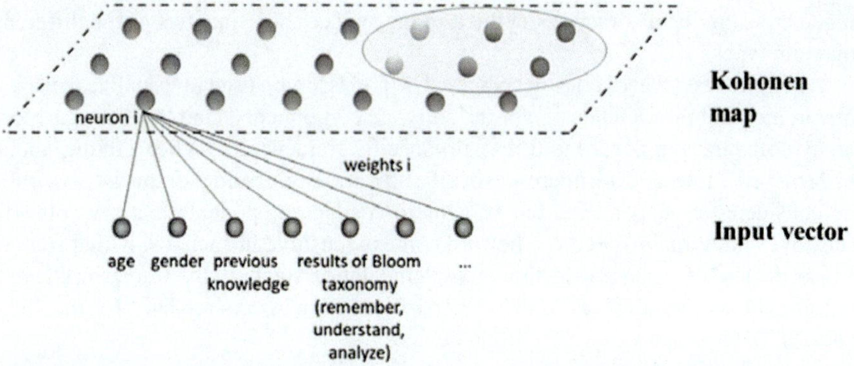


**Fig. 4** Sketch of the KFM for UC 2

outcomes of the OO SET according to the Bloom categories. On the map, every student is modeled as an output neuron. Figure 4 shows a sketch of such a KFM for our OO SET. Because the KFM algorithm is based on a stochastic process and input data have to be randomly presented in order to avoid a bias, results, i.e. the feature map, may differ from experiment to experiment. Neurons in the circle represent students who do not pass the OO SET, identifying commonalities among them will help optimize the future admission process and to sharpen the profile of the BIT program. For clarity, only the links from the input vector to a neuron on the map are displayed.

# 5 Application of OO Thinking in a Hybrid World

In this section, we reflect on the application of OO skills in the context of our current system architecture. In addition, foreseeable future IT architectures, in which some of the components will remain programmed, and AI components will be trained. In order to reflect on such a future hybrid architecture, we first take a look at the lifecycle of our current IT architecture and aspects of reusing components where OO thinking was one of the major driving forces. Second, we will provide an outlook on how OO and related concepts or design patterns could be transferred to the AI domain.

## *5.1 OO in the Context of Current System Design*

Current system architectures are usually documented with the help of unified modeling language UML [16]. Software engineers communicate OOD ideas using one of the many different UML diagram types and entire system architectures can be modeled using the abstractions defined in the syntax and semantics of the different diagram types.

At this lowest level of code design, the UML classes are used to visualize the class hierarchies and the OOU design of the software components. One level up, the hierarchy software engineers build domain-specific frameworks. When talking about patterns as a lower level micro-cosm of software engineering, frameworks could be considered as upper level but still micro-cosmic components that are context-sensitive and domain-specific. They are context-sensitive in such a way that frameworks depend on a specific technical implementation stack. In the framework layer, specific implementations of MVC or persistence frameworks are built for the JEE- and .NET-stack, as well as for mobile devices.

At this higher abstraction level (see Fig. 5), UML component diagrams can be used instead of class diagrams based on the same kind of abstract thinking in terms of encapsulating data and functionality. There are different classification styles of software architectures and many different vocabularies are used to describe the network of IT systems, components and their connections, as described in the field guide to boxology by Shaw and Clements [17]. By adding time as one of the main drivers of ever-changing software architectures different macrocosmic components (e.g. technology stacks such as Java EE or .NET or different programming languages) as well as micro-cosmic components such as frameworks may become obsolete. Development paradigms, frameworks, and domain specific applications become the victims of opportunity are influenced by the skills of IT staff in companies, which in turn are influenced by the educational portfolio of universities.

Transferring this lifecycle problem to a lower abstraction level of class design, the strategy design pattern would quickly come into the mind of an experienced Java developer. They would implicitly apply programming knowledge to an interface, not

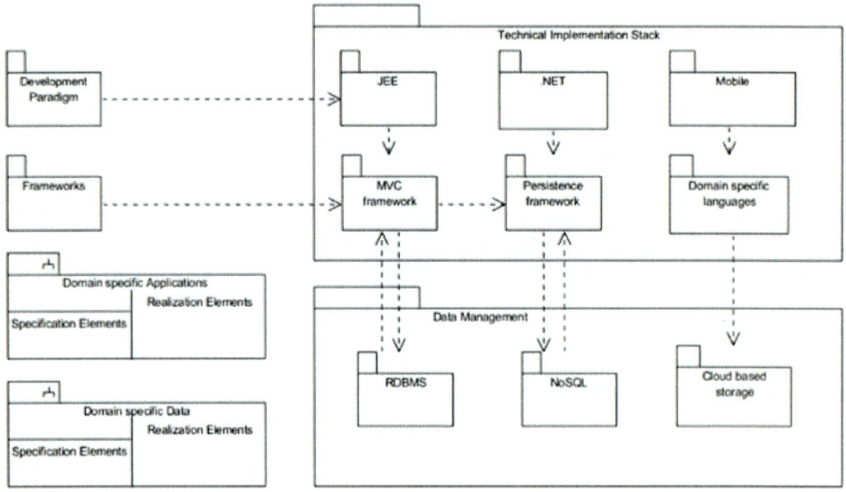


**Fig. 5** High level abstractions using layers and compositions

to an implementation, which is key to staying flexible over time and being able to seamlessly exchange a class with a different class.

Many concepts of coupling and decoupling hold true at different levels of abstraction, be it for the solution of the lower level, class-oriented software design, be it at the higher abstraction level applied to architectural lifecycles. The conceptual design skills learned from the microcosm can be transferred the macrocosm of traditional programmed systems. However, could the acquired skills also be useful in the context of AI and machine learning? Which concepts need to be evolved and which need to be kept when designing systems are based not only on programming but also on training? This adds a fundamentally different kind of behavior as macrocosm "boxes" to the eco-system which will be addressed in sections below.

## 5.2 OO Thinking in the Context of AI and Machine Learning

Given the fact that many ANN and deep learning-based components show an excellent performance in some specific use cases, while failing in others, one has to think about how to incorporate them into system architectures. Parts of systems will undergo major changes because dedicated trained AI components will become superior to their programmed counterparts. However, how can we reuse or expand them? To expand functionality in the OO-based programming paradigm one has two options to add functionality to a set of existing classes: inheritance or delegation. What will be the abstraction level at which AI based components can be reused or expanded? Will it be on the microcosmic or macrocosmic level? Could AI components that train rather than program model behavior inherit existing functionality?

For ANNs, there is a method of so-called transfer learning [18], which is comparable to inheritance. A given ANN trained in thousands of GPU hours, capable of recognizing and labeling a thousand different classes of pictures, can be re-used to learn how to label an additional set of classes. The existing ANN with its weight layers is preserved and can be retrained with much less GUP hours compared to the initial training. In fact, this new ANN somehow "expands" the original ANN to solve the classification problem of a new or expanded target domain. This could be seen as an analogy to classes that can be expanded by additional methods, forming the basis of current class hierarchies in OOA and OOD.

In addition, a considerable amount of research is being invested in studying the hierarchical classification of data and the error propagation in classification trees [19]. Hierarchies and tree-like structures can be used for supervised learning or can be learned with the help of reinforcement learning and have shown improved performance over sequential models [20, 21].

With the increase of new AI-based components, the system architectures will embrace the new capabilities of these novel components, and new architectural patterns will evolve. Architectural design thinking will most probably still use boxologies and incorporate novel abstractions. It is most likely that traditional concepts of OOA and OOD, where encapsulation, inheritance hierarchies and delegation patterns are key, will be transferred to the design of AI-enriched IT systems. As a matter of fact, van Harmelen and ten Teije [22] already proposed a set of design patterns, where they propose an informal graphical boxology notation to express their findings about different combinations of hybrid learning and reasoning systems. Thus far, they analyzed current literature in terms of existing combinations of machine learning and knowledge engineering and expressed them as compositional design patterns. If transfer learning will evolve as a kind of inheritance mechanism in ANNs, it could lead to the finding of further design patterns that combine both alternative extension mechanisms—inheritance and delegation.

## 6 Conclusion and Outlook

OO thinking and appropriate abstraction skills are crucial not only for learning current programming languages but also for understanding current IT systems from microcosmic design patterns to macrocosmic architectures. We offer prospective students two different types of self-evaluation tests to check their aptitude and abstraction skills that are fundamental for our course of studies. While the initial SET focused more on the courses of the assessment stage, the second test OO SET is not only valid for the assessment stage, but could serve as an aptitude test focusing on all technical modules from the entire course. The questions are categorized according to the different levels of the Bloom taxonomy. Further categorizations could be integrated using AI algorithms, where we presented two different use cases: performance prediction and student clustering.

However, OO thinking is not only present in current IT architectures, it will also have an influence in the future, when system architectures will be based on traditional components that are programmed as well as AI components that are based on machine learning. New architectural patterns will evolve where traditional ideas of OOA, OOD and appropriate design patterns will be transferred into a hybrid world of AI-extended hybrid IT systems. It seems that the re-use mechanisms similar to inheritance in OOD in the form of transfer learning of ANNs are still in their infancy, but the ability to perform software design with the help of different kinds of boxologies will most probably continue because it is one of the fundamental abstraction patterns of software engineers. It is well known from the Bloom taxonomy that transferring knowledge from one domain to another is considered to be the highest—which will probably take some time before the maturity level of the available architectural design patterns for the hybrid world is comparable to the existing set of patterns where components are programmed.

## References


1. Gamma, E., Helm, R., Johnson, R., Vlissides, J.: Design Patterns: Elements of Reusable Object-Oriented Software. Addison-Wesley Longman Publishing Co., Inc, Boston, MA, USA (1995)
2. Alphonce, C., Ventura, P.: Object orientation in CS1-CS2 by design. SIGCSE Bull. **34**, 70–74 (2002). https://doi.org/10.1145/637610.544437
3. Bennedssen, J., Caspersen, M.E.: Abstraction ability as an indicator of success for learning computing science? In: Proceedings of the Fourth International Workshop on Computing Education Research, pp. 15–26. ACM, New York, NY, USA (2008)
4. Bennedsen, J., Caspersen, M.E.: Abstraction ability as an indicator of success for learning object-oriented programming? SIGCSE Bull. **38**, 39–43 (2006). https://doi.org/10.1145/1138403.1138430
5. Or-Bach, R., Lavy, I.: Cognitive activities of abstraction in object orientation. ACM SIGCSE Bull. (2004). https://doi.org/10.1145/1024338.1024378
6. Kramer, J.: Is abstraction the key to computing? (2007)
7. Waite, J., Curzon, P., Marsh, W., Sentance, S.: Abstraction and common classroom activities. In: Proceedings of the 11th Workshop in Primary and Secondary Computing Education on ZZZ—WiPSCE '16 (2016)
8. Bloom, B.S.: Taxonomy of educational objectives, the classification of educational goals—cognitive domain (1956)
9. Bennedsen, J., Schulte, C.: A competence model for object interaction in introductory programming. Proc. 18th Work. Psychol. Program. Interes. Gr. 215–229 (2006)
10. Okur, M.C.: Teaching object oriented programming at the introductory level. J. Yasar Univ. **1**, 149–157 (2007)
11. Karagiannis, D. Telesko, R.: Wissensmanagement: Konzepte der künstlichen Intelligenz und des Softcomputing. Oldenbourg (2001)
12. Plagge, M.: using artificial neural networks to predict first-year traditional students second year retention rates. In: Proceedings of the 51st ACM Southeast Conference, pp. 17:1–17:5. ACM, New York, NY, USA (2013)
13. Okubo, F., Yamashita, T., Shimada, A., Ogata, H.: A neural network approach for students' performance prediction. In: Proceedings of the Seventh International Learning Analytics & Knowledge Conference, pp. 598–599. ACM, New York, NY, USA (2017)
14. Wei, X.: Student achievement prediction based on artificial neural network. In: 2011 International Conference on Internet Computing and Information Services, pp. 485–487 (2011)
15. Oja, E., Kaski, S.: Kohonen maps. Elsevier (1999)
16. OMG: UML 2 Specification. https://www.uml.org/#UML2.0
17. Shaw, M., Clements, P.C.: A field guide to boxology: preliminary classification of architectural styles for software systems. In: Proceedings of the 21st International Computer Software and Applications Conference, pp. 6–13. IEEE Computer Society, Washington, DC, USA (1997)
18. Weiss, K., Khoshgoftaar, T.M., Wang, D.: A survey of transfer learning. Springer International Publishing (2016)
19. Naik, A., Rangwala, H., States, U.: Inconsistent node flattening for improving top-down hierarchical classification
20. Jacob, A.P., Lin, Z., Sordoni, A., Bengio, Y.: Learning hierarchical structures on-the-fly with a recurrent-recursive model for sequences. In: 3rd Workshop on Representation Learning for NLP, pp. 154–158 (2018)
21. Salakhutdinov, R., Tenenbaum, J.B., Torralba, A.: Learning with hierarchical-deep models. IEEE Trans. Pattern Anal. Mach. Intell. **35**, 1958–1971 (2013). https://doi.org/10.1109/TPAMI.2012.269
22. van Harmelen, F., ten Teije, A.: A boxology of Design Patterns for Hybrid Learning and Reasoning Systems. J. Web Eng. **18**, 97–124 (2019). https://doi.org/10.13052/jwe1540-9589.18133

# Adapting the Teaching of Computational Intelligence Techniques to Improve Learning Outcomes

**Thomas Hanne and Rolf Dornberger**

**Abstract** In the Master of Science program Business Information Systems at a Swiss university, the authors have been teaching artificial intelligence (AI) methods, in particularly computational intelligence (CI) methods, for about ten years. AI and CI require the ability and readiness of a deeper understanding of algorithms, which can hardly be achieved with classical didactic concepts. Therefore, the focus is on assignments that lead the students to develop new algorithms or modify existing ones, or make them suitable for new areas of applications. This article discusses certain teaching concepts, their changes over time and experiences that have been made with a focus on improving students' learning outcomes in understanding and applying special AI/CI methods such as neural networks and evolutionary algorithms.

**Keywords** Computational intelligence · Artificial intelligence · Teaching · Learning assessment · STEM

## 1 Introduction

In research as well as in the labor market, STEM skills are strongly required—beyond all technical disciplines—also in business and society in general, and often provide above-average job and income opportunities for qualified students [1]. However, teaching STEM subjects (Sciences, Technology, Engineering, and Mathematics) provides particular challenges (see, e.g. [2]). Students often have difficulties in understanding and learning the respective subjects, which may have other reasons than real

T. Hanne (✉)
School of Business, Institute for Information Systems, FHNW University of Applied Sciences and Arts Northwestern Switzerland, Riggenbachstrasse 16, 4600 Olten, Switzerland
e-mail: thomas.hanne@fhnw.ch

R. Dornberger
School of Business, Institute for Information Systems, FHNW University of Applied Sciences and Arts Northwestern Switzerland, Peter Merian-Strasse 86, 4002 Basel, Switzerland
e-mail: rolf.dornberger@fhnw.ch



R. Dornberger (ed.), *New Trends in Business Information Systems and Technology*, Studies in Systems, Decision and Control 294,
https://doi.org/10.1007/978-3-030-48332-6_8

or imaginary deficiencies in their respective competencies (e.g. analytical thinking) to insufficient pre-arrangements or inadequate teaching methods.

Computational intelligence (CI), comprising mainly the nature-inspired artificial intelligence (AI) methods and further metaheuristics, is one of the fields of science at the cutting edge of STEM disciplines. It has become particularly important during the last ten years along with the new rise of artificial intelligence (see, e.g. [3]), which is assumed to provide some of the most important changes and disruptions in society since the introduction of computer and information technologies in the middle of the last century.

Usually, CI is not specified by an unambiguous definition but by enumerating subareas (see, e.g. [4]), in particular

- Fuzzy logic
- Neural networks
- Evolutionary computation
- Swarm intelligence.

Fuzzy logic provides theoretical insights, modeling approaches and methods to cope with problems, which can hardly be described by traditional binary logic and therefore contain elements of vagueness of knowledge in order to cope with uncertainty similar to human reasoning. (Artificial) neural networks mimic the function of real nerves and nerve nets as observed in animals and humans. It is one of the most important approaches based on machine learning in current AI developments and, nowadays, finds particular attention in the field of deep learning. Evolutionary computation is based on a simulation of biological evolutionary processes (as first described by Darwin) to find superior (optimal) solutions to complex problems. Even with moving objectives, evolutionary computation is able to continuously adapt the solution to new conditions just as biological evolution. Swarm intelligence comprises a number of similar approaches to solving complex problems, which are based on strategies found in biology, especially the behavior of animal swarms (i.e. flocks, packs, hives), which may emerge to a complex problem solving behavior, which is not shown in the individual behavior of animals.

The CI algorithms derived from these approaches are used to find good alternative solutions, optimizing candidate solutions, identifying patterns in data, and mapping input data to possible outputs. In general, CI methods work in a static context of the problem, as well as in time-dependent, changing problems, where some of these CI methods are also used for controller design. A recent development is to apply CI methods in robotics to make the robots "more intelligent", as swarm intelligence allows the self-organizing of a bunch of robots in a swarm.

These approaches have in common that they can deal with poorly structured and/or difficult-to-solve problems and mostly rely on solution concepts that are adapted from processes in nature. Therefore, such concepts are often denoted nature-inspired methods. As they are often based on incomplete or uncertain knowledge or allow for good, although not optimal solutions, e.g. by using heuristics or metaheuristics, soft computing or nature-inspired computation are other expressions to refer to this field of science.

The competencies to understand and apply CI are manifold. On the one hand, there are the algorithms, which require skills in programming and software engineering for understanding and coding the algorithms or for their adaptation or further development. On the other hand, the treated problems are often complex optimization problems that require a significant mathematical understanding to define the optimization problem with its search space and constraints correctly. Mathematics is also (for some part) necessary to understand and analyze the respective algorithms. In addition, the power of the CI methods lies in solving real-world problems, which requires profound skills to abstract the complexity of the world up to computational models, as e.g. in computational sciences or operations research.

The goal of this article is to present and discuss the authors' special teaching concepts. The changes of the teaching concepts over time and the authors' experiences are discussed with a focus on improving the students' learning outcomes in understanding and applying special AI/CI methods. In the presented case, the focus is set on the personal experiences while teaching neural networks and evolutionary algorithms and similar heuristics. However, it can be assumed that many other lecturers teaching AI methods face similar problems and will benefit from the reported measures and generalized statements.

In Sect. 2 of this chapter, we discuss selected related work. Our teaching concepts related to CI are described in Sect. 3. In Sect. 4, we discuss reasons for several changes of the course concept during the last ten years. Evaluation aspects regarding the course success are considered in Sect. 5. Conclusions are provided in Sect. 6.

## 2 Related Work

In general, there is very few published knowledge about teaching computational intelligence. It is possible to find various more or less detailed descriptions of university courses related to CI, but few insights into why they were designed as they are. In addition, little is known about the evaluation of different course designs.

Although [5] explicitly addresses CI, there is little specific insight into how to teach such topics. In [6] more specific aspects of CI teaching are considered. As one of the few concrete examples related to individual CI courses let us mention the paper by [7]. The didactic setting has some similarities with our teaching approaches, which we report on later, e.g. a student project assignment, but stronger focus is laid on traditional teaching and assessment. For instance, the project work only makes up 30% of the overall mark, whereas it is 70% in our case. While that course is mainly intended for undergraduate students, ours is offered at the Master level. Unlike the course described, our study program no longer includes any more specialized follow-up courses. Both courses have in common that they strongly support the fact that the project results should be sufficient for a scientific publication.

In [8], an elective graduate-level course on CI related to engineering applications is considered, which strongly focuses on term projects similar to our course. In addition to the standard CI approaches, the use of swarm robotics is mentioned, which is a topic also frequently offered in our course. In addition, summer internships related to CI are offered in the considered study program. Apart from some general remarks on STEM education, the didactic setting of the training under consideration is not described in detail. This also holds for evaluation results. Samanta and Turner [9] addresses a course, which also includes some CI contents, but the main focus is on mechatronics and robotics. Stachowicz [10] is another example that focuses on a single CI-related course. Most of the time, however, content is described, while details about the educational settings remain largely unclear. In [11], an overview of CI courses at different universities is provided. In most cases, however, only content is briefly described, while educational settings are not further discussed.

In some cases, such as [12], insights are reported from courses that focus on more specific content (such as a design optimization method) in a related university course. Another example is [13], which focuses on a tool based on a particular CI method (particle swarm optimization) for its use in a teaching setting. However, from our perspective, there are no particular reasons to use the specific "teaching implementations" of such methods. In contrast, we assume that it makes more sense to use regular implementations, which might be a better choice to have more freedom to use them later, e.g. when working in a company. Some other publications focus on the use of gamification or serious games related to teaching computational intelligence (e.g. [14]).

However, if we consider the much larger area of STEM disciplines, there is quite a lot of published work related to teaching aspects. We only mention a few of them, since they mostly give us only rather general insights of what might be a useful setting for a CI-related course. Often, these publications are too broad to allow for useful conclusions and practical applications. For instance, [15] discusses ideological aspects rather than concrete problems and suggestions related to the teaching of information and communication technologies (ICT).

The focus in [16] is also quite broad and includes different bachelor, master, and Ph.D. study programs in the areas Financial Management, Management of Tourism, Applied Computer Science, Information Management, and Information and Knowledge Management. The derived recommendations are quite concrete and based on abilities and skills required in industry and business. However, recommendations are not as detailed as required for the design of individual courses.

An even broader survey of study settings in Computational Science and Engineering is given in [17], which analyzes respective study programs from several universities including the content of courses, curricula, and the degrees offered. Although the topic appears to be more specific, the focus here is not on one single institution, and details of teaching concepts are mainly missing. In [18], the situation in teaching in electrical and computer engineering is considered. While the main focus of this paper is quite broad, it provides some insights that are also relevant to our teaching of CI, such as the emphasis on project–problem-based learning.

Detailed aspects of learning in mathematics are considered in [19]. For instance, they point out the learner's difficulties in the (re-)construction of general mathematical knowledge. Since more general mathematics content than in CI is considered, the aspects of problem-based learning and individual assignments are not so much taken into account.

Although teaching concepts for STEM subjects and selected AI methods (e.g. decision rules) at school and university level are quite well researched and understood, guidance to teach sophisticated CI methods (such as evolutionary computing, swarm intelligence, etc.) at university level is still lacking. That is why we present the case of reporting, evaluating, and discussing our continuous teaching approaches and share our experiences from more than ten years of lecturing CI at the University of Applied Sciences and Arts Northwestern Switzerland FHNW.

## 3 CI in the Master of Science Program in Business Information Systems

We teach CI within a course of the Master of Science (M.Sc.) program in Business Information Systems (BIS) at a school of business in Switzerland. The study program can be studied full- or part-time, which is rather an administrative distinction of an assumed study period of three or five semesters in order to obtain the required credit points. The M.Sc. BIS includes four core courses and 14 electives including the CI course to provide substantial opportunities for choice and specialization among the students. The core courses are not or only marginally related to the required skills in programming or mathematics. The only preparation in the M.Sc. BIS program provided for programming-related aspects is a short pre-course in programming that will give at least a short introduction to software development, algorithms, and programming to those students holding a bachelor degree in Business Administration or similar areas, and who have little background in these matters. Consequently, there are almost no basics from the study program, which could facilitate the learning in the CI course.

In addition, the students come from heterogeneous backgrounds in relation to their countries of origin and their culture, but also taking into account the knowledge they have acquired in their previous bachelor's studies, their apprenticeships, and work experience. Some of them have a bachelor degree in information systems, while others come from other areas such as business administration, computer science, engineering, or social sciences. Especially students with a business administration background make up a significant part of the student population with no or little background in programming and software engineering. Moreover, the mathematical background among the students is often rather weak, because many of them followed the way of apprenticeships with more practice-oriented teaching and learning instead of a stringent high school education.

Another aspect is that our university belongs to the group of "Universities of Applied Sciences", which put a stronger focus on practically applied content. While this is preferred for subsequent work in business and industry, theoretical skills (such as in mathematics) can be taught in a reduced way. The above-mentioned difficulties are particularly relevant for teaching the CI course.

A further aspect that impedes the acquisition of missing skills during the course semester is the limited amount of time among students. In addition to other learning requirements and personal requests, today, more than two thirds of the students study part-time and have a workload from their jobs of about 60%, occasionally up to 100% of full-time workload. Today, we observe that the focus of many students has shifted away from the idea of learning new topics at the university, as students want to obtain a Master's degree with minimal effort.

The course on CI was introduced in 2009, shortly after the start of our master program in Business Information Systems in 2008. In the meantime, students can begin the M.Sc. BIS program twice a year, in autumn and spring. The course on CI is based on approximately twelve lecturing blocks of four teaching hours ($4 \times 45$ min) and one presentation block during a semester of about 15 weeks [20]. The course is usually completed by students in the second, third, or fourth semester. Classroom lessons are used for traditional lectures, exercises, student presentations, and discussions. According to the European Credit Transfer and Accumulation System (ECTS), the course yields six credit points, which corresponds to approximately 180 working hours per student (including self-study time).

The lecturing part is mainly for teaching basics in CI with some focus on business applications as discussed in [21]. From the CI subareas mentioned above, we put a stronger emphasis on the subareas of evolutionary computation and swarm intelligence because we assume that these fields are still underdeveloped in various practical applications (see, e.g. [22]). The more important part of the course is based on student assignments. In these assignments, the students are expected to familiarize themselves with a particular problem (usually an optimization problem) and/or a solution approach (usually an optimization method), usually based on a given publication. In addition, the following types of tasks are typical for the assignments:

- The considered problem should be solved by applying a problem-solving method different from the original problem-solving method,
- a different type of problem (similar to the considered one) has to be solved with the original problem-solving method or another one,
- variations of the methods are explored,
- computational studies are done, which are more extensive than those reported in the literature.

For that purpose, the students (in groups or alone) are given sample reference papers related to their chosen topic and are expected to make themselves familiar with the optimization problem and/or suitable methods for solving it. Then they search for additional references, e.g. other problem types or further methods, which could be used. In the following, the students are expected to implement another method to solve a given problem, or to apply a given method to a different type of problem, or to do some modifications regarding an existing method.

In the past, we have strongly recommended that the students use the OpenCI software suite for this implementation work, which has been developed at our university in cooperation with some other institutions [23, 24]. In recent years, we have opened the choice of a suitable software platform: We also encourage the students to use a different framework or implement or modify a stand-alone code. After that, they are expected to complete computational tests with the considered settings regarding problems and solution approaches (e.g. exploring different parameter settings) and a subsequent analysis of results. In the end, they have to submit a scientific paper that should be similar to a typical conference paper, and all the artifacts elaborated during this project, such as the developed software, a short user guide, and the data used.

As discussed above, these tasks present significant difficulties for the students due to the required mathematical and software engineering background. In particular, we try to alleviate the software engineering difficulties in the following ways: In the beginning of the course, a general introduction to the programming-related aspects is given, including an introduction to the Java-based OpenCI framework. For instance, the students learn about the installation and architecture of this tool and related software such as the integrated development environment Eclipse. In addition, they learn about the OpenCI architecture and how to embed a new solution algorithm or a new type of problem into OpenCI. OpenCI is meanwhile a rather large software framework including various implementations of problem representations, solution algorithms (basic algorithms for search and optimization and machine learning), a graphical user interface, and further tools for visualization and data analysis, and additional libraries. Therefore, it is already a challenge to understand the basic ideas of this framework despite the endeavor from our side to make it as simple as possible. The maintenance of OpenCI and the integration of the students' implementation into one common version of OpenCI is always a great challenge for us lecturers.

In the software development-related part within the CI project, the students learn how to embed a new problem representation as a (Java) class and/or a new algorithm, each of them being represented as classes. While the lecturing part serves a better theoretical understanding of problems and solutions (including some related mathematics), the students receive a basic preparation for writing a scientific paper in a course on research methodology and a related project to be elaborated at the beginning of the study program. However, bringing everything together—software development, mathematics, scientific writing, and the specific contents treated in the CI course—is certainly a challenging task.

It has turned out to be useful to define milestones for the student projects such as intermediate presentations of their work. On the one hand, this allows us (i.e. the lecturers, but also other students) to give feedback and discuss problems occurring in a student group. On the other hand, the students are requested to start early with their work and work continuously during the semester, which should help to avoid situations with too much time pressure at the end of their projects. During previous semesters of the course, there were one or two mandatory presentations without grading. During the last semester of the course (spring semester, 2019) we changed that to a voluntary presentation, but with (mandatory) graded deliverables in form of short papers related to the topics. On the one hand, the grading aspect should put more emphasis on the quality of these intermediate deliverables. On the other hand, we skipped any other deliverables for grading apart from the student assignment (research paper) and the two intermediate papers.

When we started the module in 2009, it was accompanied by a regular written exam that focused on the contents of the lecture part, i.e. mostly basic CI contents. Later (in 2015), we changed this to an oral exam at the end of the semester, which included, in addition to already taught contents, a part related to the group assignments. After that (in 2017), the oral exam was skipped and replaced by individual tasks during the semester. These individual tasks were responsible for 30% of the overall grade, whereas the significance of the "big" group assignment was increased to 70%. In 2019, the individual tasks became related to the respective group assignments. The reasons for these changes were that the students are much more motivated to focus on their assigned topic and that learning here goes much deeper than the general lecture contents. As a disadvantage, however, the familiarity with a general basis of CI, as taught during the lectures, can be lost, since it is no longer necessary for the grading success of the students.

## 4 Reasons for Adapting the Course Concepts

There are several reasons why we have changed the course concept repeatedly over time and why, in particular, the group assignment was upgraded with respect to its importance in student evaluation. First, this refers to the course objectives designed to make students familiar with solving real-world problems. With regard to CI (and many other topics), this means that it is not sufficient to teach techniques such as modeling, simulation, and optimization to manage complex systems. Instead, it is necessary for students to become strongly involved with related problems and apply appropriate methods to solve them themselves without detailed supervision. Such an approach is usually denoted as problem-based learning (PBL) and usually assumes that students work in collaborative groups and learn by resolving complex, realistic problems [25]. It is supposed that this learning concept improves process skills (teamwork, project management skills, but also autonomous learning skills and other metacognitive skills), which we consider as particularly relevant at the Master level. In addition, it is assumed that PBL increases the students' motivation and engagement [25]. This assumption has been confirmed by our frequent discussions with students in the course (but also in other courses) and was also expressed in course evaluations conducted among students (see, e.g. [26] for further details).

Another reason for a stronger focus on complex assignments can be traced back to Bloom's taxonomy [27]. It distinguishes the knowledge-based aspects of learning in six levels of objectives (each represented by a characterizing noun and describing verbs, denoted here in parentheses): 1. Knowledge (to know), 2. comprehension (to understand, to demonstrate an understanding), 3. application (to apply, to solve problems in new situations), 4. analysis (to analyze), 5. synthesis (to create), and 6. evaluation (to judge). By critical examination of these levels, we found that, although CI involves complex aspects sui generis at all levels, it is difficult to evaluate the upper levels in traditional examination forms such as questions to be answered in a written exam. Based on our experience with the various learning assessment concepts in our CI courses, we are in line with the general experiences of universities: Mostly only the first three to four levels of the hierarchy can be assessed in this form, whereas questions intended for higher levels often fall back to lower levels. For instance, the evaluation of a CI method is done by repeating arguments from the literature, i.e. showing achievements on the level of "knowledge". When we changed from a written exam to an oral exam, we assumed that higher-level skills (especially levels 5 synthesis and 6 evaluation) could be better assessed in this way [26]. Although this was partly confirmed, we were still dissatisfied with the outcomes, even taking into account that the available time during oral exams was rather short for in-depth assessment and that the answers result from group experiences, which do not necessarily show a student's individual achievements. For this reason we replaced the oral exam by individual student assignments (written), which turned out to be rather easy (see grades in Table 1 for 2017 and 2018) and did not prevent collaboration among students who worked on the same assignments. As mentioned above, we changed these assignments to individual assignments related to the group assignment in 2019. Another reason for abolishing the oral exams was the provided time slot within the two weeks for examination. With an increasing number of students, it is no longer possible to realize oral exams with reasonable effort and without time conflicts with other exams.

Despite such changes in relation to individual student assessments, we still believe that the "big" group assignment is best suited to support learning objectives on all levels of Bloom's taxonomy. In particular, synthesis/evaluate (level 5) is strongly supported, as the assignment requires providing an integrated solution to a complex problem (frequently with a real-life background). Evaluation/create (level 6) is required in various ways, e.g. with respect to the decision which solution methods to use, how to adapt them, and in relation to comparisons with other approaches during computational experiments.

Another reason for repeated adaptations of course concepts is given by the accreditation regulation of the AACSB, the Association to Advance Collegiate Schools of Business, [28], where our business school is about to be accredited. The AACSB "provides quality assurance, business education intelligence, and professional development services to over 1,600 member organizations and more than 800 accredited business schools worldwide". The assurance of learning (AoL) process specified by [29] requires that a "school uses well-documented, systematic processes for determining and revising degree program learning goals; designing, delivering, and improving degree program curricula to achieve learning goals; and demonstrating that degree program learning goals have been met". Thus, the overall learning goals must be broken down to individual courses and assessed on a student's level. Since the goals are defined in terms of students' specific competences, the didactic concept from teaching to grading needs to consider them. The CI course mostly focuses on the following aspects:

**Table 1** Survey student results in CI course, 2009–2019. AS = autumn semester, SS = spring semester. Grades are measured on a scale from 1 to 6, 6 being the best achievable degree. A degree of 4 is the minimum for passing a course

| | AS 2009 | SS 2011 | SS 2012 | SS 2013 | SS 2014 | SS 2015 | SS 2016 | SS 2017 | SS 2018 | SS 2019 |
|---|---|---|---|---|---|---|---|---|---|---|
| No. of students in module | 13 | 26 | 25 | 16 | 14 | 14 | 20 | 14 | 14 | 27 |
| No. of assignments per student | 3 | 2 | 2 | 1 | 1 | 1 | 1 | 4 | 4 | 3 |
| Avg. degree for assignments | 4.87 | 4.92 | 4.9 | 4.69 | 4.81 | 4.93 | 5.01 | 4.8 | 4.5 | 4.45 |
| Type of exam | Written | Written | Written | Written | Written | Oral | Oral | Assignm. | Assignm. | Assignm. |
| Aug., degree exam | 4.62 | 4.22 | 3.77 | 4.65 | 4.91 | 4.86 | 5.06 | 5.86 | 5.69 | 5.07 |
| Avg. total degree | 4.74 | 4.57 | 4.34 | 4.67 | 4.86 | 4.91 | 5.05 | 5.12 | 4.71 | 4.64 |
| No. of student groups | 8 | 5 | 6 | 8 | 7 | 7 | 7 | 8 | 7 | 15 |
| Resulting no. of publications | 0 | 0 | 0 | 1 | 3 | 5 | 3 | 4 | 3 | |

- Our graduates can create innovative solutions and seize opportunities for given business situations,
- our students are able to (a) identify problems, risks, challenges, and opportunities for improvements or innovations, (b) create improvements and innovations.

Additionally, the CI course contributes to

- Our graduates have extraordinary methodological skills,
- our students are able to a) identify and describe a research or innovation problem, b) define/delineate and apply an appropriate research.

Following the AACSB paradigm "closing-the-loop", the teaching framework will be evaluated repeatedly. When the learning outcome of the students might be improved, the teaching concepts should be adapted. The question, however, is what changes in the teaching concept result in which effects? When do the students obtain results in terms of intended learning goals? In the next section, we present and discuss the results from the last ten years in continuation of our previous reflections concerning the design of the CI course [26, 30, 31].

## 5 Defining and Evaluating the Course Success

While the accreditation by AACSB focuses on the program evaluation by assessing the students' learning outcome with "exceed, meet, does not meet expectations", the typical forms of course evaluations at Swiss universities are the absolute marks of students (see below in this section for the grading scheme), and the feedback from the students regarding their satisfaction with the course. Regarding the latter, let us mention again that a regular course evaluation based on students' feedback is done, but mostly the number of responding students is rather small so that it does not appear to be reasonable to evaluate it quantitatively. On the other hand, we received frequent feedback in informal ways (e.g. discussions during the in-class teaching). Although this feedback cannot be evaluated in quantitative form, we had the clear impression that the students unambiguously supported a stronger emphasis on group assignments in contrast to other assessment forms (such as traditional exams). In particular, they seem to prefer the stronger real-life aspects of such assessment forms, the requisites regarding creativity and self-determination, and mostly the group work aspect as well. Only for a few students, the last requirement for this form of learning (research paper and additional documents) turned out to be difficult, because of too many other responsibilities that impede group work (esp. heavy job-related workload of part-time students). However, most of them have mastered this problem well due to modern means of communication, which do not require too many personal meetings within a group of students. Thus, a rigor evaluation of the students' feedback is not possible due to the lack of clear and statistically relevant data.

A major aspect that we consider relevant for the course success, is the quality of the research papers as a main contribution of the student assignment. In order to support the possibility of getting the student papers published, we guide the students as much as possible towards research. In addition, this improves their capabilities for the subsequent Master thesis and possibly further research. To reach sufficient maturity for publication, usually some subsequent rework or extensions of the paper may be necessary. Whenever possible (i.e. the quality appears to be good enough), we support the students to get the paper published, usually as a regular conference paper including peer review, e.g. [32–37]. Students frequently like to put additional effort into a paper, when they have the chance of getting accepted in a scientific publication. In addition, we can consider the number of published papers, another less subjective measure of the students' learning outcome, which is in alignment with our learning objectives and the AACSB accreditation.

Results from the publication efforts of course-related papers are shown in Table 1 together with other details from the various course semesters. In the early semesters of the course, the possibility of getting student papers published seemed to be rather slim, but with a more focus on these aspects, it worked out quite well. Since 2014, an increasing number of resulting papers has been accepted at various scientific conferences with regular peer reviews. Obviously, the strongest year with respect to student paper publications was 2015. In relation to the slightly weaker results in 2018, let us mention that there are still a few papers from this course in the "publishing pipeline". For 2019, papers have not yet been submitted to conferences. In general, we assume that the stronger weighting of the student assignments has contributed to the observed improvements in student paper publications.

In relation to the number of students participating in the CI course, we do not see a clear trend. However, the number of students in the Master program increased significantly during the last years, so that we assume to have an increasing number of participants in the future. The comparably high number of students (27) enrolled in the 2019 course may be an indicator of a rising trend.

With regard to the grades, we observe that those for the assignments have remained relatively stable during the specified period. The grades for the written exam declined from 2009/10 to 2012, but then have increased until 2014. The grades for the oral exams appear to be slightly better than the average grade of written exams. The assignments in 2017 and 2018 resulted in much better grades. Possible reasons are collaborations among the students and (almost) no limitation of used time or perhaps the assignments were simply too easy. In addition, the students could submit a number of such smaller individual assignments, but only the grade of the three best results were counted.

Regarding the overall grade, we can assume a rather increasing trend (i.e. better grades) with interruptions in 2012 and 2018. On the one hand, this could be due to course improvements. On the other hand, frequently observed phenomena regarding grade inflation [38] could be relevant. However, the success of student papers in conference peer reviews might indicate real improvements at least with respect to group assignments. Though the increasing trend is mainly observed for the exam part (or individual assignments from 2017), the assignment part could still be relevant for grade improvements, since its percentage for the total grade increased from 50 to 70% during the total period 2009–2019.

More detailed results from the student grading are shown in Table 2. According to the Swiss grading system, 6 is the best grade, while 4 is the minimum grade for passing a course. Part grades use decimal grades, where the final grades are half grades: 6.0, 5.5, 5.0, 4.5, 4.0, 3.5, and below.

Obviously, the failure rate in the CI course is rather small, always between 0 and 4 students from a group of 13–27 students. In this statistic, we did not count students who drop the module very early (without submitting an assignment or participating in a group assignment). Although the M.Sc. BIS program allows this retreat officially until very close to the final exam, the number of these students is usually quite small (1–2 students). We also observe that there is frequently a significant number of good and very good students (grades of 5 or higher).

Table 2 also shows the standard deviation of the students' grades. These were particularly high in 2015, 2018, and 2019, whereas they remained at a similarly lower level during other years. Since the course settings were different in 2015 and 2018/2019, while some strong differences can be observed for some years with almost identical settings and course sizes (e.g. 2017 vs. 2018), we assume that the reasons are of statistical nature. For instance, we observe frequently that the student population from year to year changes significantly, i.e. in relation to their background. This could be one of the reasons for the changes in the standard deviation of grades.

## 6 Conclusions

Many of the decisions regarding the design of the CI course and its adaptations done over time are not deeply founded on general didactic knowledge or on our specific experiences, how the best CI course might look. On several occasions, we aimed at finding better settings in a try-out-fashion instead.

Designing a CI course (but other courses as well) is similar to solving complex problems with CI methods: One usually does not know what a good design might be and experiments with different "candidate solutions" in an unknown "fitness landscape". Unfortunately, we cannot make as many trials as, for instance, an evolutionary algorithm would during hundreds or thousands of generations. The fact that (a) every year we have another group of students with different background knowledge and (b) we only teach this class of students once makes our optimization problem very complex to provide a (near) optimal solution (course curriculum).

In addition, we must stick to the feasibility of solutions taking into account various restrictions on our course design (such as from the study program, institutional requirements, etc.). However, we believe that it makes sense to experiment with course design parameters and that previous efforts have mostly resulted in observable improvements.

We will continue with our modifications to the course and hope to gain further insights into effective and efficient course designs, also from various other researchers who document their didactic settings and experience.

**Table 2** Detailed student evaluations of CI course and standard deviations of grades, 2009–2019

| Distribution of grades | AS 2009 | SS 2011 | SS 2012 | SS 2013 | SS 2014 | SS 2015 | SS 2016 | SS 2017 | SS 2018 | SS 2019 |
|---|---|---|---|---|---|---|---|---|---|---|
| 6 (rounded) | 1 | 0 | 0 | 2 | 1 | 4 | 1 | 2 | 4 | 3 |
| 5.5 (rounded) | 3 | 3 | 0 | 0 | 4 | 4 | 7 | 7 | 5 | 11 |
| 5 (rounded) | 3 | 7 | 5 | 5 | 4 | 2 | 7 | 2 | 2 | 4 |
| 4.5 (rounded) | 3 | 8 | 10 | 4 | 1 | 0 | 3 | 2 | 0 | 1 |
| 4 (rounded) | 3 | 4 | 7 | 5 | 4 | 2 | 2 | 0 | 1 | 4 |
| Below 4 (failed) | 0 | 2 | 3 | 0 | 0 | 2 | 0 | 1 | 2 | 4 |
| Stddev | 0.587 | 0.541 | 0.509 | 0.609 | 0.610 | 1.060 | 0.534 | 0.658 | 1.583 | 1.295 |

A valuable guidance for designing course curricula is provided within accreditation processes. As our school of business is in the AACSB accreditation process, the regulations for course curricula made us define how we measure the assurance of learning of our students. In the CI course, we formulate traits such as "Can the student understand the optimization problem (by identifying the problem instance and the optimization method)?" (level 2, understand, following Bloom) up to "Can the student create a computational optimization experiment to generate optimization results?" (level 6, evaluation/create, following Bloom).

## References


1. Rothwell, J.: The Hidden STEM Economy. Brookings Institution. https://www.brookings.edu/wp-content/uploads/2016/06/TheHiddenSTEMEconomy610.pdf (2013). Accessed 8 Nov 2019
2. Bybee, R.W.: The Case for STEM Education: Challenges and Opportunities. NSTA Press, the United States of America (2013)
3. Pan, Y.: Heading toward Artificial Intelligence 2.0. Engineering **2**(4), 409–413 (2016)
4. Engelbrecht, A.P.: Computational Intelligence: An Introduction. Wiley, Wiltshire (2007)
5. Feigenbaum, E.A.: Some challenges and grand challenges for computational intelligence. J. ACM **50**(1), 32–40 (2003)
6. Kim, K.J., Cho, S.B.: Game AI competitions: an open pfor computational intelligence education [Educational Forum]. IEEE Comput. Intell. Mag. **8**(3), 64–68 (2013)
7. Venayagamoorthy, G.K.K.: A successful interdisciplinary course on co[m]putational intelligence. IEEE Comput. Intell. Mag. **4**(1), 14–23 (2009)
8. Samanta, B.: Computational intelligence: a tool for multidisciplinary education and research. In: Proceedings of the 2011 ASEE Northeast Section Annual Conference. University of Hartford (2011)
9. Samanta, B., Turner, J.G.: Development of a Mechatronics and Intelligent Systems Laboratory for Teaching and Research. ASEE Comput. Educ. (CoED) J. **4**(1), 60 (2013)
10. Stachowicz, M.S.: Soft computing for advanced control applications: teacher experience. In: 10th Proceedings of the 2011 ASEE Northeast Section Annual Conference. University of Hartford (2011)
11. Minaie, A., Sanati-Mehrizy, P., Sanati-Mehrizy, A., Sanati-Mehrizy, R.: Computational intelligence course in undergraduate computer science and engineering curricula. In: Paper presented at the 120th ASEE Annual Conference & Exposition, Atlanta, 23 June 2013
12. Gerber, D.J., Flager, F.: Teaching design optioneering: a method for multidisciplinary design optimization. In: Congress on Computing in Civil Engineering. Miami, Florida, United States, 19–22 June 2011
13. Dos Santos Coelho, L., Sierakowski, C.A.: A software tool for teaching of particle swarm optimization fundamentals. Adv. Eng. Softw. **39**(11), 877–887 (2008)
14. Pustulka, E., Hanne, T., Wetzel, R., Adriaensen, B., Eggenschwiler, S., Kaba, E.: A game teaching population based optimization using teaching learning based optimization. In: Proceedings of 4th Gamification and Serious Game Symposium (GSGS´19). Neuchatel, 3–5 July 2019
15. Selwyn, N.: The use of computer technology in university teaching and learning: a critical perspective. J. Comput. Assist. Learn. **23**(2), 83–94 (2007)
16. Poulova, P., Klimova, B.: Education in computational sciences. Procedia Comput. Sci. **51**, 1996–2005 (2015)
17. Yasar, O., Landau, R.: Elements of computational science and engineering education. SIAM Rev. **45**(4), 787–805 (2003)

18. Kyriacou, G.A.: A view on electrical & computer engineering education: challenges toward convergence of different disciplines. In: Paper presented at the SEFI 40th Annual Conference, Thessaloniki, Greece, 23–26 Sept 2012 (2012)
19. Sutherland, R., Balacheff, N.: Didactical complexity of computational environments for the learning of mathematics. View Electr. Comput. Eng. Educ. Chall. Toward Converg. Differ. Discipl. **4**(1), 1–26 (1999)
20. University of Applied Sciences Northwestern Switzerland, School of Business. https://www.fhnw.ch (2019). Accessed 8 Nov 2019
21. Hanne, T., Dornberger, R.: Computational Intelligence in Logistics and Supply Chain Management. Springer International Publishing, Cham (2017)
22. Akabuilo, E., Dornberger, R., Hanne, T.: How advanced are advanced planning systems. In: The Proceedings of the International Symposium on Information Systems and Software Engineering: ISSE 2011, pp. 27–30 (2011)
23. Dornberger, R., Hanne, T., Frey, L.: The way to an open-source software for automated optimization and learning—OpenOpal. In: IEEE Congress on Evolutionary Computation. pp. 1–8. Barcelona, Spain, 18–23 July 2010 (2010)
24. Dornberger, R., Ernst, R., Frey, L., Hanne, T.: Solving optimization problems by metaheuristics using the OpenOpal-Framework—integration of travelling salesman problem and selected solvers. Work Report No. 27, University of Applied Sciences and Arts Northwestern Switzerland FHNW, School of Business (2012)
25. Allen, D.E., Donham, R.S., Bernhardt, S.A.: Problem-based learning. New Dir. Teach. Learn. (Special Issue: Evidence-Based Teaching) **2011**(128), 21–29 (2011)
26. Dornberger, R., Hanne, T.: Problem-Based Learning in teaching the module "optimization for business improvement." In: The 8th International Conference on Education, Training and Informatics (ICETI 2017), p. 5 (2017)
27. Bloom, B.S., Engelhart, M.D., Furst, E.J., Hill, W.H., Krathwohl, D.R.: Taxonomy of Educational Objectives: The Classification of Educational Goals. Handbook I: Cognitive Domain. David McKay Company. Inc., New York (1972)
28. AACBS—Association to Advance Collegiate Schools of Business: About AACSB. https://www.aacsb.edu/about (2019). Accessed 29 Apr 2019
29. AACBS—Association to Advance Collegiate Schools of Business: Accreditation Standard 8 (2013 Business Standards): Curricula Management and Assurance of Learning—An Interpretation. https://www.aacsb.edu/-/media/aacsb/publications/white-apers/aol_white_paper_standard_8.ashx?la=en (2019). Accessed 8 Sept 2019
30. Dornberger, R., Hanne, T.: E-learning and problem-based learning in teaching information systems-changing the style of teaching in the information systems programs. In: Proceedings of the 2nd International Conference on Education, Training and Informatics: ICETI, pp. 27–30. 2011
31. Dornberger, R., Hanne, T.: Problem-based learning in teaching information systems - experiences in teaching computational intelligence. In: Proceedings of the 5th International Conference on Education, Training and Informatics (ICETI 2014). Orlando, Florida, USA, 4–7 Mar 2014
32. Affolter, K., Schweizer, D., Hanne T., Dornberger, R.: Index tracking with invasive weed optimization. In: 2014 International Conference on Soft Computing and Machine Intelligence (ISCMI), pp. 110–114. New Delhi, 26–27 Sept 2014
33. Nienhold, D., Schwab, K., Dornberger, R., Hanne, T.: Effects of weight initialization in a feedforward neural network for classification using a modified genetic algorithm. In: 2015 3rd International Symposium on Computational and Business Intelligence (ISCBI), pp. 6–12. Bali, Indonesia, 7–9 Dec 2015
34. Schär, M., Brennenstuhl, M., Dornberger R.: Improving the fuzzy logic controller of a car racing competition with adjusted fuzzy sets. In: 4th International Symposium on Computational and Business Intelligence (ISCBI 2016), pp. 118–124. University of Applied Sciences and Arts Northwestern Switzerland, Olten, Switzerland, 5–7 Sept 2016

35. Graf, M., Poy, M., Bischof, S., Dornberger, R., Hanne, T.: Rescue path optimization using ant colony systems. In: 2017 IEEE Symposium Series on Computational Intelligence (SSCI 2017). Honolulu, USA, 27 Nov–1 Dec 2017
36. Pochon, Y., Dornberger, R., Zhong, V.J., Korkut, S.: Investigating the democracy behavior of swarm robots in the case of a best-of-n selection. In: 2018 IEEE Symposium Series on Computational Intelligence (SSCI 2018), pp. 743–748. Bengaluru, India, 18–21 Nov 2018
37. Lehner, J.E., Simić, R., Dornberger, R., Hanne, T.: Optimization of multi-robot sumo fight simulation by a genetic algorithm to identify dominant robot capabilities. In: 2019 IEEE Congress on Evolutionary Computation, pp. 490–496 Wellington, New Zealand, 10–13 June 2019
38. Gordon, M.E., Fay, C.H.: The effects of grading and teaching practices on students' perceptions of grading fairness. Coll. Teach. **58**(3), 93–98 (2010)

# Automatic Programming of Cellular Automata and Artificial Neural Networks Guided by Philosophy

**Patrik Christen and Olivier Del Fabbro**

**Abstract** Many computer models have been developed and successfully applied. However, in some cases, these models might be restrictive on the possible solutions or their solutions might be difficult to interpret. To overcome this problem, we outline a new approach, the so-called allagmatic method, that automatically programs and executes models with as little limitations as possible while maintaining human interpretability. Earlier we described a metamodel and its building blocks according to the philosophical concepts of structure and operation. They are entity, milieu, and update function that together abstractly describe a computer model. By automatically combining these building blocks in an evolutionary computation, interpretability might be increased by the relationship to the metamodel, and models might be translated into more interpretable models via the metamodel. We propose generic and object-oriented programming to implement the entities and their milieus as dynamic and generic arrays and the update function as a method. We show two experiments where a simple cellular automaton and an artificial neural network are automatically programmed, compiled, and executed. A target state is successfully reached in both cases. We conclude that the allagmatic method can create and execute cellular automaton and artificial neural network models in an automated manner with the guidance of philosophy.

**Keywords** Automatic programming · Meta-modelling · Allagmatic method · Structure · Operation · Complex system · Cellular automaton · Artificial neural network

---

Both authors contributed equally to this work.

---

P. Christen (✉)
Institute for Information Systems, School of Business, FHNW University of Applied Sciences and Arts Northwestern Switzerland, Riggenbachstrasse 16, 4600 Olten, Switzerland
e-mail: patrik.christen@fhnw.ch

O. Del Fabbro
Chair for Philosophy, Department of Humanities, Social and Political Sciences, ETH Zurich, Clausiusstrasse 49, 8092 Zurich, Switzerland
e-mail: olivier.delfabbro@phil.gess.ethz.ch

 
R. Dornberger (ed.), *New Trends in Business Information Systems and Technology*, Studies in Systems, Decision and Control 294,
https://doi.org/10.1007/978-3-030-48332-6_9

# 1 Introduction

Computer modelling has become an important approach in many scientific disciplines that study their respective systems. It allows to experiment digitally with the modelled system and as such to explore and test possible explanations for its functioning. This is especially important and useful for the understanding of complex systems, where the system's behaviour emerges from the interaction of local elements. Most of the computer models that have been proposed so far have their specific applications or type of problems for which they are most appropriate. E.g. cellular automata are particularly suitable for simulating complex behaviour in general [1] and development and growth processes in particular as they occur in materials [2] and biology [3]. Due to their capability to model complex behaviour in general, cellular automata have been also applied in many other and very different fields, e.g. theoretical biology [4], medicine [5, 6], and quantum mechanics [7]. Agent-based models, on the other hand, are currently mostly used in the social sciences studying the behaviour of agents such as pedestrians moving in a subway station [8]. With the availability of large training data sets and computing power, artificial neural networks and other artificial intelligence methods are widely used today in many applications, from the classification of skin cancer [9] to the solution of the quantum many-body problem [10]. Although current artificial intelligence methods seem to be general enough to apply them to many applications, they are still limited to certain problems and impose requirements on the available data and the interpretability of the solutions. Evolutionary algorithms are another type of model that has been successfully used for problems where the data might be incomplete, or an optimisation with multiple objectives is required [11]. They have been also suitable for searching for model parameters in combination with other computer models [12].

This indicates that there are specific computer models for certain problems and applications rather than a metamodel that is able to tackle any problem and in any application without limitations. It also means that by choosing a particular computer model, we might limit the possible solutions and their interpretability. E.g. if a cellular automaton is used, the solution will be represented by the way the cellular states are updated over time with a transition function. If an artificial neural network is used, the solution will be represented by the weights that are generally difficult if not impossible to interpret. This is referred to as the interpretability problem of current methods of artificial intelligence and is especially critical in applications such as medicine. It seems, however, that these methods have a large solution space and thus can potentially come up with creative solutions. Models providing more interpretable solutions, in contrast, seem to have a smaller solution space limited by the model itself. An alternative way to choose models or to build them is therefore proposed in the present study, which seeks to overcome these restrictions on possible solutions and their interpretability.

We argue that such an alternative option could be the automatic programming of computer models based on a metamodel where as little limitations as possible are imposed on the model creation. The input data would guide the concretisation of a suitable model from the metamodel. Possible solutions or models could be programmed from certain code blocks that are meaningful to humans and combined together in an evolutionary computation. In extreme cases, i.e. without limitations, program code would be written automatically by choosing symbols that are valid in the respective programming language and then models would evolve based on choosing and putting together these symbols to create program code. One can imagine that this would generate program code that would almost always fail when compiling. In addition, even if it were to compile or the compilation problem could be reduced with programming languages such as LISP or interpretative languages, it is still hard to imagine that this would create readable program code and thus interpretable for humans. Instead of allowing any valid symbol, one might limit the choice to certain model building blocks that can be chosen, adapted, and combined in an evolutionary computation. If, in addition, these model building blocks are defined in such a way that they make sense for humans, the evolutionary computation would potentially be able to develop novel solutions or models that can be interpreted by humans and where only the input data imposes restrictions.

However, this is a large problem in which, in a first step, the questions need to be addressed how to define model building blocks for automatic programming of computer models, and how to implement and combine them in a running computer program. The first question has been recently addressed in detail in another study by the authors [13]. The purpose of the present study is to answer the second question based on philosophical concepts and model building blocks from the previous study. Philosophers have thought for centuries about the structure and behaviour of objects and phenomena of many kinds. We proposed in the earlier study to proceed in the same way, particularly by using concepts developed by the French philosopher Gilbert Simondon. Via paradigms and schemes borrowed from cybernetics [14, 15], his philosophy describes the functioning of what could be called complex systems today, especially technical objects such as engines but also natural processes such as crystallisations [16, 17]. Therefore, his method seems closely related to computer modelling of complex systems that wants to achieve the same. With a slight deviation in the use of these concepts, we have integrated Simondon's concepts directly into the functioning of a metamodel for complex systems, where it uses philosophical concepts as a guideline for computation. In this sense, one could speak of philosophical computation. The aims of this study are therefore to propose an approach on how the previously defined model building blocks could be implemented and combined into computer models, and to provide first experiments of automatically programmed cellular automata and artificial neural networks from the metamodel.

## 2 The Metamodel

In this section, we describe the metamodel for complex systems and its model building blocks as developed in the earlier study [13]. These building blocks were developed according to the philosophical concepts structure, operation, and system as defined and described by Simondon [16].

## 2.1 Structure

Structure represents the spatial or topological dimension of the system, e.g. agents, cells, nodes, lattices, or grids. In this configuration, no processes and no dynamics are active. Structure represents the topological configuration of the system in its most basic arrangement.

## 2.2 Operation

Operation represents the system's temporal dimension. It considers the behaviour of the system, e.g. it looks at how cells, agents, or nodes control and affect each other. Moreover, on a more fundamental level, it also describes how structures dilate topologically over time, e.g. how single cells, agents, or nodes are formed initially. Similar to drawing a straight line on a sheet of paper, the spatial dimension of the line is formed at the same time as the temporal operation of drawing moves on [16]. In summary, operation defines how agents are generated temporally and, once formed, how they behave in their specific environment.

## 2.3 System

However, in reality, no system is composed by operations or structures alone, i.e. every system has a spatial *and* a temporal dimension. Hence, a *system* is defined by the product of structure and operation, without any concrete parametrisation. Once parametrisation begins, structure and operation find themselves in a so-called *metastable system*. This means that in metastability more and more parameters such as initial conditions of the parameter states and the dynamics of the modelled system are defined in order to have a computable model. In this sense, metastability represents a transitional regime, which is at the same time partly virtual and actual. Hence, structure and operation are initially defined in a virtual regime and while more and more parameters are included, such as initial conditions and dynamic update functions, the model itself becomes more and more concrete in metastability and finally computes in actuality (Fig. 1).

## 2.4 Model Building Blocks

These philosophical concepts were built into model building blocks of a generic computer model that are fundamental to every computer model. They are therefore

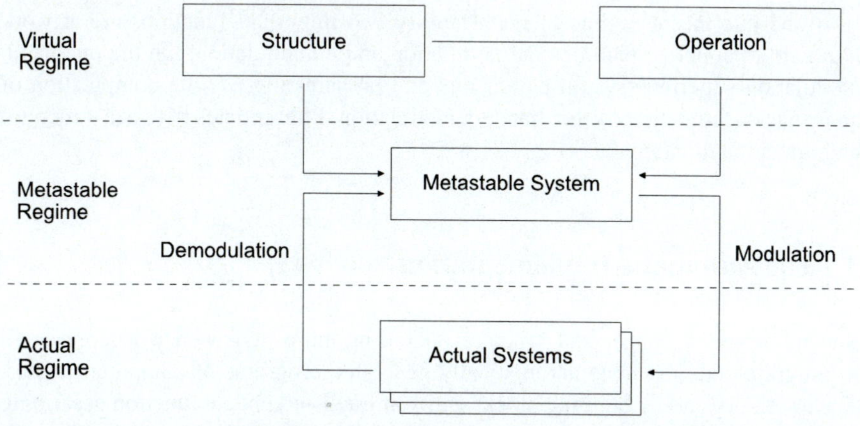


**Fig. 1** Gilbert Simondon's philosophical concepts applied to meta-modelling of complex systems

independent from the concrete models such as cellular automata and artificial neural networks and can be regarded as components of the metamodel.

On an abstract level, computer models have at least one structure and at least one operation that are described in a so-called virtual regime. The virtual regime can be regarded as abstract descriptions of the spatial and temporal dimensions of a model (Fig. 1).

Formally, structure has been described by a $p$-tuple $e$ of $p$ basic *entities* such as cells, nodes, agents, or elements forming a topology such as a lattice, grid, or network. The environment, neighbourhood, or *milieu* of an entity describing its interaction with neighbouring entities is defined in a $q$-tuple $m_i$ consisting of $q$ neighbouring entities for the $i$-th entity $e_i$ in $e$. The milieus of all entities in $e$ can furthermore be described by the matrix $\mathbf{M}$, which is structured as an adjacency matrix [18].

Operation, on the other hand, can be described with an *update function*

$$\phi : s^{q+1} \rightarrow s, \tag{1}$$

where the states of each entity $e_i$ are updated over time. An entity can have a certain state, $e_i \in s$, where $s$ defines the set of possible states. The new state of an entity $e_i^{(t+1)}$ at the next time step $t + 1$ is determined based on the states of the entity $e_i^{(t)}$ itself at the current time step $t$ and the states of the neighbouring entities in $m_i^{(t)}$ at the current time step $t$.

The model building blocks entity, milieu, and update function are the basic components of the metamodel. They were described in formal and abstract form. As they represent structure and operation, combining them can form a system, which can be interpreted as a model. If concrete parameters are fed to the system, a metastable system is created, which is then capable of acting in the actual regime. It is the initialisation of the model with given parameters ($\phi$, $\mathbf{M}$, $e^{(t=0)}$, $s$) while the execution of the model afterwards occurs in the actual regime (Fig. 1).

In the interjacent regime of metastability two important functions are at work in regard to concrete parameters: modulation and demodulation. On the one hand, modulation superimposes all parameters and prepares thereby the computation of an actual system. On the other hand, demodulation differentiates between structural and operational parameters (Fig. 1).

# 3 The Automatic Implementation

The philosophical concepts of structure, operation, and system need to be represented in program code to create automatically generated programs of computer models. The developed model building blocks entity, milieu, and update function describing the metamodel on an abstract level are in the following used to automatically map and implement the philosophical concepts into program code.

## 3.1 Representation

Structure, and therefore the entity tuple $e$ and the milieu matrix $\mathbf{M}$ are well represented with arrays or similar data containers. Generic programming is used to generically define their data type, and dynamic array containers are used to dynamically define their size. In this sense, the metamodel is independent of the type as well as the size of input data provided by the system to be modelled. It allows creating a model of the system that can be concretised in terms of data type and size by feeding concrete parameters into it to form the metastable system. We used the `vector` template in C++ for the present implementation but other dynamic array containers such as the `ArrayList` class in Java would also be possible to use.

Operation and therefore the update function $\phi$ describes the temporal dimension defining how the entities $e^{(t)}$ at time point $t$ change their states to the next time point $e^{(t+1)}$ at $t+1$ based on their current states $e^{(t)}$ and the states of the entities in their respective milieu as described in $\mathbf{M}$. Mathematical functions can be represented by functions and methods in procedural and object-oriented programming, respectively. Abstract methods could be used to define these methods in the metamodel, or classes for specific implementations could be used. For simplicity reasons, we implemented specific classes for cellular automaton and artificial neural network models in the present implementation. The parameters are defined in a generic and dynamic way given by the structure implementation. The method body, on the other hand, is specific to the application and therefore is implemented in specific classes that can be used by the metamodel to create concrete systems or models.

### 3.2 Implementation

In the present study, we used an object-oriented programming language to implement the first experiments with cellular automata and artificial neural networks. The reason for this choice was twofold: first, classes allow the abstract and complementary description of structure and operation in the virtual regime and objects allow implementing the metastable regime through initialisation. Second, they often provide dynamic and generic types. C++ was used for the implementation, but any other object-oriented programming language providing these features would have been suitable as well.

The entity tuple $e$ and the milieu matrix $\mathbf{M}$ were implemented with the `vector` template as data members of a class `System`. $e$ contains objects of the class `Entity` with the member data `state` while $\mathbf{M}$ contains either Booleans to represent relationships only or floating point numbers to represent weighted relationships between entities. The update function $\phi$ was implemented with a method in a specific class. In these first experiments, the classes `CA` and `ANN` were created, each implementing a respective `updateFunction`.

### 3.3 Automation

The present implementation allows to automatically program computer models by specifying the concrete parameters in different ways concretising entities and their milieus as well as choosing $\phi$ from different classes. It is therefore possible to provide all concrete parameters and compute the structural and operational evolution of the system or to provide them only partially varying unknowns in an evolutionary computation. In the present study, as a first step, we varied a part of the update function $\phi$ through an evolutionary computation in the cellular automaton and through a learning rule in the artificial neural network to find a given state of the system.

Automatic programming was achieved through source code generation and writing it to a file followed by programmatically compiling and running the code using `system()` in C++. The metamodel allows creating different models in the source code generation where a string containing the source code is determined. Variables, model building blocks, and specific update functions are encoded as separate string variables concatenated by the main program. The advantage of a metamodel is evident because the different model building blocks are compatible with each other independently from a specific model implementation and are therefore more easily defined in this way of creating source code. In preliminary tests, we also used the Java Compiler API as well as running it based on the concept of reflection. It led to a rather complicated programming and a code prone to errors. In addition, it is a time-consuming way to solve the current problem. We therefore favoured the less complicated and much faster C++ implementation.

### 3.4 Computation

We have seen that with the transition from the metastable to the actual regime, concrete parameters concretise more and more the actual system. Hereby, further parameters can be categorised, i.e. within structure, the number of entities $p$ and their set of possible states $s$ and the concrete states $e$ define in total the initial conditions $e^{(t=0)}$. On the other hand, within operation, specific dynamic rules, boundary conditions, and the size of the milieu $q$ define in total the update function $\phi$. If all these single parameters are known, then we have a complete system. However, if there are certain unknowns, the system is incomplete. Therefore, one is allowed to conclude that as soon as unknown parameters occur, demodulation processes arise, in which structural and operational parameters can be differentiated. On the contrary, as we have seen above, if concrete parameters are brought together, modulation processes are at work, since structural and operational parameters are more and more superimposed (Fig. 1).

To search for these unknown parameters, several steps have now to be taken. First, a problem has to be defined, i.e. one or several unknown parameters have to be defined as such, which then become the subject of investigation. This also means that while several concrete parameters are unknown, several others are known. Therefore, in a second step, new actual systems can be created by modulation starting from these known parameters and by giving the unknown parameters concrete values, e.g. in an evolutionary computation giving the asked parameters random values. It is possible to automatically implement these modulation processes with the present approach. In the main program, the given structures are declared and then used to instantiate a respective `System` object via the constructor, creating the virtual regime. A first system is then created by initialising known parameters and setting the asked parameter with a random value, which is the transition from the virtual to the metastable regime. The `updateFunction` method is then repeated until the system leads to acceptable results in the given iteration or until a maximum number of time steps is reached. In a third step, demodulation processes are required to check whether the unknown parameters have been found. Thus, every time a certain number of time steps have been computed, the concrete parameters of the newly obtained system are disaggregated via demodulation. In a fourth step, the new parameters are then compared to the ones that were initially searched for. If there is no correlation or the result is not satisfying, certain values within the initially unknown parameters are changed in order to compute further modulation processes, which then are subjected again to demodulation and comparison. As a result, if one searches for certain unknowns, demodulation and modulation processes are applied interchangeably. The comparison of new and old parameters is required to evaluate whether a system reveals an acceptable result. This is achieved by calculating the match or overlap with the target, quantified as a number between 0 and 1, which is similar to the concepts of fitness in evolutionary computations or loss function in artificial neural networks. The closer it gets to 1, the more significative the new system is and the closer it is to the unknown parameters searched for. The search of these unknowns can be and is in the present work achieved with an evolutionary computation in a cellular automaton experiment and with a perceptron learning rule in the artificial neural network experiment. However, it is important to notice the difference that here the model is created automatically and from abstract concepts, while in the usual case the model is given and implemented and executed manually.

## 4 The First Experiments

In the following, the basic building blocks of the metamodel are concretised in a cellular automaton and an artificial neural network.

### 4.1 Cellular Automaton

A cellular automaton consists of a discrete cellular state space $\mathcal{L}$ on which the dynamics unfolds, a local value space $\Sigma$ that defines the set of possible states of an individual cell, boundary conditions, and a dynamic update rule $\phi$ defining the dynamics and thus temporal behaviour [19]. In the present study, we considered a two-state one-dimensional cellular automaton. It is made of cells that are all identical, and periodic boundary conditions were assumed [20]. In the one-dimensional case, this boundary condition connects both ends of the lattice $\mathcal{L}$ together forming a ring. A Boolean state was furthermore used where the cell $c_i$ at position $i$ in $\mathcal{L}$ can have one of two possible states, $c_i \in \Sigma = \{0, 1\}$. Each cell at time point $t$ is updated by a dynamic update function or local transition function

$$c_i^{(t+1)} = \phi(c_{i-1}^{(t)}, c_i^{(t)}, c_{i+1}^{(t)}), \tag{2}$$

depending on its current state $c_i^{(t)}$ as well as the states of its neighbouring cells $c_{i-1}^{(t)}$ and $c_{i+1}^{(t)}$. Therefore, there are $n_\Sigma = 2$ possible cell states and three input parameters, which leads to $2^3 = 8$ possible update rules in a two-state cellular automaton with a neighbourhood defined by the nearest neighbours.

Structure is therefore represented by the fact that there is a lattice of cells and that spatially a further cell is situated in the next time generation at $t + 1$, e.g. below the considered cells at $t$. Operation, on the other hand, is represented by the fact that within the considered neighbourhood in the lattice, each cell is initially formed and thereby directly linked to its adjacent cell as well as that these cells inform the state of the cell at $t + 1$. Only in combination, structure and operation are able to form a system. This system can then be concretised into a metastable and finally actual system or in this case a real computable cellular automaton, by feeding concrete parameters into the system and executing it.

In this first experiment, the dynamic rules were set randomly to automatically generate systems or metastable cellular automata with various update rules. The idea therefore is to generate metastable systems and automatically program them with various update rules until one of them produces a given target output system. This target system was provided, and its specific values were assumed to be the output of Wolfram's rule 110 at $t = 15$ [21] while the update rule was searched with an evolutionary computation assuming only one possible solution at a time that is randomly assigned without any selection criterion. With $\phi(c_{i-1}^{(t)}, c_i^{(t)}, c_{i+1}^{(t)})$, rule 110 is the following update function:

$$\begin{aligned} &\phi(0,0,0)=0, \quad \phi(0,0,1)=1, \quad \phi(0,1,0)=1, \quad \phi(1,0,0)=0, \\ &\phi(0,1,1)=1, \quad \phi(1,0,1)=1, \quad \phi(1,1,0)=1, \quad \phi(1,1,1)=0. \end{aligned} \tag{3}$$

Starting with the initial configuration of

$$e^{(t=0)} = 0000000000000001000000000000000, \tag{4}$$

the output or target is

$$e^{\text{target}} = 1101011001111101000000000000000. \tag{5}$$

Although the used cellular automaton model is very simple, rule 110 shows exceptional and interesting behaviour that is complex and could be considered on the edge of order and chaos [22]. It is even computationally universal, i.e. capable of running any given program or algorithm [23].

In a system where the structural parameters, i.e. the number of entities $p$ and their concrete states $e$ are known, the defined problem is to search for the operational parameters, i.e. a specific dynamic rule.

Structures of type Boolean were implemented representing two states $\Sigma = \{0, 1\}$ and 31 entities were modelled. Operation was implemented as a truth table of three input parameters representing the nearest neighbours and the considered cell itself determining one output according to the specified dynamic rules.

The experiment was repeated several times, in each case revealing Wolfram's rule 110 as a solution after less than 1000 attempts of building and automatically programming a metastable and actual system with randomly generated rules.

## 4.2 Artificial Neural Network

This model consists of a network of artificial neurons called perceptrons, where typically input neurons are fed with data and connected with middle or hidden layers of neurons, which in turn are connected to other hidden layers before they are finally connected to the output neurons that provide the result. Different network topologies are therefore possible. Each perceptron $j$ consists of an input function $in_j$ that calculates a weighted sum of its incoming activation signals $a_i$ from perceptrons $i$. The value of the input function is then used in the activation function $g$ to calculate the outgoing activation signal $a_j$ of the perceptron $j$. The output or result of an artificial neural network is thus computed by calculating the activation of each individual perceptron from layer to layer until the output neurons are reached. This type of network is called multilayer feedforward artificial neural network. The weights of the network are then adjusted through a learning rule.

In our experiment, we automatically programmed a multilayer feedforward artificial neural network based on the metamodel with entities, milieus, and an update function. A neuron or perceptron is represented by an object of the class `Entity` and the milieus by the adjacency matrix **M** also storing the weights. The entity tuple $e$ therefore contains all the perceptrons with their current states and the milieu matrix **M** the neighbourhood or topology of the network. Comparable to the cellular automaton experiment, we used the same target system $e^{\text{target}}$ and 15 layers, each with 31 perceptrons. The update function consists of the input function $in_j$ and the activation function $g$ calculating the activation signal $a_j = g(in_j)$. In the input function

$$in_j = \sum_{i=0}^{n} \omega_{i,j} \cdot a_i, \tag{6}$$

the $n$ incoming activation signals $a_i$ of neuron $j$ are weighted through a respective weight $\omega_{i,j}$ and summed up. Please note that $a_0$ is a bias weight. The activation function was defined as a threshold function

$$g(in_j) = \begin{cases} 0 \text{ if } 0.5 > in_j \\ 1 \text{ if } in_j \geq 0.5 \end{cases} \tag{7}$$

and the perceptron learning rule

$$\omega_{i,j} \leftarrow \omega_{i,j} + r(y - a_j)a_i \tag{8}$$

was used, where $r$ is the learning rate, $y$ the target or desired activation signal, $a_j$ the current outgoing activation signal, and $a_i$ the current incoming activation signals [24]. It is implemented as a method in the `ANN` class.

Here, too, the experiment was repeated several times, generally requiring less than 100 000 attempts to build and automatically program a metastable and actual system. The exact solution was not found in any of the tested cases but a match of approximately 90% was achieved comparing the solution with target entity states.

## 5 Discussion

The aim of the present study was to make use of a previously defined metamodel of complex systems for automatic programming of computer models and to generate and automatically program cellular automata and artificial neural networks. Model building blocks of this metamodel were defined based on the concepts of structure and operation in a previous study [13]. It might not be surprising that structure is formally described with mathematical tuples and has thus been implemented with some kind of array or list data structure. Philosophy, however, allows novel and creative guidance. Hence, the philosophical concepts borrowed from Simondon imply a system that consists of some local elements with connections to each other. On this basis, we defined the model building blocks entity and milieu. In line with this, from a temporal perspective, structure can be of different sizes and entity states can be of different types. Therefore, the data structure has to be implemented dynamically and generically. In addition, due to the philosophical definition of operation, it is not surprising to formally describe it with a mathematical function and thus to implement it with a method. Here, the usefulness of philosophy is to allow the operation to change structure. According to the two philosophical concepts of structure and operation, a system is created by superimposing them. It is important to highlight that structure and operation are always interconnected. While structure provides the spatial dimension for the operation to occur, operation, on the other hand, forms the evolving structure and thus defines the connections between the entities. Hence, input and output parameters of the update function have to be implemented and defined, which is part of the model building block operation. This function is application specific, i.e. it represents the functioning of the concrete system. Hence, philosophical concepts such as structure and operation can guide the definition of a metamodel for complex systems. Because of their abstractness, they suggest generic and object-oriented programming for their implementation in the virtual regime.

Moreover, starting from the abstract definition of model building blocks, additional concrete models are formed in a metastable state. Hereby, philosophy provides a framework on how to concretise the abstract model building blocks into a concrete computer model. Concrete parameters are fed into the metastable system. These concrete parameters are the initial conditions and an update function, which in turn inform the metastable system. This guides the implementation by using the concrete parameters to initialise a `System` object, which then represents the metastable system.

Since structure and operation are complementarily interrelated, it is important to emphasise that in the virtual regime, structure and operation have neither a categorical nor a hierarchical relationship. Therefore, every structure is operated, and every operation is structured. While concrete parameters are being fed into the system, thereby transitioning from the virtual to the metastable regime, the virtual regime itself is not being altered. Hence, it is the starting point in order to create constantly new metastable models within the metastable regime. By accepting the virtual regime as underlying governor, computation itself is constantly shifting from the metastable to the actual regime producing new systems, new models. This also means that modelling itself happens only within the field of concrete parameters. It is not re-entering the virtual regime of abstractness. Spoken philosophically: The starting point as much as the result of every computation is always a model or some kind of image. Even if the majority of concrete parameters is unknown, it should be possible to compute different possible outcomes, simply by giving the unknowns random values, e.g. in an evolutionary computation.

Because of the complementary relation of structure and operation and its unfolding in metastability, computation of actual systems is not reduced to the meta-level of the virtual regime. Our purpose was to create a *method*, which is able to create models of complex systems. In this sense, our method can be seen as a tool rather than a theory. We call this method the allagmatic method, because of its relatedness to Simondon's work [16]. For Simondon, structure and operation are generic concepts, which are able to describe all sorts of systems in reality. Allagmatic is derived from the Greek verb allatein, meaning change, transition, or transformation. Hence, structure and operation are not only complementary, they also influence each other reciprocally. Based on this relationship between abstract and concrete computation, the allagmatic method is highly adaptable. Since the virtual regime, except from its spatial and temporal constitution is not fixed to any specific type of model or image, it can potentially undermine all types of computer models. As we have shown, the metamodel allows to produce different types of models, such as cellular automata and artificial neural networks, if not novel kinds of models that are still undiscovered.

However, with the immense adaptive behaviour comes also a high degree of freedom, which has to be controlled in order to use the allagmatic method in a pragmatic way. This control is achieved by defining initial values based on the problem to be solved. Yet, since computation always happens within the metastable and actual regime, three problems can occur: (a) having a totally incomplete system, i.e. all parameters are unknown, (b) having a semi-complete system, i.e. certain parameters are unknown and certain are known, and (c) having a totally complete system, i.e. all parameters are fully known. Even if at first sight a problem of kind (a) seems meaningless, it might be fruitful if one considers the possibility of the allagmatic method to create totally new types of models. Here, it would be possible to explore the possible unfolding of the computational universe [21]. A problem of type (b) was the initial scenario of the presented experiments; not only could it be widened by increasing the number of unknowns, it could also be modelled by a different kind of model such as an agent-based model. Problems of type (c), where all parameters are fully known, can be looked at as being subjected to possible predictions. Starting from fully known parameters, future possible outcomes can be computed and evaluated. Notice that it is also possible for problems to transmute, i.e. if starting from a problem (b) by knowing only certain parameters and with the help of computation the unknown parameters are found, one ends up with the situation of problem (c) where all parameters are known and where from now on predictions of the behaviour of the system can be made.

We also provided first examples of formal descriptions as well as automatic implementations of a two-state one-dimensional cellular automaton and a multilayer feedforward artificial neural network using the allagmatic method. What is usually defined as a cellular automaton cell is built from the abstract entity model building block. These cells form a lattice, which is determined by the entity tuple *e*. Typically, this lattice would be implemented with an array and the milieu would not be stored for each cell as this is always defined by the neighbours and thus given from the grid structure in a cellular automaton. Here it becomes evident that the traditional way of implementing a specific model is more efficient but at the same time it is also more restrictive with respect to the topology of connections between entities. Our approach does not have this limitation but comes with a less efficient implementation. This is not an issue in the present study since the aim is to find a way to build and automatically program computer models with as few limitations as possible.

In addition to cellular automata and artificial neural networks, the metamodel defined in the allagmatic method arguably also allows to build other computer models such as agent-based models. Agent-based models are closely related to cellular automata and as such have a similar structure and operation. Even more so, they have agents and agent behaviours, which are well represented by the model building blocks entity and update function, respectively. Agents also communicate with each other from which the behaviour of the whole system emerges. This is well represented by a model building block, the milieu and output of the update function. Maybe even more importantly, the generality of the model building blocks possibly allows creating novel computer models still not known to us. The created models therefore are no longer necessarily of one known model type, they might only show certain features of them. They blur the line between distinct computer models and thus are able to produce novelty. In addition, these newly generated computer models can possibly also be interpreted by humans. They consist of the model building blocks that can be related to the philosophical concepts, which provide a description of the role of each model building block and how they are related to each other.

## 6 Conclusion and Outlook

We conclude that automatic programming of computer models can be achieved by a previously developed metamodel [13] that was guided by philosophical concepts such as structure and operation, particularly on an abstract level. Other concepts such as metastable system and concrete parameters are also useful for building concrete computer models and for executing them. Both the abstract and the concrete definitions provide guidance for mathematical description and automatic programming of computer models.

While we provided exemplified descriptions and automatic programming experiments of simple cellular automaton and artificial neural network models, further studies are required to formalise and implement other models, especially agent-based models and artificial intelligence methods. Because of the generality of the building blocks in the metamodel of the allagmatic method, it is likely that the formalisation and automatic programming of models other than cellular automata and artificial neural networks are possible. This will allow to not only automatically program existing and novel computer models, it will also create models that are interpretable and thus understandable to humans. Such novel but still interpretable computer models will finally help us to more deeply explore and understand the computational universe.

**Acknowledgements** This work was supported by the Hasler Foundation under Grant 18067.

## References


1. Wolfram, S.: Cellular automata as models of complexity. Nature **311**, 419–424 (1984). https://doi.org/10.1038/311419a0
2. Ding, R., Guo, Z.X.: Microstructural modelling of dynamic recrystallisation using an extended cellular automaton approach. Comput. Mater. Sci. **23**(1–4), 209–218 (2002). https://doi.org/10.1016/S0927-0256(01)00211-7
3. Jiao, Y., Torquato, S.: Emergent behaviors from a cellular automaton model for invasive tumor growth in heterogeneous microenvironments. PLOS Comput. Biol. **7**(12) (2011). https://doi.org/10.1371/journal.pcbi.1002314
4. Keller, F., Christen, P.: Bone as a complex system: computing bone remodelling across biological hierarchies. In: Proceedings of the 8th World Congress of Biomechanics, O1389 (2018)
5. Ohs, N., Keller, F., Blank, O., Lee, Y.-W.W., Cheng, C.-Y.J., Arbenz, P., Müller, R., Christen, P.: Towards in silico prognosis using big data. Curr. Dir. Biomed. Eng. **2**(1), 57–60 (2016). https://doi.org/10.1515/cdbme-2016-0016
6. Christen, P.: Moving beyond the genome with computer modeling. Pers. Med. **15**(3), 145–148 (2018). https://doi.org/10.2217/pme-2017-0066
7. Hooft, G.T.: The Cellular Automaton Interpretation of Quantum Mechanics. Springer, Cham Heidelberg New York Dordrecht London (2016)
8. Chen, X., Li, H.Y., Miao, J.R., Jiang, S.X., Jiang, X.: A multiagent-based model for pedestrian simulation in subway stations. Simul. Model. Pract. Theory **71**, 134–148 (2017). https://doi.org/10.1016/j.simpat.2016.12.001
9. Esteva, A., Kuprel, B., Novoa, R.A., Ko, J., Swetter, S.M., Blau, H.M., Thrun, S.: Dermatologist-level classification of skin cancer with deep neural networks. Nature **542**, 115–118 (2017). https://doi.org/10.1038/nature21056
10. Carleo, G., Troyer, M.: Solving the quantum many-body problem with artificial neural networks. Science **355**, 602–605 (2017). https://doi.org/10.1126/science.aag2302
11. Bezerra, L.C.T., Lopez-Ibanez, M., Stutzle, T.: Automatic component-wise design of multi-objective evolutionary algorithms. IEEE Trans. Evol. Comput. **20**(3), 403–417 (2016). https://doi.org/10.1109/Tevc.2015.2474158
12. Bidlo, M.: On routine evolution of complex cellular automata. IEEE Trans. Evol. Comput. **20**(5), 742–754 (2016). https://doi.org/10.1109/Tevc.2016.2516242
13. Christen, P., Del Fabbro, O.: Cybernetical concepts for cellular automaton and artificial neural network modelling and implementation. 2019 IEEE International Conference on Systems, Man and Cybernetics (SMC), 4124–4130 (2019). https://doi.org/10.1109/smc.2019.8913839 arXiv:2001:02037
14. Wiener, N.: Cybernetics: Or Control and Communication in the Animal and the Machine. The MIT Press, Cambridge (1961)
15. Ashby, W.R.: An Introduction to Cybernetics. Chapman & Hall, London (1956)
16. Simondon, G.: L'individuation à la lumière des notions de forme et d'information. Editions Jérôme Millon, Grenoble (2013)
17. Simondon, G.: On the Mode of Existence of Technical Objects. University of Minnesota Press, Minneapolis London (2017)

18. Cormen, T.H., Leiserson, C.E., Rivest, R.L., Stein, C.: Introduction to Algorithms. The MIT Press, Cambridge London (2009)
19. Ilachinski, A.: Cellular Automata: A Discrete Universe. World Scientific, Singapore (2001)
20. Mainzer, K., Chua, L.: The Universe as Automaton: From Simplicity and Symmetry to Complexity. Springer, Heidelberg Dordrecht London New York (2012)
21. Wolfram, S.: A New Kind of Science. Wolfram Media, Champaign (2002)
22. Kaufmann, S.A.: The Origins of Order: Self-Organization and Selection in Evolution. Oxford University Press, New York (1993)
23. Cook, M.: Universality in elementary cellular automata. Complex Syst. **15**(1), 1–40 (2004)
24. Russel, S.J., Norvig, P.: Artificial Intelligence: A modern Approach. Prentice Hall, Upper Saddle River (2010)

# Facial Recognition and Pathfinding on the Humanoid Robot Pepper as a Starting Point for Social Interaction

**Achim Dannecker and Daniel Hertig**

**Abstract** Interaction between humanoid robots and humans is a complex process. Speech, gestures, and recognition of communication partners are important aspects in a well-defined interaction. To seem more natural, a humanoid robot should not be stationary. It should be able to be part of a crowd and wander around a specific area. Therefore, pathfinding is important to give a humanoid robot the ability to connect with people in more than one place. In addition, the recognition of communication partners is the backbone of social interaction. This chapter demonstrates how OpenCV, a well-known computer vision library, supports the robot Pepper in the recognition of communication partners and in addition, how this is the starting point to different types of small talk as the basis for a prototypical interaction process of humanoid robots and humans. Additionally, the navigation functions that allow the robot to move autonomously and enable a better human–robot interaction will be discussed.

**Keywords** Robotics · Humanoid robots · Human–robot interaction · Facial recognition · Pathfinding

## 1 Introduction

In Fong et al. [1] it has already been stated, that social robots will play a significant role in our world. According to them, robots will work in close proximity to humans in different areas such as health care, office assistance, serving as tour guides, rehabilitation, etc. Especially in the health care sector, there are many assistive robots that help with surgery, rehabilitation, elderly care, and medication delivery [2–4]. Or, for example, tour guides that focus on getting people from one place to another

A. Dannecker (✉) · D. Hertig
School of Business, Institute for Information Systems, FHNW University of Applied Sciences and Arts Northwestern Switzerland, Peter Merian-Strasse 86, 4002 Basel, Switzerland
e-mail: achim.dannecker@fhnw.ch

D. Hertig
e-mail: daniel.hertig@fhnw.ch



R. Dornberger (ed.), *New Trends in Business Information Systems and Technology*, Studies in Systems, Decision and Control 294,
https://doi.org/10.1007/978-3-030-48332-6_10

autonomously and in some cases with explanations of interesting points on the way [5–7]. The ability to interact with people in an entertaining, engaging, or anthropomorphic manner is an important part of the robots' functionality [8]. In order to establish a social interaction between a human and a robot, humans must get the impression, that robots have beliefs, desires, and intentions [9]. In addition, social robots should have greeting, face recognition, user identification, and dialog skills [10, 11].

Humans are a deeply social species, and it is important to apply a social model when observing and interacting with autonomous robots [12]. Autonomous humanoid robots can perceive their world, make decisions on their own, and perform coordinated actions to carry out different tasks [12].

The goal of this chapter is to show how the facial recognition capabilities of a robot in combination with its pathfinding ability can be improved for a better social interaction of the humanoid robot with people around it. The robot used in the research is Pepper of SoftBank Robotics [13]. OpenCV [14], a well-known Computer Vision Library was used for facial recognition.

In paragraphs two and three, the facial recognition and pathfinding abilities of Pepper and their limits are discussed. In paragraph four, the importance of these abilities for a human–robot interaction, based on the outcome of qualitative research within an elderly care institution, is pointed out before the conclusion is drawn in the last paragraph.

## 2 Facial Recognition

As a prerequisite for advanced human–robot interaction, it is necessary for the robot to recognize people it has previously met [15]. Without this ability, it will be almost impossible to build a relationship between a human and a robot.

Since the facial recognition solution supplied on Pepper is not sufficient for use in a real-world business case, we implemented our own facial recognition solution based on OpenCV [14] on the robot. To improve the recognition performance, an experiment regarding the camera settings of Pepper was conducted. In this experiment, four different parameters of the camera settings where changed to find the setting with the best recognition performance and to analyze whether the default settings perform well. For the facial recognition, all of the three recognition algorithms provided by OpenCV [14] were used and compared in the experiment.

Systems that support facial recognition are technologies that can identify or verify a person from a digital image or a video frame. There are multiple methods and algorithms for how facial recognition systems work. What they usually have in common is comparing selected facial features from given images with faces within a database. These methods and algorithms can uniquely identify a person by analyzing patterns based on the person's facial textures and shapes.

## 2.1 Algorithms

OpenCV provides three different algorithms for facial recognition. Eigenfaces [16, 17] which is based on Principal Component Analysis (PCA), Fisherfaces [18] which is based on Fishers Linear Discriminant (FLD) [19] and Local Binary Pattern Histograms (LBPH) [20] which is based on local feature extraction.

Turk and Pentland were the first to use Eigenfaces for face classification [16]. This algorithm builds on the fact that each face has features that make it unique and differentiate it from other faces. The Eigenfaces method tries to collect those features, encodes them in an efficient way and afterwards compares them to new faces to enable recognition.

Fisherfaces is an improvement of Eigenfaces that was introduced by Belhumeur et al. [18]. Instead of looking at the principal components of all faces in the training set, Fisherfaces concentrates on the features of a face that discriminate it from other faces. Thus, the impact of illumination changes is not as big as before because the algorithm only looks at one face at a time and the features of a single face cannot become as dominant as in the Eigenface algorithm [18].

LBPH [20] is a texture descriptor that looks for patterns that form a feature vector, which can be used to classify a face. To do so, each training image is divided into blocks. Each block contains nine pixels that build a grid of $3 \times 3$ pixels with a value from 0 to 255 according to their grey level. In a next step, the pixel in the center of the grid is compared to its neighboring pixels. If the value of the compared pixel is equal to or bigger than the value of the center pixel, it is assigned a 1, if the value is lower than the one of the center pixel the assigned value is 0. After all neighboring pixels have been compared and converted into a binary value, they are read clockwise or counter-clockwise to build a binary representation of a number. After converting the binary representation back to decimal, the resulting number will be assigned as the new value of the center pixel. This procedure is repeated for each pixel in a block. In a next step, the values of a block are converted into a histogram, which leads to a representation of the image in one histogram per block. In a final step, all block histograms are combined to a histogram for an image, which contains all the features needed to classify an image. When the algorithm is confronted with a new face, a histogram for the face is created. Then, the histogram is compared to the histograms with which the algorithm was trained, and the face is assigned to the histogram with the shortest distance to its own.

As mentioned in [21] Fisherfaces and Eigenfaces are known to be more sensitive on illumination changes in comparison to LBPH.

## 2.2 OpenCV

OpenCV is a well-known library for computer vision which also includes facial recognition classes [21]. Before facial recognition can be applied, the facial recognition algorithms need to be trained to have data that allows them to classify unknown images in the recognition process. When recognizing a face, the algorithms compare the image of the face to be recognized with the images with which they were trained and output a distance value and a class name to which the algorithm assigned the face. The distance value is a Euclidian distance and specifies how similar the recognized face is to a trained face. The lower the distance, the closer the recognized face relates to the trained face. Once the distance is calculated, the OpenCV algorithms assign the test image to the class of the training image with the lowest distance. If desired, a distance threshold can be applied so that a face is classified only if the calculated distance is below a certain value. If no threshold is set, the algorithms will always classify the test image into a class, even if the distance is rather large.

## 2.3 Experiment

The experiment collected data on the different settings of the robot's camera. The experiment was conducted in an office at the University. For the experiment, the humanoid robot Pepper in combination with a computer and a LAN connection were used. The LAN connection was chosen because it is much faster in transferring data compared to WiFi in combination with Pepper. To have remote access to the file system of the robot, PuTTY and the Remote System Explorer plugin for Eclipse were used. One part of the program was written as Python scripts in Eclipse, another part was realized with Choregraphe, which is the development environment delivered with the Pepper robot and also includes some additional Python code. As mentioned in the introduction, OpenCV was used for the facial recognition in the experiment instead of the standard solution delivered on Pepper. To store the data collected in the experiment, a SQLite database was used.

For the experiment, the robot was placed in a marked spot in front of a white wall, so that all images taken in the course of the experiment were recorded in a similar setting. The experimental subjects were placed in a marked spot in front of the white wall, so that they were in the center of the robot's field of sight. The distance between the subject's feet and the robot's wheels was around 75 cm which approximately reflects the distance between the robot and a human in a real-life interaction. All images were taken in daylight without artificial light (Fig. 1).

In a first step, a series of 44 images was taken of all four subjects involved in the experiment. This image set reflects 44 different settings of the parameters brightness (BRT) [22], contrast (CON) [23], saturation (SAT) [24] and resolution (RES) [25], through which the robot iterated in predefined intervals, as can be seen in Table 1. One of the parameters was always changed, while the others stayed at the default

**Fig. 1** Experimental setup

**Table 1** Interval camera parameters

| | Start | End | Interval | DF |
|---|---|---|---|---|
| BRT | 0 | 240 | 20 | 0 |
| CON | 0 | 120 | 10 | 32 |
| SAT | 0 | 240 | 20 | 64 |
| RES | 160 × 120 | 2560 × 1920 | – | 640 × 480 |

value (DF). At the time the image was taken, the average luminance was measured to allow the settings to be optimized later according to the measured luminance. The captured images, including the metadata of the various parameters and the subject, were stored in a SQLite database.

This first sets of images were used as training sets for the facial recognition algorithms. A few days after taking the first set of images, the experimental subjects returned to take another set of 44 images with the same settings as for the first one, which were later used as test sets to perform facial recognition. In the first recognition tests, it became clear that no face could be recognized with the highest brightness settings, and also in some cases with the lowest contrast settings, so these images were excluded and the test data reduced to 41 resp. 40 images. Again, the image set was stored in the SQLite database. As soon as the two sets of images were taken, the involvement of the subjects in the experiment was terminated and the collecting of data started.

In the first phase of data collection, the three different facial recognition algorithms of OpenCV were trained with a training set of one of the subjects. Each image of the training set represented a class for the recognition process. For the training, triplets of three images (classes) with different settings from the same parameter were built. The first triplet started at the beginning of the range of a parameter, e.g. BRT0 and then continued in the intervals of the parameter, in this case BRT20 and BRT40. After the training of the first triplet, the facial recognition was performed on the test images of the same subject as the training set. The results for the facial recognition by all three algorithms and the training set used were stored in the SQLite database. After the recognition, a next iteration started, altering the training set by one image by increasing the parameter values by one step so that the second triplet would be BRT20, BRT40, and BRT60. These iterations were repeated until the end of the range of the first parameter was reached and then resumed at the beginning of the range of the next parameter. After all parameters completed the iteration, phase one of the data collection for that subject ended. To end the first phase of data collection, the process just described was applied to all remaining subjects. The general procedure was repeated for the second phase of data collection. The difference with the first phase was that the training set consisted of a combination of triplets of all subjects through which it was iterated, and the test set consisted of the images of all subjects as well. After the completion of the two data collection phases, 9,396 data sets were collected that were used for data analysis.

The goal of the data analysis was to research whether the camera's default settings perform well in facial recognition with real-world lighting conditions, or if not, which camera settings would perform better. For this purpose the results of the data collection phases were analyzed. First, the data was filtered by subjects, then all parameter settings were analyzed and the occurrence percentage of each setting was calculated.

The next step was to identify the three best settings per parameter and algorithm. This was carried out individually for phase one and phase two of the data gathering, so that in the end there were two rankings per experimental subject, one for each phase. These rankings were then compared to find the best parameter setting across all phases. To get the weighted ranking $R_w$, Eq. 2.1 was applied where $R_p$ is the ranking of a parameter setting, $O_p$ is the number of occurrences of a parameter setting and $O_{\max}$ is the maximum of occurrences possible. In a last step of the analysis, the best parameter settings from the experiment were compared to the results of the default parameter settings to see which settings performed better.

$$R_w = \frac{\sum R_p}{\sum O_p} \times (O_{\max} - O_p) \tag{2.1}$$

## 2.4 Results

Differences resulted between the three tested algorithms as can be seen in Table 2 and Figs. 2, 3 and 4. Figures 2, 3 and 4 each show four images that represent the best setting for each algorithm and parameter. LBPH had the biggest variance between the default (DF) and the best experimental setting. Fisherfaces (FF) and Eigenfaces (EF) had remarkably smaller variances but still performed better than the default values. The Fisherfaces and Eigenfaces algorithms matched the brightness parameter, while the LBPH algorithm had different results for all parameters except the resolution. Eigenfaces is also the only algorithm with which a default value other than the resolution had the best result. For the resolution parameter, the default value of 640 × 480 was the best setting for all algorithms.

Depending on the distance threshold, which indicates when an algorithm can recognize a subject, the different algorithms result in different numbers of correctly recognized subjects. The range of the threshold differs for the three algorithms [16–20]. For all three algorithms, the "best" parameter settings for the images were used for the pictures tested with a training set. One training set included all images that were taken with the different types of settings. Another training set contained only images with the "best" parameter set for the specific algorithm. The following tables therefore show two different results. One with the general set of pictures (Quality

**Table 2** Best settings for parameters

| | DF | LBPH | FF | EF |
|---|---|---|---|---|
| BRT | 0 | 60 | 40 | 40/80 |
| CON | 32 | 120 | 20 | 60 |
| SAT | 64 | 40 | 240 | 64 |
| RES | 640 × 480 | 640 × 480 | 640 × 480 | 640 × 480 |

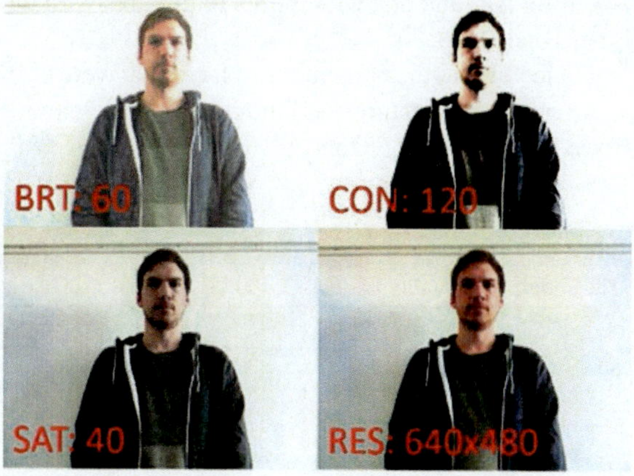


**Fig. 2** Best LBPH settings

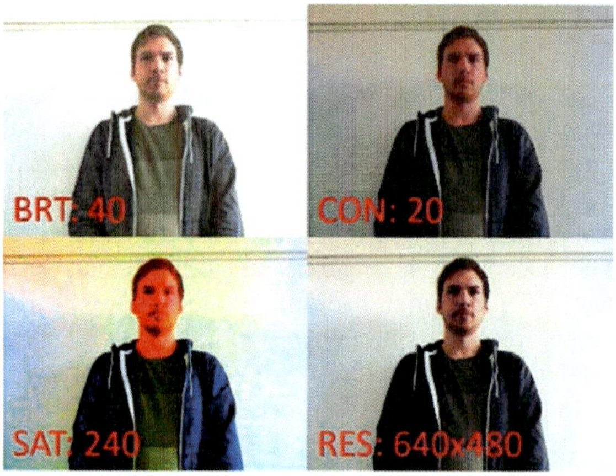


**Fig. 3** Best Fisherfaces settings

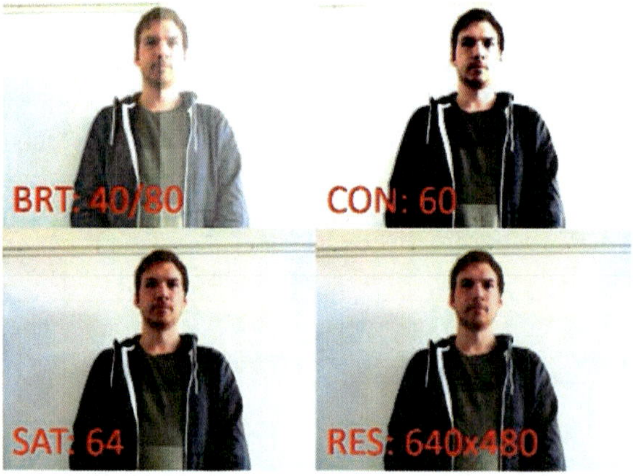


**Fig. 4** Best Eigenfaces settings

All, Total All) and one with the set of pictures with the "best" parameter set (Quality Best, Total Best).

Table 3 shows the percentage of faces that were correctly recognized. In particular, starting with the pictures and the related setting above, Table 3 shows, for example, based on a threshold 55 that 89% of the recognized subjects were identified correctly,

**Table 3** Number of correctly recognized subjects for LBPH

| Threshold | Quality all (%) | Total all (%) | Quality best (%) | Total best (%) |
|---|---|---|---|---|
| Avg | 70 | 91 | 75 | 92 |
| 60 | 79 | 84 | 82 | 85 |
| 55 | 88 | 82 | 89 | 83 |
| 50 | 86 | 93 | 75 | 92 |

**Table 4** Number of correctly recognized subjects for Fisherfaces

| Threshold | Quality all (%) | Total all (%) | Quality best (%) | Total best (%) |
|---|---|---|---|---|
| Avg | 66 | 92 | 65 | 93 |
| 1000 | 70 | 88 | 71 | 87 |
| 500 | 71 | 89 | 69 | 88 |
| 250 | 74 | 90 | 76 | 89 |

**Table 5** Number of correctly recognized subjects for Eigenfaces

| Threshold | Quality all (%) | Total all (%) | Quality best (%) | Total best (%) |
|---|---|---|---|---|
| Avg | 66 | 93 | 65 | 87 |
| 1000 | 68 | 91 | 68 | 85 |
| 500 | 60 | 89 | 57 | 91 |
| 250 | 78 | 67 | 94 | 68 |

which represents 83% of the sample. 17% of the subjects were not recognized and 11% were classified falsely.

Since the recognition with Fisherfaces generally results in higher distance values, different thresholds are used. In general, the recognition rate was not so good and the number of falsely recognized subjects is higher (Table 4).

For Eigenfaces, the thresholds are the same as for Fisherfaces. In general, the recognition rate was not so good, except with threshold 250 (Table 5).

The same set of tests was made with the default face recognition of Pepper. Pepper's default face recognition resulted in a correct recognition rate of 50% (best case) based on a total of 89% of the complete sample in which it recognized a face.

## 3 Pathfinding

Since autonomous navigation of the robot is a major concern in a business case with human interaction, we also conducted smaller experiments in a nursing home for the elderly to research how well the built-in navigation capabilities of the robot Pepper perform and what improvements should be implemented in the future.

Pathfinding or pathing in most cases is implemented with the calculation of the shortest path. It is often based on Dijkstra's algorithm for finding a shortest (cheapest, fastest, etc.) path on a weighted graph [26]. First, the robot autonomously creates a map of a defined area. The area is often defined with a starting point and a radius the robot should scan. During the scan, the robot searches for open spaces and obstacles in the defined area.

In addition to facial recognition, the autonomous navigation of a robot in an interaction with a human is also of high importance. At the moment, Pepper robots

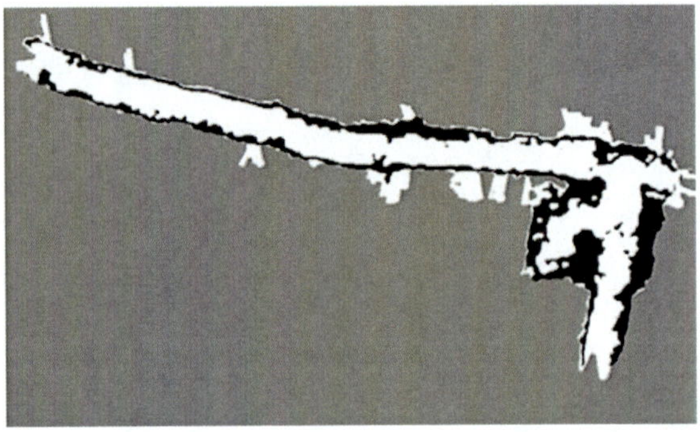

**Fig. 5** Map created by Pepper

are mainly used in a stationary manner where they provide information [27, 28]. To enable a human–robot interaction that feels more natural, the robot needs to be able to move in a way that does not interfere with its human collaborator [29, 30]. For navigation, an existing sample application [31] has been adapted to our needs. With the modified navigation application, a nursing home was visited to see how the application performs. On the first visit, Pepper was placed in the middle of the hallway it was supposed to scan, and after defining the radius for the exploration (20 meters), it started to explore the hallway. To prevent Pepper from entering the rooms of patients who were asleep, one of the authors always stood on the doorstep and was therefore recognized as an obstacle. In the first attempt, the radius chosen was too small, which led Pepper to stop the exploration prematurely. The radius was adjusted to 30 meters and the second attempt went better and included the whole hallway including an area called “Marketplace” where patients can meet each other (see Fig. 5). While exploring, Pepper successfully maneuvered around static and moving obstacles thanks to its laser and sonar sensors [32, 33]. In a last exploring attempt, Pepper explored the same hallway twice and displayed the scanned hallways parallel to each other on its tablet that showed the created map.

On the second day, the goal was to test how well Pepper could localize itself on a previously created map and navigate to predefined points on the map. To help Pepper localize, the user must estimate the position of Pepper on the map that is displayed on its tablet. After the user guessed the robot’s position, Pepper continued with a scan of its surroundings to localize itself on the map. In the first attempts, localization difficulties occurred as Pepper could not distinguish between the main hallway and the access ways to other rooms which were similar in width to the hallway. After adjusting Pepper’s starting position, these problems could be eliminated and a first patrol to different points on the map was possible. In general, the navigation to points on the map was quite accurate. When inaccuracies in the navigation occurred, they were mostly related to the localization of the robot on the map. The two most common error scenarios were that Pepper thought it had already reached a point because in terms of its localization, it was at the specified point on the map and the robot was unable to navigate to the point because there were large obstacles, e.g. a wall, in its way due to a wrong localization. The best results were achieved when Pepper was starting at the "Marketplace", probably because it had a very distinguishable layout and allowed a good initial localization.

Although the localization and patrolling worked in some cases, it became clear that there is still a lot of room for improvement in the navigation capabilities of Pepper since a real-life application in an elderly care home would be hard to realize with the existing solution.

## 4 Social Interactions

To bring both aspects—facial recognition and pathfinding—together, there is a wide range of possibilities related to social interactions. This section will give an overview of possible further social interaction dialogues that can be part of a nursing home for the elderly. These possible dialogues are the result of several workshops together with a nursing home for the elderly, caregivers, nurses, and the people who live there.

### *4.1 Searching for People*

People are often searched e.g. when grandchildren visit their grandmother and she is not in her room. Due to its pathfinding abilities, a humanoid robot can wander around and is aware (because of facial recognition) of who is in which place. The robot would be able to give assistance in finding a person. This is also useful if an elderly person has an appointment, e.g. physiotherapy or an event such as dancing, and they are missing. The robot could give hints in advance that there is an appointment with the physiotherapist in 10 min and offer to escort the person to their appointment.

### *4.2 Menu Plan*

The menu plan for lunch and dinner is an important aspect in nursing homes. At the moment, caregivers are asked repeatedly what is on the menu during the course of the day. If the robot would know the menu plan, it could give respective information to people. Additionally, if the robot knew what the preferred meal of an elderly person was, it would be able to talk about it with someone if it recognized the person via facial recognition, e.g.: "You know, today there is fish for lunch. You like fish, are you happy?"

### 4.3 Taking Orders

Often the residents of a nursing home order something to drink etc. The robot could take orders from persons and, in addition, if it recognized a person and knew their preferred drink, it could motivate the person: "Do you want some peppermint tea, I know you like it and already two hours have passed since you have ordered your last drink." If there are other activities in an old people's care center such as a dancing event etc. the robot could accept registration for these events.

### 4.4 Sport Small Talk

If the robot knew what an elderly person's preferred sports team was, it could do research on the internet (via keywords, e.g. soccer, Basel, Switzerland, < last 3 days) and start a dialogue, e.g.: "You know your preferred team the FC Basel won last weekend against the BSC Young Boys from Bern 2:0, it was a great game. Fabian Frei scored the two goals in minutes 75 and 83. There was also a red card for…"

There are plenty of other possibilities for dialogues a humanoid robot can hold if it is able to navigate, recognize people and has some background information about them.

## 5 Conclusion

Analyzing the collected data, it became clear that the default settings of the robot's camera do not perform at their best when facial recognition is done in real-life lighting conditions.

The experiment showed that OpenCV improves the rate of facial recognition up to 27% and also reduces the misidentified subjects. The results show that the different algorithms bring up different results in quality and quantity. Nevertheless, the differences are minor compared to the differences to the default settings of the robot Pepper. For each algorithm there is a "best" setting for the images. These settings differ except for resolution. Interestingly, the resolution $640 \times 480$ had the highest recognition rate for all three algorithms. The difference between the default settings and the "best" setting due to the different algorithms is remarkable. In all cases the use of OpenCV led to better results than the standard tools that are used on the robot Pepper. The different algorithms have special ways to handle images with different properties. How to determine the best algorithm with the best setting depends on the focus of the business case. Is it more important to recognize more subjects correctly but less in total or recognize less correctly but more in total? There is no general "best" configuration in terms of algorithm and threshold. What can be said is that there is a general best configuration per algorithm w.r.t. brightness, contrast, saturation and resolution, excluding the threshold. This paper can be used as an initial decision helper, when it comes to the setup or improvement of the ability of Pepper to recognize people.

In our experiment at the nursing home it became clear that the navigation abilities of the robot Pepper are not precise enough to allow an autonomous navigation on a previously explored map. On the one hand, the exploration works very well and results in maps that can be used for navigation. The navigation on the map, on the other hand, has several weaknesses, e.g. the localization of the robot and the precision of reaching points marked on the map.

In terms of navigation and as a next research step, the on board abilities of Pepper will be compared to the tools that are provided by ROS [34] in the future to evaluate which platform allows better pathfinding. In addition, there is a vast amount of possibilities to combine facial recognition, pathfinding, and small social dialogues to improve the interaction between humans and humanoid robots.

## References

1. Fong, T., Nourbakhsh, I., Dautenhahn, K.: A survey of socially interactive robots. Robot. Autonom. Syst. **42**, 143–166 (2003). https://doi.org/10.1016/S0921-8890(02)00372-X
2. Hu, J., Edsinger, A., Lim, Y.-J., Donaldson, N., Solano, M., Solochek, A., Marchessault, R.: An advanced medical robotic system augmenting healthcare capabilities—robotic nursing assistant. In: 2011 IEEE International Conference on Robotics and Automation, pp. 6264–6269 (2011)
3. Nejat, G., Sun, Y., Nies, M.: Assistive robots in health care settings. Home Health Care Manage. Pract. **21**, 177–187 (2009). https://doi.org/10.1177/1084822308325695
4. Broekens, J., Heerink, M., Rosendal, H.: Assistive social robots in elderly care: a review. Gerontechnology **8** (2009)
5. Lim, M., Aylett, R.: Intelligent mobile tour guide. In: Symposium on Narrative AI and Intelligent Serious Games for Education, AISB '07 (2007)
6. Han, B., Kim, Y., Cho, K., Yang, H.S.: Museum tour guide robot with augmented reality. In: 2010 16th International Conference on Virtual Systems and Multimedia, pp. 223–229 (2010)
7. Cho, S.T., Jung, S.: Control of a robotic vehicle for entertainment using vision for touring University Campus. In: 2014 14th International Conference on Control, Automation and Systems (ICCAS 2014), pp. 1415–1417 (2014)
8. Sidner, C.L., Dzikovska, M.: Human-robot interaction: engagement between humans and robots for hosting activities. In: Proceedings. Fourth IEEE International Conference on Multimodal Interfaces, pp. 123–128 (2002)
9. Breazeal, C., Scassellati, B.: How to build robots that make friends and influence people. In: Proceedings 1999 IEEE/RSJ International Conference on Intelligent Robots and Systems. Human and Environment Friendly Robots with High Intelligence and Emotional Quotients (Cat. No.99CH36289), vol. 2, pp. 858–863 (1999)
10. Gorostiza, J.F., Barber, R., Khamis, A.M., Malfaz, M., Pacheco, R., Rivas, R., Corrales, A., Delgado, E., Salichs, M.A.: Multimodal human-robot interaction framework for a personal robot. In: ROMAN 2006—The 15th IEEE International Symposium on Robot and Human Interactive Communication, pp. 39–44 (2006)
11. Fischer, K.: Interpersonal variation in understanding robots as social actors. In: 2011 6th ACM/IEEE International Conference on Human-Robot Interaction (HRI), pp. 53–60 (2011)
12. Breazeal, C.: Toward sociable robots. Robot. Autonom. Syst. **42**, 167–175 (2003). https://doi.org/10.1016/S0921-8890(02)00373-1

13. Pepper—Documentation—Aldebaran 2.5.11.14a documentation. http://doc.aldebaran.com/2-5/home_pepper.html
14. OpenCV: OpenCV library. https://opencv.org/
15. Vossen, P., Baez, S., Bajčetić, L., Kraaijeveld, B.: Leolani: a reference machine with a theory of mind for social communication. arXiv:1806.01526 [cs]. (2018)
16. Turk, M.A., Pentland, A.P.: Face recognition using eigenfaces. In: IEEE Computer Society Conference on Computer Vision and Pattern Recognition, 1991. Proceedings CVPR'91., pp. 586–591. IEEE (1991)
17. Sirovich, L., Kirby, M.: Low-dimensional procedure for the characterization of human faces. Josa a. **4**, 519–524 (1987)
18. Belhumeur, P.N., Hespanha, J.P., Kriegman, D.J.: Eigenfaces vs. fisherfaces: Recognition using class specific linear projection. Yale University New Haven United States (1997)
19. Fisher, R.A.: The use of multiple measurements in taxonomic problems. Annals Eugenics. **7**, 179–188 (1936). https://doi.org/10.1111/j.1469-1809.1936.tb02137.x
20. Ahonen, T., Hadid, A., Pietikäinen, M.: Face recognition with local binary patterns. In: Pajdla, T., Matas, J. (eds.) Computer Vision—ECCV 2004, pp. 469–481. Springer, Berlin Heidelberg, Berlin, Heidelberg (2004)
21. OpenCV: Face Recognition with OpenCV—OpenCV 2.4.13.7 documentation. https://docs.opencv.org/2.4/modules/contrib/doc/facerec/facerec_tutorial.html
22. Definition of bright. https://www.merriam-webster.com/dictionary/bright
23. Definition of contrast. https://www.merriam-webster.com/dictionary/contrast
24. Saturation|Definition of Saturation by Merriam-Webster. https://www.merriam-webster.com/dictionary/saturation
25. Definition of resolution. https://www.merriam-webster.com/dictionary/resolution
26. Wang, H., Yu, Y., Yuan, Q.: Application of Dijkstra algorithm in robot path-planning. In: 2011 Second International Conference on Mechanic Automation and Control Engineering, pp. 1067–1069 (2011)
27. Niemelä, M., Arvola, A., Aaltonen, I.: Monitoring the acceptance of a social service robot in a shopping mall: first results. In: Proceedings of the Companion of the 2017 ACM/IEEE International Conference on Human-Robot Interaction, pp. 225–226. ACM, New York, NY, USA (2017)
28. Aaltonen, I., Arvola, A., Heikkilä, P., Lammi, H.: Hello pepper, may i tickle you?: children's and adults' responses to an entertainment robot at a shopping mall. In: Proceedings of the Companion of the 2017 ACM/IEEE International Conference on Human-Robot Interaction, pp. 53–54. ACM, New York, NY, USA (2017)
29. Khambhaita, H., Alami, R.: A human-robot cooperative navigation planner. In: Proceedings of the Companion of the 2017 ACM/IEEE International Conference on Human-Robot Interaction—HRI '17, pp. 161–162. ACM Press, Vienna, Austria (2017)
30. Kretzschmar, H., Spies, M., Sprunk, C., Burgard, W.: Socially compliant mobile robot navigation via inverse reinforcement learning. Int. J. Robot. Res. **35**, 1289–1307 (2016). https://doi.org/10.1177/0278364915619772
31. GitHub—aldebaran/naoqi_navigation_samples: contains samples for NAOqi navigation and exploration features. The samples are packaged into Choregraphe applications. https://github.com/aldebaran/naoqi_navigation_samples
32. Lasers—Aldebaran 2.5.11.14a documentation. http://doc.aldebaran.com/2-5/family/pepper_technical/laser_pep.html
33. Sonars—Aldebaran 2.5.11.14a documentation. http://doc.aldebaran.com/2-5/family/pepper_technical/sonar_pep.html
34. Documentation—ROS Wiki. https://wiki.ros.org/

# Digital Business Transformation

# Social Robots in Organizational Contexts: The Role of Culture and Future Research Needs

**Theresa Schmiedel, Janine Jäger, and Vivienne Jia Zhong**

**Abstract** The integration of social robots in organizations is on the rise. In the future, an increase in the application of physically embodied robots who socially interact and collaborate with humans is expected. In this chapter, we outline the organizational contexts in which current research examines social robots and point out cultural challenges regarding their implementation in the workplace. We suggest that the successful integration of social robots in organizations requires a cultural fit between values embedded in social robots and values lived in the organizational context in which the robot is deployed. In addition, we propose an agenda for future research that addresses the associated cultural challenges of introducing social robots into organizational contexts.

**Keywords** Social robot · Organizational culture · Embedded values

## 1 Introduction

A new generation of robots has emerged. Unlike conventional industrial robots that perform repetitive tasks in a delimited area in factories, this relatively new type of robot is characterized by sociality, which means that it has the ability to express emotions, communicate, and learn from other agents [1]. Compared to industrial robots, social robots no longer function as mere "machines" that perform some predefined task such as assembling car parts. They are able to cooperate and interact

T. Schmiedel (✉) · J. Jäger · V. J. Zhong
School of Business, Institute for Information Systems, FHNW University of Applied Sciences and Arts Northwestern Switzerland, Peter Merian-Strasse 86, 4002 Basel, Switzerland
e-mail: theresa.schmiedel@fhnw.ch

J. Jäger
e-mail: janine.jager@fhnw.ch

V. J. Zhong
e-mail: viviennejia.zhong@fhnw.ch



R. Dornberger (ed.), *New Trends in Business Information Systems and Technology*, Studies in Systems, Decision and Control 294,
https://doi.org/10.1007/978-3-030-48332-6_11

autonomously with human partners to achieve mutual goals based on their cognitive abilities.

A variety of application areas for social robots is currently evolving. For example, social robots may assist patients in "checking in" at the doctors, support customers in hotel lobbies, or they can help students with their learning activities. This new kind of interaction with robots comes with challenges such as the integration of social robots in existing organizational contexts. This integration has to be tailored to the specific needs of each organization and be aligned with the organizational culture, so that the social robot can unfold its potential and contribute to increased organizational performance [2].

Particularly, the integration of social robots into organizational cultures, i.e. the acceptance among employees to interact with the social robot, will play a key role in social robot adoption. Social robots will increasingly collaborate with humans in the workplace, where they will take over autonomous roles in work processes. Hence, social robots will be part of the organizational culture and both human and robot behavior will mutually shape each other [3]. Thus, the concept of culture is an essential research field in robot design and robot integration in organizations.

The purpose of this chapter is to highlight the role of culture in integrating social robots into organizational contexts. Therefore, we first give an overview of organizational application areas that current social robot research focuses on. Second, we conceptualize the role of culture in integrating social robots into work environments.

The remainder of this chapter is organized as follows. Section 2 specifies the concept of social robots and introduces main organizational application areas. Section 3 specifies the concept of culture and outlines how far current social robot research has addressed cultural aspects. Section 4 conceptualizes the role of culture regarding the adoption of social robots in organizations and suggests a research agenda. Section 5 concludes the chapter with a summary and an outlook.

## 2 Social Robots in Organizational Contexts

### *2.1 Social Robots*

Although the history of robots might be traced back to the 3rd Century, B.C.E, when the first robotic bird—the Pigeon—was made using stream power [4], social robotics represent a relatively new study field. Over the past two decades, several attempts have been made to define the concept of a social robot. A review of definitions of social robots reveals that a variety of facets have been taken into account. Scholars from the field have not reached consensus on the typology of social robots. Social robots are generally understood as machines, applications, or autonomous agents [1]. Although the ability of social interaction is considered the main function of a social robot, some researchers also explicitly mention other functional skills such as task execution, assistance, and problem solving (e.g. [5, 6]). Furthermore, beyond the interaction with humans, other researchers additionally include the interaction with the environment or other robots as one of the qualification criteria of a social robot [7]. Moreover, research has defined various attributes that characterize a social robot. While most definitions emphasize sociality and communicativeness, for some researchers, the physical embodiment or the possession of cognitive functions are very important elements of a social robot (e.g. [8]).

As the definition of the term social robot is still evolving, synthetized from the existing definitions, as listed in [7], we argue that a social robot in its simplest form is:

> A physically embodied robot that possesses a certain degree of autonomy and a certain level of cognitive functions. A social robot engages in social interactions with humans and adheres to social norms.

The physical embodiment is emphasized to distinguish social robots from virtually embodied agents (e.g. embodied conversational agents), which can also perform social interaction with its counter partner. Autonomy generally refers to the ability of the robot to perform processes and operations on its own [9]. Cognitive functions range from basic functions such as perception and memory to high-level functions such as speech and decision making [10]. Both autonomy and cognitive functions are required to perform social interactions, which are defined by Gist as "the reciprocal influences that human beings exert on each other through interstimulation and response" (cited in [11], p. 14).

The term social robot is often used synonymously with service robots. Unarguably, there is an overlap of the capabilities between social robots and service robots, which are robots that perform "useful tasks for humans" (see [12], p. 1). Both might be partially autonomous, possess some sort of cognitive functions, and carry out social interactions with their counterpart such as avoidance of collisions, communication through audio, visual, and written signals, collaboration in assembly tasks, etc.

To distinguish social robots from service robots, we suggest using the characteristics and the extent of social interactions of their main task as indicators. While the main task of a service robot is to perform a certain activity that physically supports humans, such as lifting, assembling, or cleaning [12], the main task of a social robot is to interact socially with people [13], for example through having a conversation, playing a game, answering questions, providing guidance, or engaging in learning activities. Despite these differences, however, we can expect that the capabilities of social and service robots become increasingly aligned over time. Even today, we can already identify robots that integrate both physical and verbal interaction with humans (e.g. Lio, which is designed for stationary healthcare support [14]).

### 2.2 *Organizational Application Areas of Social Robots*

The ability to interact socially with the user and engage with the user on an emotional level opens up new possibilities for the application of social robots in organizations.

In the domains of healthcare, education, and tourism, the use of social robots has been intensively studied. The following sections give an overview of these application areas.

### 2.2.1 Healthcare

Several trends drive the use of social robots in healthcare: On the one hand, the demand for care has been increasing due to the globally ageing population [15] and the rising prevalence of people with disabilities [16]. On the other hand, the cost of care has continuously grown while the shortage of healthcare professionals is still considerable [17].

Social robots find a broad range of applications in healthcare. Research has mostly focused on two target groups: elderly people and children. When it comes to elderly people, researchers have investigated the use of social robots for various application scenarios, such as the treatment of mental health (e.g. dementia) [18, 19], promotion of psychological well-being [20, 21], support of physiotherapy [22] and companionship [23].

Research targeting children has mostly focused on involving social robots in the treatment of autistic spectrum disorder [18, 24, 25], in which a social robot can take on various roles to achieve the therapeutic goals [26]. In addition, social robots are used in health elucidation (e.g. quizzes on diabetes knowledge) and well-being (e.g. distraction during vaccination) of children [27].

### 2.2.2 Education

A further popular application domain of social robots is education. In this context, social robots can take on the role of a teacher, teaching assistant, novice, companion, and peer [28]. Most research is aimed at children and the education of adolescents [28].

Language learning and teaching is increasingly using social robots. A meta-review [29] showed that social robots were predominantly involved in vocabulary learning. However, a few studies also reported how robots can help improve reading skills, grammar learning, and sign language [29].

In further application scenarios, social robots acted, for example, as teaching assistants to support learning activities in small groups [30]. In [31], the authors reported that the combination of the robot and the think-aloud strategy achieved positive effects on children's learning outcomes.

#### 2.2.3 Tourism

A growing number of research has investigated the use of robots in tourism including organizations such as hotels, museums, and shopping malls, where social robots often play the role of an information provider.

In the hospitality and catering sector, the robot is preferably seen as a receptionist and concierge, performing tasks related to logistics and goods delivery, taking customer orders, providing information, and processing payments [32, 33]. Furthermore, from the hotel manager's perspective, registering guests for breakfast, connecting the guest to the hotel desk via video conference, and guiding the guests to the desired location are considered useful applications [34].

Additionally, application scenarios in shopping malls are similar to hospitality, with a primary focus on information provision, guidance [35–38], and entertainment [39]. Social robots also have the potential to be used at the airport to accompany passengers during their flight transfer [40].

### *2.3 Need for Cultural Integration of Social Robots in Organizational Contexts*

Many application scenarios for social robots are still in a research and testing phase and scholars are working intensively on the technical realization and improvement of the use of social robots. However, studies have also shown that the integration of social robots in given contexts is also culturally challenging, since people do not necessarily accept social robots as interaction partners, especially in tasks that require intensive relationship building (e.g. in education or healthcare) [41].

## 3 Culture as a Concept in Social Robot Research

### *3.1 Culture*

Culture is a concept that is difficult to grasp and researchers have defined it in many different ways. Basically, many researchers agree that the essential elements that define culture are values [42, 43], which are concepts of the desirable—in other terms, what a group deems proper to want or pursue [44]. For example, a group may appreciate the value reliability and, thus, act accordingly through being there for each other, or a group may appreciate the value individualism and find it preferable that everyone can do as they like.

This specification leads to another important aspect of culture: the concept always refers to a certain social system, be it a group of friends, an organization or a geographical region [45]. Researchers refer to this lens as levels of culture, such as national culture, professional culture, organizational culture, and group culture [46].

Considering these specifications of culture, organizations have to understand which cultural levels and associated values are already in place or are required for a successful integration of technologies [47]. As culture has a strong influence on (work) behavior and thus on organizational performance [48], it should also be considered as an important factor in social robot research.

## 3.2 *Culture as Part of Social Robot Studies*

Little research is addressing cultural aspects of social robots and human–robot collaboration because the vast majority of robotics studies come from the fields of engineering and computer sciences and focus on technological robot development and performance. When robotics studies investigate culture, they mainly focus on differences in national culture to optimize the functionalities and design of robots [49–51]. Next, a brief overview of the study of culture in relation to robots in the three organizational application areas described above will be given.

### 3.2.1 Healthcare

Only a few robotics studies focus on the concept of culture in healthcare, even though cultural competence is essential in this context [52]. A robot should adapt to the specific cultural context in which it assists people, such as healthcare workers, elderly people in domestic environments, or patients in hospitals. Especially patients and the elderly in need of care are in a vulnerable state and deployed care robots should represent the cultural values that are needed in this context. Further, a person's attitude towards health and self-care strongly depends on their cultural background [53]—another cultural aspect social robot designers should take into account.

The few culture-related studies of robots in healthcare either address the implementation of surgical robots in different national cultures [54], or they focus on how people perceive and interact with robots at a national or cross-national level in different application scenarios [55–58]. These studies identify the cultural competence of robots as highly relevant for their application in healthcare and acknowledge cultural differences with regard to the investigated concepts. However, these studies mostly address cultural values in relation to national culture and do not examine further cultural layers that predominate in healthcare, such as organizational values and patient values.

### 3.2.2 Education

There is also little research on the cultural requirements of social robot implementation in education or the cultural values embedded in educational robots.

Few studies on robots in the education sector, focusing on culture, investigate either attitudes towards robots or user acceptance [59–61] as well as child–robot interaction across national cultures [62]. Some studies investigate the difference between Eastern and Western cultures and their openness and willingness to use robots in educational contexts [63].

The underlying assumption is that culture has a strong influence on the acceptance and use of robots, as confirmed by these studies. Overall, most research conducted on social robots in education focuses on identifying and evaluating learning scenarios and applications without highlighting the culture or underlying values.

### 3.2.3 Tourism

With regard to tourism, some studies on robots focus on their integration into museums to innovate cultural heritage [64, 65]. Further research investigates the cultural differences in attitudes towards robots in Eastern and Western national cultures [66]. Some researchers examine for example the national evolution of robot culture in Japan [67].

Ideas for robot applications in tourism are manifold, however, the concept of culture is not yet being studied intensely at many levels other than national culture. Regardless of the application scenario, the consideration of cultural values is essential to successfully integrate social robots into organizations together with an appropriate design of these robots and the respective human–robot interactions.

## *3.3 Lack of Focus on Culture in Social Robot Research*

Overall research on social robots hardly focuses on organizational culture, even though existing studies and conceptual research emphasize its importance [52]. Nevertheless, we can identify a strong need for research that addresses not only the national, but all different levels of culture, and considers cultural values that should be embedded in social robots for their successful integration in organizations [68]. In fact, not only the strategies and goals of an organization determine the success or failure of technology integration, but also the acceptance of the users who determine whether the integration of an artefact into the work routines makes sense [69, 70].

# 4 The Role of Culture in Integrating Social Robots in Organizations

The integration of robots in organizations influences organizational structures and workflows as well as roles and responsibilities. Following prior research on the implementation of novel approaches in organizations [71], we argue that organizations have to understand their cultural readiness for the integration of social robots in order to design a work environment that enables meaningful human–robot collaboration. In other words, we recommend that organizations are aware of which of their cultural values support or hinder the implementation of social robots.

We take a value-based perspective to conceptualize the cultural integration of social robots [68]. In fact, research has called for a value-sensitive focus on social robots. Particularly, the value-sensitive design approach has become popular in recent years. It emphasizes the need to integrate human values in technology through a bottom-up approach [72]. Accordingly, we focus on the type of values to consider for the cultural integration of social robots into organizational contexts.

## 4.1 Conceptualization

Regarding the cultural integration of social robots into organizations, we identify two key sets of values based on previous research [71]. First of all, we need to consider the organizational culture through the *values that employees live by in the organization*. In other words, we need to have a good understanding of how far the organizational culture supports the integration of social robots or how far the given culture needs to be influenced so that social robots can actually support the organizational performance. Second, we need to consider the *values that are embedded in the specific social robot* that an organization intends to integrate. Examples for such values could be fun, modesty, control, and companionship.

The overlap of these two types of values (organizational values and social robot values) is what we call the *cultural fit* (following [73]). A high cultural fit exists when the organizational culture embraces values that are also embedded in social robots. A low cultural fit means that the social robot values oppose the organizational culture. In the latter case, cultural resistance regarding the application of the social robot can be expected.

Accordingly, we suggest that a cultural fit between the values embedded in social robots and the values lived in the organizational context has a positive influence on *social robot performance*, i.e. the degree to which the social robot supports the organization in achieving its goals. More specifically, we argue that social robots which culturally fit the organization provide the organization with performance benefits such as more efficient and effective business processes. For example, social robots may provide information more quickly, more accurately, and more cost-efficiently than

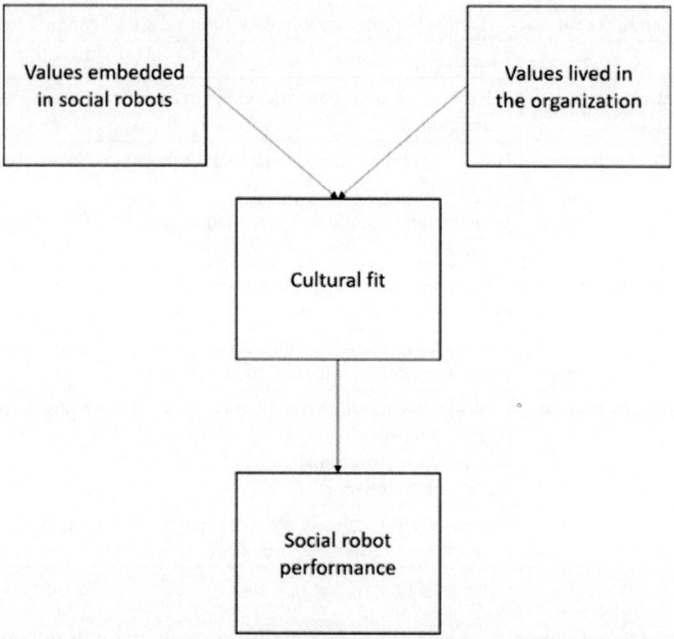


**Fig. 1** Conceptualization of the role of cultural values in social robot integration

employees. We call such performance benefits social robot performance. Figure 1 illustrates the relations described above.

## 4.2 Research Agenda

Following the above conceptualization, we propose various fields in the intersection of social robots and culture that future research can follow up on. In particular, we identify three fields of research that future studies can work on: social robot values, organizational values, and social robot performance.

For each of these fields, we identify research questions and potential methods that researchers can apply to address these questions. Table 1 provides an overview of the research agenda we propose in the field of social robotics.

As to the field of social robot values, future research may at first study existing social robots and find out which values are already embedded in them. Interviews with social robot designers and users can help identify the as-is integration of particular values. Since currently embedded values may be highly diverse and have various labels depending on people's perceptions, we consider it relevant to categorize the identified values to be able to compare values embedded in social robots on a more abstract level. Focus groups with categorization exercises and discussions would be a

**Table 1** Agenda for future research at the intersection of culture and social robots

| Field | Research questions | Potential methods |
|---|---|---|
| Social robot values | What kind of values are embedded in current social robots? | Expert interviews |
| | How can values embedded in social robots be classified? | Focus groups |
| | Which values should be embedded in social robots? | Delphi study |
| Organizational values | Which values positively influence the integration of social robots in organizations? | Survey |
| | How can we assess the cultural readiness of an organization to integrate social robots? | Scale development |
| Social robot performance | How can we assess social robot performance (e.g. the contribution of social robots to increased organizational efficiency/effectiveness)? | Scale development |
| | Which cultural contexts are most beneficial for high social robot performance? | Survey |

suitable way to collect data for this purpose. Finally, we consider it most important to examine which values should be integrated into social robots for a particular industry context. We recommend conducting a comprehensive Delphi study with experts from various backgrounds (e.g. designers, users; academia, practice).

Regarding organizational values, future research could examine which values have a positive effect on the integration of social robots in organizations. For example, based on a survey, researchers could analyze what kind of organizational values support the integration of a social robot among employees. Furthermore, a scale/measurement instrument could be developed to examine the cultural readiness of an organization to integrate social robots. Such a scale could measure how far an organizational culture already embraces those values embedded in a particular social robot to anticipate potential difficulties and required cultural changes for a successful social robot implementation.

Regarding social robot performance, future research may also develop a scale to measure the performance of social robots. This scale could assess how far the social robot contributes to the efficiency and effectiveness of organizational processes (e.g. what factors determine efficient and effective processes supported by a social robot, and how can we measure them?). Finally, researchers could examine what kind of cultural contexts support the performance of social robots the most. A survey could be used that builds on the developed scale for social robot performance and on an assessment of the organizational culture.

The examples show a broad range of future research possibilities at the intersection of social robots and cultural values, and we call for studies following up on these examples.

## 5 Conclusion

Our paper addresses the role of culture in the application of social robots. We can observe a lack of research tackling cultural challenges with regard to the integration of social robots into organizational contexts. Based on this finding, we propose a conceptualization of the role of cultural values in social robot integration and we derive an agenda for future research to follow up on the research need we identified.

We encourage fellow researchers to take a value-based perspective and address the call for research in this area [74]. A better understanding of the cultural requirements to successfully integrate social robots into organizations will help practitioners to achieve more powerful performance benefits through the usage of social robots, it will help robot designers develop social robots with integrated cultural values, and it will help employees and customers to see social robots as part of their cultural groups.

## References

1. Hegel, F., Muhl, C., Wrede, B., Hielscher-Fastabend, M., Sagerer, G.: Understanding social robots. In: Proceedings of the 2009 Second International Conferences on Advances in Computer-Human Interactions, pp. 169–174. IEEE Computer Society, Washington, DC, USA (2009)
2. Duffy, B.R., Joue, G., Bourke, J.: Issues in assessing performance of social robots. In: Presented at the WSEAS International Conference on Robotics, Skiathos Island (2002)
3. Šabanović, S.: Robots in society, society in robots. Int. J. Soc. Robot. **2**(4), 439–450 (2010). https://doi.org/10.1007/s12369-010-0066-7
4. Schaefer, K.E.: The Perception and Measurement of Human-Robot Trust, Dissertation, University of Central Florida (2013)
5. Castro-González, Á., Malfaz, M., Salichs, M.A.: Learning the selection of actions for an autonomous social robot by reinforcement learning based on motivations. Int. J. Soc. Robot. **3**(4), 427–441 (2011). https://doi.org/10.1007/s12369-011-0113-z
6. Duffy, B.R.: Anthropomorphism and the social robot. Robot. Auton. Syst. **42**(3), 177–190 (2003). https://doi.org/10.1016/S0921-8890(02)00374-3
7. Sarrica, M., Brondi, S., Fortunati, L.: How many facets does a "social robot" have? a review of scientific and popular definitions online. Inf. Technol. People (2019). https://doi.org/10.1108/ITP-04-2018-0203
8. Li, H., Cabibihan, J.-J., Tan, Y.K.: Towards an effective design of social robots. Int. J. Soc. Robot. **3**(4), 333–335 (2011). https://doi.org/10.1007/s12369-011-0121-z
9. Beer, J.M., Fisk, A.D., Rogers, W.A.: Toward a framework for levels of robot autonomy in human-robot interaction. J. Hum. Robot. Interact. **3**(2), 74–99 (2014). https://doi.org/10.5898/JHRI.3.2.Beer
10. Glisky, E.L.: Changes in cognitive function in human aging. In: Riddle, D.R. (ed.) Brain Aging: Models, Methods, and Mechanisms. CRC Press/Taylor & Francis, Boca Raton (FL) (2007)
11. Carpentier, N.: Differentiating between access, interaction and participation. Conjunct. Transdiscipl. J. Cult. Particip. **2**(2), 7–28 (2015)
12. International Federation of Robotics: Introduction into Service Robots. https://ifr.org/service-robots
13. Hegel, F., Muhl, C., Wrede, B., Hielscher-Fastabend, M., Sagerer, G.: Understanding social robots. In: Presented at the Advances in Computer-Human Interactions, 2009. ACHI'09. Second International Conference February (2009)

14. Minor, L.: Es darf nicht passieren, dass Lio jemanden umfährt , https://www.tagesanzeiger.ch/zuerich/region/seine-geduld-ist-endlos/story/17626511, (2019)
15. World Health Organization: World Report on Ageing and Health 2015. World Health Organization, Geneva, Switzerland (2015)
16. World Health Organization: World Report on Disability. World Health Organization, Geneva, Switzerland (2011)
17. Riek, L.D.: Healthcare robotics. Commun. ACM **60**(11), 68–78 (2017)
18. Rabbitt, S.M., Kazdin, A.E., Scassellati, B.: Integrating socially assistive robotics into mental healthcare interventions: Applications and recommendations for expanded use. Clin. Psychol. Rev. **35**, 35–46 (2015). https://doi.org/10.1016/j.cpr.2014.07.001
19. Whelan, S., Murphy, K., Barrett, E., Krusche, C., Santorelli, A., Casey, D.: Factors affecting the acceptability of social robots by older adults including people with dementia or cognitive impairment: a literature review. Int. J. Soc. Robot. **10**(5), 643–668 (2018). https://doi.org/10.1007/s12369-018-0471-x
20. Chen, S.-C., Jones, C., Moyle, W.: Social robots for depression in older adults: a systematic review. J. Nurs. Scholarsh. **50**(6), 612–622 (2018). https://doi.org/10.1111/jnu.12423
21. Kachouie, R., Sedighadeli, S., Abkenar, A.B.: The role of socially assistive robots in elderly wellbeing: a systematic review. In: Rau, P.-L.P. (ed.) Cross-Cultural Design, pp. 669–682. Springer International Publishing, Cham (2017)
22. Lagrange, F.J., Jacq, F.: Developing an innovative oral unit dose robot dispenser: patient care performance and industrial perspectives. Le Pharmacien Hospitalier et Clinicien. **49**(2), e1–e14 (2014). https://doi.org/10.1016/j.phclin.2014.03.001
23. Abdi, J., Al-Hindawi, A., Ng, T., Vizcaychipi, M.P.: Scoping review on the use of socially assistive robot technology in elderly care. BMJ Open **8**(2), e018815 (2018). https://doi.org/10.1136/bmjopen-2017-018815
24. Pennisi, P., Tonacci, A., Tartarisco, G., Billeci, L., Ruta, L., Gangemi, S., Pioggia, G.: Autism and social robotics: a systematic review. Autism Res. **9**(2), 165–183 (2016). https://doi.org/10.1002/aur.1527
25. Robinson, N.L., Cottier, T.V., Kavanagh, D.J.: Psychosocial health interventions by social robots: systematic review of randomized controlled trials. J. Med. Internet Res. **21**(5), e13203 (2019). https://doi.org/10.2196/13203
26. Cabibihan, J.-J., Javed, H., Ang, M., Aljunied, S.M.: Why robots? a survey on the roles and benefits of social robots in the therapy of children with autism. Int. J. Soc. Robot. **5**(4), 593–618 (2013). https://doi.org/10.1007/s12369-013-0202-2
27. Beran, T.N., Ramirez-Serrano, A., Vanderkooi, O.G., Kuhn, S.: Reducing children's pain and distress towards flu vaccinations: a novel and effective application of humanoid robotics. Vaccine **31**(25), 2772–2777 (2013). https://doi.org/10.1016/j.vaccine.2013.03.056
28. Belpaeme, T., Kennedy, J., Ramachandran, A., Scassellati, B., Tanaka, F.: Social robots for education. A Rev. Sci. Robot. **3**(21), eaat5954 (2018). https://doi.org/10.1126/scirobotics.aat5954
29. van den Berghe, R., Verhagen, J., Oudgenoeg-Paz, O., van der Ven, S., Leseman, P.: Social robots for language learning: a review. Rev. Educ. Res. **89**(2), 259–295 (2019). https://doi.org/10.3102/0034654318821286
30. Rosenberg-Kima, R., Koren, Y., Yachini, M., Gordon, G.: Human-Robot-Collaboration (HRC): social robots as teaching assistants for training activities in small groups. In: 2019 14th ACM/IEEE International Conference on Human-Robot Interaction (HRI), pp. 522–523, Daegu, South Korea (2019)
31. Ramachandran, A., Huang, C.-M., Gartland, E., Scassellati, B.: Thinking aloud with a tutoring robot to enhance learning. In: Hri '18: Proceedings of the 2018 Acm/Ieee International Conference on Human-Robot Interaction, pp. 59–68. Assoc Computing Machinery, New York (2018)
32. Ivanov, S., Webster, C., Garenko, A.: Young Russian adults' attitudes towards the potential use of robots in hotels. Technol. Soc. 55(November 2018), 24–32 (2018). https://doi.org/10.1016/j.techsoc.2018.06.004

33. Lai, C.-J., Tsai, C.-P.: Design of Introducing Service Robot into Catering Services. In: Proceedings of the 2018 International Conference on Service Robotics Technologies (ICSRT '18). ACM, New York, NY, USA (2018)
34. Pinillos, R., Marcos, S., Feliz, R., Zalama, E., Gomez-Garcia-Bermejo, J.: Long-term assessment of a service robot in a hotel environment. Robotics and Autonomous Systems. 79(May 2016), 40–57 (2016). https://doi.org/10.1016/j.robot.2016.01.014
35. Brodeur, T., Cater, A., Vaz, J.C., Oh, P.: Directory navigation with robotic assistance. In: 2018 IEEE 8th Annual Computing and Communication Workshop and Conference (CCWC). IEEE, New York (2018)
36. Chen, Y., Wu, F., Shuai, W., Chen, X.: Robots serve humans in public places-KeJia robot as a shopping assistant. Int. J. Adv. Robot. Syst. **14**(3) (2017). https://doi.org/10.1177/1729881417703569
37. Satake, S., Hayashi, K., Nakatani, K., Kanda, T.: Field trial of an information-providing robot in a shopping mall. In: 2015 Ieee/Rsj International Conference on Intelligent Robots and Systems (iros), pp. 1832–1839. IEEE, New York (2015)
38. Sabelli, A.M., Kanda, T.: Robovie as a mascot: a qualitative study for long-term presence of robots in a shopping mall. Int. J. Soc. Robot. **8**(2), 211–221 (2016). https://doi.org/10.1007/s12369-015-0332-9
39. Foster, M.E., Alami, R., Gestranius, O., Lemon, O., Niemelä, M., Odobez, J.-M., Pandey, A.K.: The MuMMER Project: Engaging Human-Robot Interaction in Real-World Public Spaces. In: Agah, A., Cabibihan, J.-J., Howard, A.M., Salichs, M.A., He, H. (eds.) Social Robotics, pp. 753–763. Springer International Publishing, Cham (2016)
40. Triebel, R., Arras, K., Alami, R., Beyer, L., Breuers, S., Chatila, R., Chetouani, M., Cremers, D., Evers, V., Fiore, M., Hung, H., Ramírez, O.A.I., Joosse, M., Khambhaita, H., Kucner, T., Leibe, B., Lilienthal, A.J., Linder, T., Lohse, M., Magnusson, M., Okal, B., Palmieri, L., Rafi, U., van Rooij, M., Zhang, L.: SPENCER: a socially aware service robot for passenger guidance and help in busy airports. In: Wettergreen, D.S., Barfoot, T.D. (eds.) Field and Service Robotics: Results of the 10th International Conference, pp. 607–622. Springer International Publishing, Cham (2016)
41. Commission, European: Public Attitutes Towards Robots. European Commission, Brussels (2012)
42. Hofstede, G., Hofstede, G.J.: Cultures and Organizations: Software of the Mind. McGraw-Hill, New York (2005)
43. Schein, E.H.: Organizational Culture and Leadership. Wiley, San Francisco (2010)
44. Kluckhohn, C.: 2. values and value-orientations in the theory of action: an exploration in definition and classification. In: Toward a General Theory of Action. Harvard University Press, Cambridge (1951)
45. Tajfel, H.: Social categorization. In: Moscovici, S. (ed.) Introduction à la psychologie sociale, pp. 272–302. Larousse, Paris (1973)
46. Karahanna, E., Evaristo, J., Srite, M.: Levels of culture and individual behavior: an investigative perspective. J. Global Inf. Manage. **13**(2), 1–20 (2005). https://doi.org/10.4018/jgim.2005040101
47. Spiegel, M., Schmiedel, T., Brocke, J.V.: What makes change harder—or easier: how embedded values fit organizational culture. MIT Sloan Manage. Rev. **58**(3), 88–89 (2017)
48. Kotter, J.P., Heskett, J.L.: Corporate Culture and Performance. The Free Press, New York (1992)
49. Trovato, G., Lucho, C., Paredes, R.: She's electric—the influence of body proportions on perceived gender of robots across cultures. Robotics **7**(3), 50 (2018). https://doi.org/10.3390/robotics7030050
50. Kanngiesser, P., Itakura, S., Zhou, Y., Kanda, T., Ishiguro, H., Hood, B.: The role of social eye-gaze in children's and adults' ownership attributions to robotic agents in three cultures. Interact. Stud. **16**(1), 1–28 (2015). https://doi.org/10.1075/is.16.1.01kan
51. Kamaruddin, N., Wahab, A., Quek, C.: Cultural dependency analysis for understanding speech emotion. Expert Syst. Appl. **39**(5), 5115–5133 (2012). https://doi.org/10.1016/j.eswa.2011.11.028

52. Bruno, B., Chong, N.Y., Kamide, H., Kanoria, S., Lee, J., Lim, Y., Pandey, A.K., Papadopoulos, C., Papadopoulos, I., Pecora, F., Saffiotti, A., Sgorbissa, A.: Paving the way for culturally competent robots: a position paper. In: 2017 26th IEEE International Symposium on Robot and Human Interactive Communication (RO-MAN), pp. 553–560 (2017)
53. Papadopoulos, I., Koulouglioti, C.: The influence of culture on attitudes towards Humanoid and animal-like robots: an integrative review. J. Nurs. Scholarsh. **50**(8), 653–665 (2018). https://doi.org/10.1111/jnu.12422
54. Cunningham, S., Chellali, A., Jaffre, I., Classe, J., Cao, C.G.L.: Effects of experience and workplace culture in human-robot team interaction in robotic surgery: a case study. Int. J. Soc. Robot. **5**(1), 75–88 (2013). https://doi.org/10.1007/s12369-012-0170-y
55. Rudovic, O., Lee, J., Mascarell-Maricic, L., Schuller, B.W., Picard, R.W.: Measuring engagement in robot-assisted autism therapy: a cross-cultural study. Front. Robot. AI. **4**, 1–17 (2017). https://doi.org/10.3389/frobt.2017.00036
56. Hashim, R., Mahamood, S.F.: Humanoid robots for skill augmentation of gifted children: teachers' perceptions and islamic implications. Procedia Comput. Sci. **42**, 345–350 (2014). https://doi.org/10.1016/j.procs.2014.11.072
57. Bidin, S.A.H., Lokman, A.M., Mohd, W.A.R.W., Tsuchiya, T.: Initial intervention study of Kansei robotic implementation for elderly. Procedia Comput. Sci. **105**, 87–92 (2017). https://doi.org/10.1016/j.procs.2017.01.205
58. Cortellessa, G., Scopelliti, M., Tiberio, L., Svedberg, G.K., Loutfi, A., Pecora, F.: A cross-cultural evaluation of domestic assistive robots. In: AAAI Fall Symposium: AI in Eldercare: New Solutions to Old Problems, pp. 24–31, 7–9 Nov, Arlington, Virginia (2008)
59. Reich-Stiebert, N., Eyssel, F.: Learning with educational companion robots? toward attitudes on education robots, predictors of attitudes, and application potentials for education robots. Int. J. Social Robot. **7**(5), 875–888 (2015). https://doi.org/10.1007/s12369-015-0308-9
60. Conti D., Cattani A., Nuovo S.D., Nuovo, A.D.: A cross-cultural study of acceptance and use of robotics by future psychology practitioners. In: 2015 24th IEEE International Symposium on Robot and Human Interactive Communication (RO-MAN). pp. 555–560 (2015)
61. Han, J., Hyun, E., Kim, M., Cho, H., Kanda, T., Nomura, T: The cross-cultural acceptance of tutoring robots with augmented reality services. Int. J. Digit. Content Technol. Appl. **3**(2), 95–102 (2009). https://doi.org/10.4156/jdcta.vol3.issue2.han
62. Shahid, S., Krahmer, E., Swerts, M.: Child-robot interaction across cultures: how does playing a game with a social robot compare to playing a game alone or with a friend? Comput. Hum. Behav. **40**, 86–100 (2014). https://doi.org/10.1016/j.chb.2014.07.043
63. Choi, J.-H., Lee, J.-Y., Han, J.-H.: Comparison of cultural acceptability for educational robots between Europe and Korea. J. Inf. Process. Syst. **4**(3), 97–102 (2008). https://doi.org/10.3745/JIPS.2008.4.3.97
64. Pang, W.-C., Wong, C.-Y., Seet, G.: Exploring the use of robots for museum settings and for learning heritage languages and cultures at the chinese heritage centre. Presence-Teleoperators Virtual Environ. **26**(4), 420–435 (2017). https://doi.org/10.1162/PRES_a_00306
65. Germak, C., Lupetti, M.L., Giuliano, L., Ng, M.E.K.: Robots and cultural heritage: new museum experiences. J. Sci. Technol. Arts. **7**(2), 47–57 (2015)
66. Shiomi, M., Hagita, N.: Social acceptance toward a childcare support robot system: web-based cultural differences investigation and a field study in Japan. Adv. Robot. **31**(14), 727–738 (2017). https://doi.org/10.1080/01691864.2017.1345322
67. Kovacic, M.: The making of national robot history in Japan: monozukuri, enculturation and cultural lineage of robots. Crit. Asian Stud. **50**(4), 572–590 (2018). https://doi.org/10.1080/14672715.2018.1512003
68. van Wynsberghe, A.: Designing robots for care: care centered value-sensitive design. Sci. Eng. Ethics **19**(2), 407–433 (2013). https://doi.org/10.1007/s11948-011-9343-6
69. Hasse, C.: How robots challenge institutional practices. Learn. Cult. Soc. Interact. (2018, in press). https://doi.org/10.1016/j.lcsi.2018.04.003
70. Mutlu, B., Forlizzi, J.: Robots in organizations: the role of workflow, social, and environmental factors in human-robot interaction. In: Proceedings of the 3rd ACM/IEEE International Conference on Human Robot Interaction, pp. 287–294. ACM (2008)

71. Schmiedel, T., vom Brocke, J., Recker, J.C.: Cultural fitness for business process management : what is it and what is it worth? The World Financial Review. Nov, 21–24 (2012)
72. Davis, J., Nathan, L.P.: Value sensitive design: applications, adaptations, and critiques. In: Handbook of ethics, values, and technological design: sources, theory, values and application domains, pp. 11–40 (2015)
73. vom Brocke, J., Sinnl, T.: Culture in business process management: a literature review. Bus. Process Manage. J. **17**(2), 357–378 (2011). https://doi.org/10.1108/14637151111122383
74. van den Hoven, J.: ICT and value sensitive design. In: Goujon, P., Lavelle, S., Duquenoy, P., Kimppa, K., Laurent, V. (eds.) The Information Society: Innovation, Legitimacy, Ethics and Democracy In honor of Professor Jacques Berleur s.j. pp. 67–72. Springer US (2007)

# Digital Transformation for Sustainability: A Necessary Technical and Mental Revolution

**Michael von Kutzschenbach and Claus-Heinrich Daub**

**Abstract** Digital technology and innovations are transforming our economy and society as never before. However, business leaders and managers are at a point where they need to rethink the way business can leverage digital transformation in order to support engagement with sustainability challenges. As organizations operate in the context of society, many factors such as social structures, political decision-making, general economic trends, and technological development, influence their operations. To be successful, an organization must be aware of these changes and at the same time be able to engage with them. Based on this viewpoint, we outline an integrated framework linking different levels of "digitalness" with necessary changes in managerial practice to support organizational inquiry. More research is necessary to better understand how the practice of management needs to transform in order to leverage digital technologies for an on-going, learning-based engagement strategy. This requires awareness of one's own thinking, suspension of one's prior beliefs, and engagement in organizational learning and experimentation with stakeholders.

**Keywords** Digital transformation · Engagement strategy · Sustainability · Organizational learning · Inquiring system

## 1 Introduction

We live in exciting times, to paraphrase an old Chinese saying. The original idea behind the statement had both positive and negative implications and the current state of the world confirms the wisdom of this insight. From the business perspective, two

M. von Kutzschenbach (✉)
Institute of Management, School of Business, FHNW University of Applied Sciences and Arts Northwestern Switzerland, Peter Merian-Strasse 86, 4002 Basel, Switzerland
e-mail: michael.vonkutzschenbach@fhnw.ch

C.-H. Daub
Institute of Management, School of Business, FHNW University of Applied Sciences and Arts Northwestern Switzerland, Bahnhofstrasse 6, 5210 Windisch, Switzerland
e-mail: clausheinrich.daub@fhnw.ch



R. Dornberger (ed.), *New Trends in Business Information Systems and Technology*, Studies in Systems, Decision and Control 294,
https://doi.org/10.1007/978-3-030-48332-6_12

issues have become dominant in the strategic thinking of organizations—stakeholder demands for greater business responsiveness to environmental and social issues and advances in digital technology [1].

The first is concern for the environment. This is more generally interpreted as including not only the climate aspects but also the effects of "business as usual" on the planetary ecosystem [2]. Added to this are concerns about the effects of business operations on society [3]. Providing wages for services in a capitalistic society clearly improves overall welfare but there are systemic distortions that shift the benefits to a few, often at the expense of others cf. [4]. This issue has a strong ethical component and vocal stakeholders argue that businesses must take responsibility for past abuses and strive to eliminate them in their future work. Without credible efforts to address these issues organizations can lose their "license to operate" in society.

Important and significant improvements in environmental performance have been achieved in the decades following the publication of Rachel Carson's *Silent Spring* in 1962 [5]. However, these have generally not come from industry-initiated actions. Regulations and laws were needed to assure compliance through mostly financial penalties. Although greater efforts have made environmental protection possible, these improvements most often represent "harvesting low hanging fruit." These are actions taken by companies because it is good for business and for their profile but without radically changing the way they do their business. Significantly more effort and creativity will be needed to attain higher levels of environmental performance [6].

The societal narrative has evolved from demanding non-polluting business operations to achieving a state of "sustainability". In reality, this is an impossible goal because it ultimately represents a violation of the Second Law of Thermodynamics [7, 8]. Beyond this, the original environmental challenge has evolved to include the social and ecosystem ethical aspects, each backed by active and vocal stakeholders from all parts of society [9].

In short, the external and internal challenges of being successful in this new "environment" is a classic "wicked problem," as originally described by Rittel and Webber [10]. Engagement with this class of problems requires quite different managerial mindsets and approaches in order to be effective. To date, there is little evidence that the magnitude of the "sustainability" challenges facing businesses is fully grasped by the business community.

While environmental and social issues are generally viewed in a negative light as a threat, digital technologies have been widely adopted by business as a means for improving operations and directly contributing to economic performance. Nevertheless, digital technologies also offer the potential to support the organization's efforts to address the first issue. Since their beginnings in military technology [11], digital technologies have grown at an exponential pace and have gone through several evolutionary stages.

Digital technologies present a company with a set of issues that are both promising and challenging [12]. Improved information gathering, processing and dissemination is an obvious benefit. Organizational disruption is one of the more challenging consequences. Both the sustainability challenge and developments in digital technology are largely driven by forces external to the organization. However, the adoption and management of digital technologies are under direct managerial control, which allows greater freedom in determining their application. However, a company's responses to the sustainability issue are not as direct and focused.

The purpose of this chapter is to present a framework that links sustainability demands with the promise of digital technologies. As with almost all technologies, there are both positive and negative consequences. An all too human tendency is, unfortunately, to concentrate on the advantages and not fully recognize the potential disadvantages. Digital technologies are no different. Additionally, the immediate benefits of, for example, automating routine procedures have tendencies to reinforce the belief that digital technologies are a panacea for all problems. The ubiquity and success of digital technologies in business suggest that these technologies can significantly support an organization's efforts to engage with the wicked problem of sustainability. However, this path is not an easy one; it implies significant changes ranging from managerial thinking and organizational culture to the larger society in which they are embedded.

This chapter highlights the importance of fostering organizational learning and management innovation within and beyond the organization to harness the power of digital technologies. It addresses how digital transformation to higher levels of organizational integration supports the engagement and learning processes required in order to cope with the wicked problem of sustainability at both the business and societal levels.

The remainder of this chapter is organized in three sections. The following includes a discussion of the complexity of sustainability challenges and provides a framework for understanding digital transformation as a process. In the subsequent section, an integrative framework for a sustainability-oriented inquiring process with the support of digital technologies will be developed. Finally, the integration of new digital technologies with organizational support for double-loop learning and innovative experimentation to foster digital transformation for the better will be discussed.

## 2 A Necessary Revolution—from Technical Fixes to Fundamental Solutions

Building on the introduction of the sustainability challenge and providing a framework for understanding different levels of "digitalness," this section argues for a more holistic approach to digital transformation. When applying digital transformation to the development of sustainability approaches, it is useful to have a shared understanding of how the different terms are used in this chapter.

The first step highlights the complexity of sustainability challenges for human decision-making capability. The second step frames the digital evolution as characterized by three levels, digitization, digitalization, and digital transformation. Because digital transformation is not an end in itself but rather an ongoing evolutionary process of innovation, the necessary organizational learning processes for a more holistic approach to digital transformation for sustainability are discussed.

## 2.1 The Sustainability Challenge

The effects of human activities on the planet have been observed for centuries, so awareness of the consequences of these activities is not new (cf. [13]). However, industrialization and economic pressures for more and faster growth have outstripped nature's ability to absorb the by-products of this growth. Carbon dioxide emissions to the atmosphere is the most common example but the list of hazardous chemicals and wastes is long [14]. Information and communication technologies, including websites and social media, have contributed significantly to raising public awareness of these issues and providing a means for stakeholder voices to be heard in settings that were previously inaccessible.

The word "sustainability" is used in discussions about a healthier world that enables more positive ways of living. The report *Our Common Future*, published by the UN World Commission on Environment and Development in 1987, defined sustainable development as "development that meets the needs of the present without compromising the ability of future generations to meet their own needs" ([15], n.p.) For the business world, the UN report has established sustainability as a keyword for a variety of standards and expectations that society uses to judge the performance of companies.

Over time, the concept of sustainability has been socially constructed and re-constructed through social, political, and economic processes [16, 17]. While the UN definition and the subsequent "Sustainable Development Goals" (SDGs) are inspirational objectives, the task of translating them into a business strategy is not obvious. Since sustainability is a wicked problem, there are no "final end-states" or "fixed targets" that we can strive for. According to Bagheri and Hjorth, the concept of sustainability is a "moving target, which is continuously evolving as we understand more about our socio-environmental system" ([18], p. 84). Thus, the sustainability challenge represents a classic "wicked problem" (cf. [10, 19]). That is, it has no definitive formulation. Multiple stakeholders have different values and objectives on the topic that lead to different views/definitions of the state of the world. Wicked problems have no "stopping rule." Consequently, sustainability is dynamic and the time horizon is indeterminate. Furthermore, every wicked problem is a symptom of another problem; this refers to the fact that it can usually be traced back to a deeper, more fundamental problem. Transferred to the sustainability challenge, all issues relating to sustainability are consequences of the nested and dynamic nature of the multiple systems involved [20].

Thus, the complexity of sustainability challenges human decision-making capability. This often leads to managerial prescriptions and recommendations based on simplifying disciplinary assumptions to reduce the complexity of the challenges that organizations face. Artificially constraining the problem definition and developing strategies for the reduced problem statement has two outcomes. In the short term, a technological solution may allow temporary mitigation of the problem. For example, technical advances have significantly improved the fuel efficiency of internal combustion engines through reducing automotive emissions. However, there is rarely a single outcome of a technological development. In the automotive case, increased fuel efficiency has actually led to increased car usage due to lower fuel expenses, which is a negative outcome in the larger environmental picture.

Digital technologies offer the prospect of contributing positively to the efforts of organizations and society to become more "sustainable." However, the application must be moderated by the awareness of possible unintended negative consequences. Essentially, the success of initial digital transformations in standard business and organizational applications has led to the belief that this approach might also be useful in dealing with sustainability. That is, an exclusive reliance on technology to solve a problem that is comprised of both technological and non-technological components. This limits the effectiveness of both organizational initiatives for learning and management innovation as well as digital technologies for fostering a more sustainable society.

## 2.2 *Classifying the Different Levels of Digitalness*

Before examining the impact that digital technology might have on enabling a transition towards a digitized sustainable society, it is important to understand exactly what "digital" means in digital transformation. Unruh and Kiron [21] proposed a framework (see Fig. 1) to describe the digital evolution. They argue, "two principal forces are driving digitalization. One is digital technology itself and its associated services and gadgets. Another is the invisible hand of the market responding to the evolving wants of consumers" ([21], n.p.) In their framework, they characterize the digital evolution as consisting of three phases starting with simple digitization followed by digitalization and leading to a digital transformation. Note that digital transformation is not an end state, but an ongoing evolutionary innovation process.

Figure 1 illustrates the different levels of "digitalness" (cf. [21]). Given the impact of digital technologies, new ways of thinking about businesses and relationships will be needed in order to achieve greater performance.

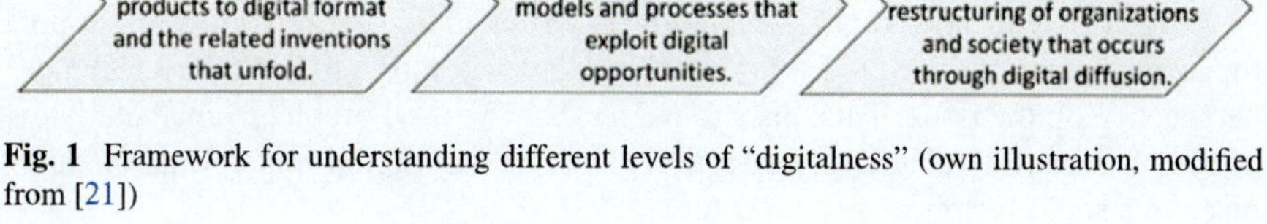


**Fig. 1** Framework for understanding different levels of "digitalness" (own illustration, modified from [21])

### 2.3 The Need for a More Holistic Approach to Digital Transformation

While business interests are currently the primary driver of digital transformation, the trends will expand beyond the organizational boundaries. Examples of this include the "Internet of Things" and cloud computing [22]. Consequently, innovation activities initiated by business interests will spill over into the larger social world. Business organizations therefore have a special responsibility to promote the best way for this disruptive transformation to support sustainability.

Previous research has shown that organizational learning is a critical capability for an organization's innovation outcome (cf. [23, 24]). Senge [4] popularized the term "Learning Organization" with the publication of his book *The Fifth Discipline: The Art and Practice of the Learning Organization.* In talking about "Learning Organizations" he described organizations in which "people continually expand their capability to create the results they truly desire, where new and expansive patterns of thinking are nurtured, where collective aspiration is set free, and where people are continually learning how to learn together" ([4], p. 3).

The literature contrast two fundamental organizational learning processes: (1) "single-loop learning" (SLL) and (2) "double-loop learning" (DLL). Organizational SSL is characterized by problem solving within the organization's "theory in use" [25] or "dominant logic" [26]. DLL expands the process of detecting and solving problems by critically reflecting on the existing organizational mindset. This involves questioning the values and norms and the underlying belief systems and assumptions. This can lead to a change in the existing "theory in use" [25, 27] and results in a new business logic for a company's theory of success.

When applying digital transformations to developing approaches to sustainability, SLL consists of introducing incremental changes within a dominant design. The objective is therefore to improve efficiency through the new digital technology ("Digitization") and to ensure competitiveness in the existing business ecosystem ("Digitalization"). In contrast, DLL involves deeper inquiry and changes in the existing managerial mindset and consequently the organization system. The goal of DLL is to promote a mindset towards a learning-orientation with the objective of achieving a continuous systemic change for sustainability ("Digital Transformation").

The next section presents an integrated framework describing the linkages between different levels of "digitalness" and necessary changes in managerial practice for organizational inquiry. The desired outcome of this is to identify high leverage system intervention points (cf. [28]) to be able to more effectively guide the organization and society towards a more "sustainable" world. Currently, most research is concerned with how digital technologies foster more sustainability or support the achievement of the SDGs (cf. [29–31]). However, most studies take a technical perspective on the issue. This tends to neglect the "wicked problem" characteristics of the sustainability challenge and ignores the fact that digital transformation takes place in a socio-technical system context.

# 3 A Framework Using Digital Transformation for the Better

This section presents a framework for understanding how digital transformation for the better requires changes in managerial and organizational mindsets. The objective is a systemic change at the organizational and ultimately societal level. These are complex and interconnected nested systems. The key to effecting meaningful systemic change is to find effective high leverage intervention points in the system under investigation [28]. Leverage point analysis is the third component of our proposed methodology to support the transition to a higher level of sustainability.

The first step in this process is the use of digital technologies, which are evolving in power and sophistication. The second step is the recognition that engaging the wicked problem of sustainability requires significant changes in mindsets, with the objective of bringing about meaningful systemic change. The third step, leverage point analysis, links the power of digital technologies to the need to develop new paradigms to effectively engage sustainability challenges.

## *3.1 The Framework for Understanding the Importance of Fostering Organizational Inquiry*

The optimistic hope that digital technologies will be a panacea for the sustainability challenge is based on society's belief in and reliance on the efficacy of technological solutions to problems. Historically, technology has succeeded brilliantly in many difficult contexts. Nevertheless, the nature of the current situation is far more complex than anything humans have faced before. Indeed, much of the problem stems from the fact that the problem context is as much social as it is technical. Strictly speaking, digital technologies are technical systems—cause and effect are understood, or at least understandable in principle through experimentation and the application of scientific methods. A strong and essentially unified theoretical base supports technology.

The social world is not as "simple". Chief among the complexities is the presence of the human agent and all the social and psychological attributes that they contribute to a society. Characteristics that complicate the work towards a more sustainable world include multiple stakeholders with differing ideologies, different goals and capabilities to achieve them, and well-documented cognitive biases in problem solving (see for example, the work of Tversky and Kahneman, summarized in [32]) and decision-making. However, one of the most striking barriers is the managerial mindset with respect to engaging with problems. In particular, Western civilization settles for solving problems [33]. The common approaches to dealing with sustainability is an example of this mindset to rely primarily on technology to "fix it". Digital technologies are just another advanced tool to finally solve the problem by achieving sustainability.

Because of the nature of wicked problems, this is not possible. The fundamental challenge for managers (and society) is to adopt a different mindset, or problem engagement paradigm, that emphasizes a process-based learning approach. This recognizes that this problem cannot be solved in the traditional sense. Rather it requires continuous interaction to improve the situation. It needs to be inspired by a collective vision of a better world, recognizing that there can be significant differences as to what that better world might look like. Despite these very difficult issues, there can be broad agreement on many points of improvement. Business organizations have a slightly simpler task because they are designed to meet a specified objective in their market. Still, their challenge is also wicked [34] and managers need to adapt their practices to meet the needs of their place in society as well as their "license to operate".

Figure 2 links the different "levels of digitalness" with the need for "a more holistic approach to digital transformation" in organizations. A key element in this transformation process is the addition of innovation as the driver that can shape the technology to support the engagement with sustainability. Innovation is very much an individual characteristic but it can be supported in an organizational context. Innovators are seen as problematic disruptors of the organization's status quo [35]. If they are identified as threatening then the likelihood that these individuals will leave the organization increases. A more productive approach to fostering innovators is that the organization offers a culture of openness, support and encouragement for all employees to express their visions and ideas to others in the company.

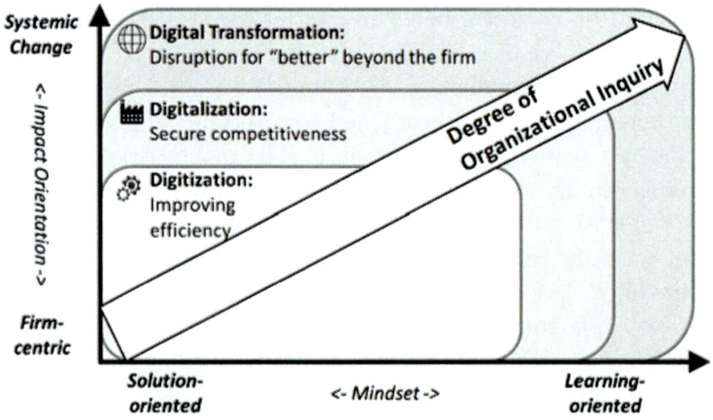


**Fig. 2** The framework for understanding the importance of fostering organizational inquiry to better address the transformation towards a digitalized sustainability society

**Table 1** Leverage points in order of their increasing power to change systems behavior. (*Adapted* from [28])

| Places to intervene in a system |
|---|
| 12. Constants, parameters, numbers (such as subsidies, taxes, standards) |
| 11. The size of buffers and other stabilizing stocks, relative to their flows |
| 10. The structure of material stocks and flows (such as transport networks, population age structure) |
| 9. The lengths of delays, relative to the rate of system change |
| 8. The strength of negative feedback loops, relative to the impacts they are trying to correct against |
| 7. The gain around driving positive feedback loops |
| 6. The structure of information flows (who does and does not have access to what kinds of information) |
| 5. The rules of the system (such as incentives, punishments, constraints) |
| 4. The power to add, change, evolve, or self-organize system structure |
| 3. The goals of the system |
| 2. The mindset or paradigm out of which the system—its goals, structure, rules, delays, parameters—arises |
| 1. The power to transcend paradigm |

### *3.2 Linking Leverage Points with Different Levels of Digitalness*

The mechanical system of a lever and a fulcrum is a metaphor for a systemic leverage point. In the mechanical system, a properly placed fulcrum enables the human user to move masses that would usually go beyond the ability of the individual. The fulcrum can be located in several possible positions with respect to the lever and the mass. Nevertheless, there is usually one "best" position that the user finds through trial and error experimentation.

This idea transfers easily to organizational and societal systems. The system is a mass and the user (manager) wants to find the best position for the fulcrum in order to obtain a desired change. Meadows [28], in the context of systemic change, identified a hierarchy of potential leverage points. These identified places that allow managers to most effectively intervene in a system to make the desired changes. Table 1 presents Meadows' leverage points. They are listed in order of their increasing power to change systems behavior. The organization's level of "digitalness" affects the kind of leverage that can be supported by the available technologies.

It is important to recognize that low leverage initiatives are relatively easy to implement but do not generate lasting change. They have the advantage of returning

| Level of Digitalness | Leverage Points |
|---|---|
| **Digitization** Improving efficiency | 12. Constants, parameters, numbers (such as subsidies, taxes, standards)<br>11. The size of buffers and other stabilizing stocks, relative to their flows<br>10. The structure of material stocks and flows (such as transport networks, population age structure)<br>9. The lengths of delays, relative to the rate of system change |
| **Digitalization** Secure competitiveness | 8. The strength of negative feedback loops, relative to the impacts they are trying to correct against<br>7. The gain around driving positive feedback loops<br>6. The structure of information flows (who does and does not have access to what kinds of information)<br>5. The rules of the system (such as incentives, punishments, constraints) |
| **Digital Transformation** Disrupting for good | 4. The power to add, change, evolve, or self-organize system structure<br>3. The goals of the system<br>2. The mindset or paradigm out of which the system—its goals, structure, rules, delays, parameters—arises<br>1. The power to transcend paradigm |

**Fig. 3** Linkages between different levels of "digitalness" and leverage points

quick results. Higher leverage initiatives are more difficult to identify and implement but lead to more lasting changes. The problem for managers, both organizational and societal, is how to harness the power of digital technologies to find higher leverage points in the system to implement meaningful policies that generate less ***un***sustainable behavior.

Figure 3 combines the different levels of "digitalness" with Meadows' leverage points. The boundaries are relative because the transition from one level to another is usually gradual. However, higher levels of "digitalness" are correlated with greater leverage.

In the context of managerial innovation for sustainability, "digitization" and "digitalization" are examples of applying single-loop learning to incorporate digital technology. However, the challenge of digital transformation lies in the tension between an organization's current dominant logic and emerging new logic. Addressing the challenges of digital innovation for sustainability, digitization and digitalization initiatives are useful but limited and if extended too far, can become counterproductive. Following Ehrenfeld's claim that "[*r*]*educing unsustainability is not the same as creating sustainability*" ([8], p. 4), it can be argued that optimizing an unsustainable system will not improve the health of the overall system, no matter how digital the existing business system is.

In addition, a new managerial mindset needs to be established that focuses more on learning and experimentation and not just on problem solving. Tackling wicked sustainability challenges with isolated efforts is inadequate. Conventional managerial models for improvement (pay for performance, rewards and punishments, benefit–cost relationships, fear as the primary motivator) are insufficient to deal with transformational or "adaptive" changes. Leveraging digital technology can help organizations learn better through supporting collaboration with many stakeholders. Each stakeholder needs to be involved in an iterative learning and enquiring process.

The outstanding issue with digital transformation is how to leverage improved shared understanding of a desired (digital) future and the possibilities to manage its resulting unintended consequences [36]. Ongoing dialogue, transparency, accountability, and a mutual willingness to accept risks are essential if organizations are to successfully address the sustainability challenge. This implies a dramatic change in most organizations.

## 4 Conclusion—Digital Transformation for the Better: An on-Going Learning-Based Engagement Strategy

The aim of this chapter was to increase awareness of the great potential digital technology has to address many challenges that face our world. However, accepting the challenge of sustainability is a wicked problem that requires an approach to engagement that is different from that used to solve technical problems. Perhaps the most significant aspect of digital transformation for greater sustainability is how management can innovate and transform their management practice in order to conduct an effective dialogue on a better digitally supported sustainable future. A key element in this transformation process is the addition of innovation as the driver that can adapt the technology of digitalization to support the engagement with sustainability challenges.

A characteristic of complex systems is that there is never a single outcome for a single intervention. Thus, the adoption of a technological "quick fix" for a complex problem will usually relieves the problem symptom, at least in the shorter term, but can have several unintended consequences. One of these unintended consequences is the reduction of the system's ability to diagnose and develop more fundamental solutions to the problem. These types of consequences stem from the interconnectedness of the system components and their dynamic interactions. We have argued that organizations can leverage advanced digital technologies to support managerial learning in addressing a firm's sustainability challenge. This can lead to innovative and potentially disruptive changes in organizational behavior and performance.

There are additional side effects from the new ways of thinking exhibited by organizational managers. These effects can potentially affect the societal system in ways that also contribute to improved sustainability. The dynamic process logic is that as the firm successfully leverages digital technologies and transforms accordingly, it will also achieve positive results. Two desirable consequences may arise. First, other companies will observe the improved performance and will likewise be inspired or challenged by their stakeholders to change their behavior similarly. This can lead to industry-wide (and beyond) improved performances. The second outcome could be that other societal actors such as governance and institutions might also be inspired and contribute to changing the societal culture in the direction of increased sustainability.

Systems thinking emphasizes that decision makers must be patient. Major system changes take time but advanced digital tools such as simulation modeling can provide the means to identify sources of delays, potential road blocks, and enable more effective communication with stakeholders.

Current discussions on how digital technology can help achieve the Sustainable Development Goals (SDGs) are important steps in the right direction. However, they are not sufficient. A wicked problem cannot be "solved" like a predictable technical problem. Thus, there is a strong need for an on-going, learning-based engagement strategy that leverages digital technology-enabled insights from advanced simulation and scenario building with organizational learning and innovation through experimentation and adaptive management.

More research is clearly needed. The possibilities offered by predictive analytics and big data can be used to draw attention to the consequences of decisions, both intended and unintended, in a dynamic and complex world. This can help overcome unrealistic beliefs about digital transformation and sustainability. Additionally, understanding how to integrate emergent technologies such as virtual and augmented reality with strategic thinking and learning to improve the state of the world is essential. These technologies can support conversations for thought experiments and scenario thinking that deepen the learning and understanding of a desired sustainable future. This technology-mediated form of learning affords the opportunity to experience the results that are separated in space and time from actions. This virtual world allows experimenting with decision-making and eliminates the risks of serious, unintended consequences.

Beyond solely technical improvements and digital innovations, building an innovation-friendly culture within business ecosystems seems to be most beneficial. These organizational development trends demand new strategies and abilities, as well as leadership mindsets, to engage with stakeholders to co-create innovations for sustainability.

The information technology sector faces a challenge in the effort to improve sustainability. By leveraging digital transformation for sustainability, both management and the IT sector must re-conceptualize digitization and digitalization as a means to an end, not as an end in itself. Digital technologies are tools that have tremendous potential to support individual companies and society as a whole as they continue to develop in ways that leverage human innovation and creativity for development while respecting environmental limits.

## References

1. WBGU: Unsere gemeinsame digitale Zukunft. Hauptgutachten. https://www.wbgu.de/de/publikationen/publikation/unsere-gemeinsame-digitale-zukunft (2019). Accessed 3 Sept 2019
2. Steffen, W., Richardson, K., Rockström, J., Cornell, S.E., Fetzer, I., Bennett, E.M., Biggs, R., Carpenter, S.R., de Vries, W., de Wit, C.A, Folke, C., Gerten, D., Heinke, J., Mace, G.M., Persson, L.M., Ramanathan, V., Reyers, B., Sörlin, S.: Sustainability. Planetary boundaries:

guiding human development on a changing planet. Science **347**(6223), 736–746 (2015). https://doi.org/10.1126/science.1259855
3. Moffat, K., Lacey, J., Zhang, A., Leipold, S.: The social license to operate: a critical review. For. Int. J. For. Res. **89**(5), 477–488 (2016). https://doi.org/10.1093/forestry/cpv044
4. Senge, P.M.: The Fifth Discipline: The Art and Practice of the Learning Organization. Doubleday, New York (2006)
5. Carson, R.: Silent spring: the classic that launched the environmental movement. A Mariner Book, Boston (1962/2002)
6. Dowling, G., Moran, P.: Corporate reputations: built in or bolted on? Calif. Manag. Rev. **54**(2), 25–42 (2012)
7. Huesemann, M.H.: The limits of technological solutions to sustainable development. Clean Technol. Enviro. Policy **5**, 21–34 (2002). https://doi.org/10.1007/s10098-002-0173-8
8. Ehrenfeld, J.R.: Beyond Sustainability: Why an All-consuming Campaign to Reduce Sustainability Fails. http://www.cirps.it/CIRPS/wp-content/uploads/2017/11/25_03_BeyondSustain.pdf (2006). Accessed 2 Oct 2019
9. Ehrenfeld, J.R.: Sustainability by Design: A Subversive Strategy for Transforming Our Consumer Culture. Yale University Press, New Haven (2008)
10. Rittel, H.W.J., Webber, M.M.: Dilemmas in a general theory of planning. Policy Sci. **4**(2), 155–169 (1973). https://doi.org/10.1007/BF01405730
11. Naughton, J.: The evolution of the internet: from military experiment to general purpose technology. J. Cyber Policy **1**(1), 5–28 (2016). https://doi.org/10.1080/23738871.2016.1157619
12. Helbing, D.: Societal, economic, ethical and legal challenges of the digital revolution: from big data to deep learning, artificial intelligence, and manipulative technologies. In: Helbing, D, (ed.) Towards Digital Enlightenment: Essays on the Dark and Light Sides of the Digital Revolution, pp. 47–72. Springer, Cham (2019). https://doi.org/10.1007/978-3-319-90869-4
13. Meadows, D.H., Meadows, D.L., Randers, J., Behrens III, W.W.: The Limits To Growth: A Report for the Club of Rome's Project on the Predicament of Mankind. Universe Books, New York (1972)
14. UNDP: Chemical and Waste Management for Sustainable Development: Results from UNPD's Work to Protect Human Health and the Environment from POPs. https://sustainabledevelopment.un.org/content/documents/1963Chemicals%20and%20Waste%20Management.pdf (2015). Accessed 29 July 2019
15. UN WCED: Report of the World Commission on Environment and Development: Our Common Future. http://www.un-documents.net/our-common-future.pdf (1987). Accessed 29 July 2019
16. Hoffman, A.J.: Institutional evolution and change: environmentalism and the U.S. chemical industry. Acad. Manag. J. **42**(4), 351–371 (1999). https://doi.org/10.2307/257008
17. Luthe, T., von Kutzschenbach, M.: Building common ground in mental models of sustainability. Sustainability **9**(5), 247–254 (2016). https://doi.org/10.1089/sus.2016.29068.tl
18. Bagheri, A., Hjorth, P.: Planning for sustainable development: a paradigm shift towards a process-based approach. Sustain. Dev. **15**, 83–96 (2007). https://doi.org/10.1002/sd.310
19. Peterson, H.C.: Transformational supply chains and the 'wicked problem' of sustainability: aligning knowledge, innovation, entrepreneurship, and leadership. J. Chain Netw. Sci. **9**(2), 71–82 (2009). https://doi.org/10.3920/JCNS2009.x178
20. Brønn, C., Brønn, P.S.: Sustainability: A wicked problem needing new perspectives. In: Borland, H., Lindgreen, A., Vanhamme, J., Maon, F., Ambrosini, V., Palacios-Florencio, B. (eds.) Business Strategies for Sustainability, pp. 3–18. Routledge, London (2019)
21. Unruh, G., Kiron, D.: Digital Transformation on Purpose. https://sloanreview.mit.edu/article/digital-transformation-on-purpose/ (2017). Accessed 22 July 2019
22. Boes, A., Kämpf, T., Lühr, T.: Von der « grossen Industrie » zum « Informationsraum » : Informatisierung und der Umbruch in den Unternehmen in historischer Perspektive. In: Doering-Manteuffel, A., Raphael, L., Schlemmer, T. (eds.) Vorgeschichte der Gegenwart. Dimensionen des Strukturbruchs nach dem Boom. Vandenhoeck & Ruprecht, Göttingen (2016)

23. March, J.G.: Exploration and exploitation in organizational learning. Organ. Sci. **2**(1), 71–87 (1991). https://doi.org/10.1287/orsc.2.1.71
24. Lynn, G.S., Mazzuca, M., Morone, J.G., Paulson, A.: Learning is the Critical Success Factor in Developing Truly New Products. Res. Technol. Manag. **41**(3), 45–51 (1998). https://doi.org/10.1080/08956308.1998.11671209
25. Argyris, C., Schön, D.A.: Organizational Learning II, Theory. Method and Practice. Addison-Wesley Publishing Company, MA (1996)
26. Prahalad, C.K., Bettis, R.A.: The dominant logic: A new linkage between diversity and performance. Strateg. Manag. J. **7**(6), 485–501 (1986). https://doi.org/10.1002/smj.4250070602
27. Argyris, C.: Double loop learning in organizations. Harvard Bus. Rev. **55**(5), 115–124 (1977)
28. Meadows, D.H.: Leverage points: places to intervene in a system. In: The Sustainability Institute (1999). http://donellameadows.org/wp-content/userfiles/Leverage_Points.pdf (1999). Accesses 3 Oct 2019
29. Vision: Uniting to deliver technology for the global goals. 2030Vision Full Report 2017. https://www.unglobalcompact.org/docs/publications/2017/2030Vision_Report.pdf (2017). Accessed 22 July 2019
30. Ericsson, The Earth Institute (Columbia University): ICT and SDGs: How Information and Communications Technology Can Accelerate Action on the Sustainable Development Goals. https://www.ericsson.com/49e8f7/assets/local/news/2016/05/ict-sdg.pdf (2018). Accessed 22 July 2019
31. GeSi.: Enabling the Global Goals: Evidence of digital solutions‘ impact on achieving the Sustainable Development Goals (SDGs). https://gesi.org/download-report/b8kdVlMhL0uKyPnLFwqiJ0AyECmFa4eX.pdf (2018). Accessed 22 July 2019
32. Kahneman, D.: Thinking, Fast and Slow. Penguin Books, London (2011)
33. Ackoff, R.: Ackoff's Best: His Classic Writings on Management. Wiley, New York (1999)
34. Camillus, J.C.: Strategy as a wicked problem. In: Harvard Business Review May, pp. 99–106 (2008)
35. Christensen, C.M., McDonald, R., Altman, E.J., Palmer, J.E.: Disruptive innovation: an intellectual history and directions for future research. J. Manag. Studies. **55**(7), 1043–1078 (2018). https://doi.org/10.1111/joms.12349
36. von Kutzschenbach, M., Brønn, C.: Education for managing digital transformation: a feedback systems approach. J Syst Cybernet Informat. **15**(2), 14–19 (2017)

# Visualization of Patterns for Hybrid Learning and Reasoning with Human Involvement

**Hans Friedrich Witschel, Charuta Pande, Andreas Martin, Emanuele Laurenzi, and Knut Hinkelmann**

**Abstract** "Boxology" is the graphical representation of patterns that are commonly observed in hybrid learning and reasoning systems. Since some hybrid systems also involve humans-in-the-loop, a need to identify patterns including humans is foreseen. With the help of use cases that involve humans-in-the-loop, this chapter provides a discussion on the typical roles performed by humans in hybrid systems and how they influence machine learning and/or knowledge engineering activities. As a result, it introduces a new element in boxology to represent a human and identify two abstract patterns for such human-in-the-loop scenarios.

**Keywords** Machine learning · Knowledge engineering · Knowledge representation · Boxology · Hybrid intelligence

## 1 Introduction

Artificial intelligence with its different methods has a long tradition and is now celebrating a veritable revival, mainly triggered by advances in the area of machine learning. Currently, the machine learning method of AI is regarded as the panacea of

H. F. Witschel (✉) · C. Pande · A. Martin · E. Laurenzi · K. Hinkelmann
School of Business, FHNW University of Applied Sciences and Arts Northwestern Switzerland, 4600 Olten, Switzerland
e-mail: hansfriedrich.witschel@fhnw.ch

C. Pande
e-mail: charuta.pande@fhnw.ch

A. Martin
e-mail: andreas.martin@fhnw.ch

E. Laurenzi
e-mail: emanuele.laurenzi@fhnw.ch

K. Hinkelmann
e-mail: knut.hinkelmann@fhnw.ch


R. Dornberger (ed.), *New Trends in Business Information Systems and Technology*, Studies in Systems, Decision and Control 294,
https://doi.org/10.1007/978-3-030-48332-6_13

digitalization. Significant advancements in learning algorithms, access to computing power and the availability of vast amounts of labeled data as the basis for training have recently resulted in impressive applications and have significantly reduced the learning effort [2].

However, learning algorithms also have weaknesses that are being addressed by combining machine learning with knowledge engineering or representation methods [6]. Some approaches additionally involve humans. This chapter provides a discussion of two contributions [12, 17] that we made to the AAAI[1] Spring Symposium on Combining Machine Learning with Knowledge Engineering,[2] with the goal of finding certain commonalities in the interaction between humans, machine learning and knowledge engineering methods. This is done with the help of the recently developed "Boxology of Design Patterns for Hybrid Learning and Reasoning Systems" [6], which provides an abstraction mechanism for all approaches that combine machine learning with knowledge engineering (and which henceforth will be referred to simply as "the boxology").

As pointed out in the future work section of [6], there is a need to extend the boxology such that it can also describe systems with a human in the loop. As this is a common theme of our symposium contributions, they are used to derive a corresponding boxology extension. Describing our approaches with the extended boxology will make it possible to reflect on the commonalities and possible general design patterns of such human-in-the-loop approaches. Although the boxology approach has already been validated by its application to a large number of research papers in [6], it does not cover the approaches involving humans [12, 17] involve human feedback or reasoning within the process of a machine learning-knowledge engineering combination. For instance, a machine might learn something and output it in a human-interpretable form and a human expert may use it to improve a knowledge representation that is then used for deductive inferencing. This is not foreseen by the boxology approach in its current form.

Therefore this chapter takes a closer look at how the boxology could be applied to the two approaches mentioned above and where it fails to capture the human involvement. We then go on to propose an extension to the boxology, aiming to also describe these missing aspects via the introduction of a new box type for humans. By describing our contributions with the extended boxology, we attempt to anticipate some general patterns of how humans-in-the-loop can be used to improve machine learning-knowledge engineering combinations.

As a detailed discussion of all of our symposium contributions is beyond the scope of this chapter, the following sections shortly summarize the boxology and then elaborate on how it was applied to two of the symposium contributions [12, 17].

[1]Association for Advancement of Artificial Intelligence.

[2]https://www.aaai-make.info.

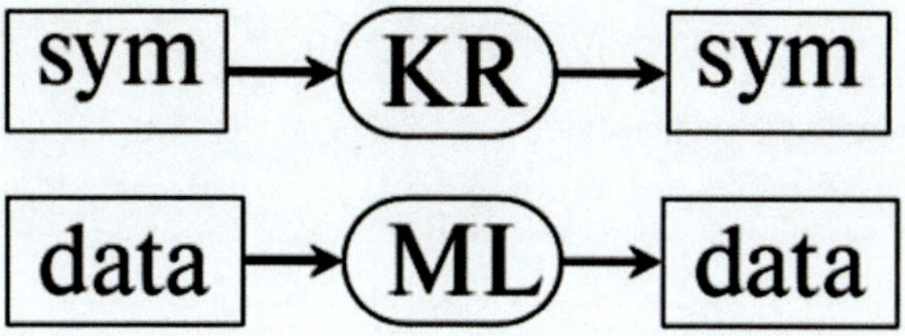


**Fig. 1** "Classical" patterns of input and output of a KR and ML system, respectively, as described in [6]

## 2 Summary of the Boxology Approach

The boxology [6] is a graphical representation of the abstraction of how to combine machine learning (ML) and knowledge representation[3] (KR). The abstraction is done via a set of design patterns that are expressed as combinations of very few basic elements.

The elements are divided into two categories, which are based on two different possibilities of categorizing work in ML and KR, namely task-based and representation-based distinctions:

- Task-based distinction: tasks can be either induction, i.e. deriving a generalization from specific observations or deduction, i.e. inferring specific conclusions from a generalized model. Since ML tends to be inductive and KR deductive, the boxology introduces ovals with the captions ML and KR to refer to inductive and deductive algorithmic components.
- Representation-based distinction: the boxology classifies the input or output of the algorithmic components into ones represented as (human-interpretable) symbols or models vs. ones represented as model-free data e.g. in the form of numerical variables or function parameters. Symbolic inputs/outputs are referred to with boxes labeled "sym" whereas the model-free representations correspond to boxes labeled "data".

Figure 1 shows the "classical" input-output behavior of KR and ML systems—where KR infers symbolic output from symbolic input using a human-engineered model and ML induces an often subsymbolic model from data. Here, a subsymbolic model usually applies a statistical learning technique and is represented by an artificial neural network.

The more interesting patterns described in [6] are those in which ML approaches learn on symbolic inputs (e.g. "graph completion", i.e. the (inductive) prediction of new connections in a knowledge graph) or produce a symbolic output learned from model-free input data (e.g. decision tree learning). Patterns can also easily become more complex, as shown in Fig. 2. In [6], this pattern is described as "learning an intermediate abstraction for learning".

[3]From now on "representation" instead of "engineering" is used because this is the terminology used by [6].

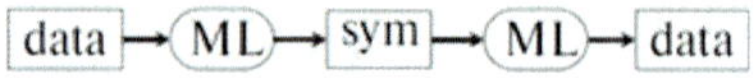


**Fig. 2** A more complex pattern from [6]

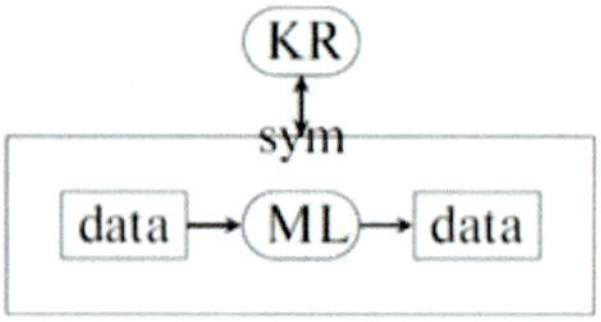


**Fig. 3** Another complex pattern from [6]

The pattern in Fig. 2 shows that combination of KR and ML can simply consist in using a symbolic *representation* at some stage of an ML process. But of course it is also possible to combine induction and deduction tasks as shown in Fig. 3—described as "meta-reasoning for control" in [6]. That pattern is used when a symbolic reasoner controls the learning process of a learning agent—one can think of e.g. rules that control how long the agent needs to learn something.

## 3 Human-in-the-Loop Patterns

The human-in-the-loop hybrid systems consist of all three components—ML, KR and humans. Since the boxology does not have a graphical representation for a human, we introduce a new "Human" box. The following sections provide a brief overview of the two approaches in [12, 17] and use the new "Human" component to illustrate the approaches with an extended boxology and to identify common human-in-the-loop patterns.

### *3.1 Learning Similarity Functions*

The first approach is the application of machine learning and knowledge representation to the problem of designing useful similarity functions, e.g. for case-based reasoning (CBR), as described in [17]. In short, the idea of this approach is to apply human knowledge to design the basic function and use machine learning to learn subsymbolic components of this function, such as weights. In CBR, humans are usually the ones who design a function and the idea of learning weights is already quite old [15].

What is new is the involvement of similarity function designers *after* the initial similarity function has been put into practice. As suggested in the CBR cycle by Aamodt and Plaza [1], case-based reasoning starts with a *retrieval* phase where, given an input or query case, the system searches a case base of historical cases for cases that are similar to the query. As pointed out in [3], there are different forms of knowledge involved in CBR—this first step involves mainly *similarity knowledge*, which is now the focus. CBR similarity functions are often composed of *local* similarity measures—which measure similarity only for one case attribute—and which in turn are combined into a *global* similarity measure, usually via a weighted sum [16].

Thus, the initial design of a global similarity function involves (a) specifying the attributes of cases that will be needed to assess similarity, (b) defining local similarity measures for these attributes and (c) specifying the weights for their combination. As pointed out in [15, 17], humans are usually not very good at this latter task.

Thus, the approach in [17] foresees that humans will not only be involved in setting up the initial function, but also in adapting the function later:

- By giving feedback on similarities computed with the initial similarity measure. For instance, a human may choose (implicitly or explicitly) to exclude a retrieved case that they consider dissimilar from the query case or, as in [15], correct the similarity-induced ranking of retrieved cases as a "similarity teacher".
- By inspecting examples of *failed differentiation* that occurred during the usage of the function: it could happen that two cases with nearly identical values of current attributes—hence exhibiting a high similarity according to the currently used similarity function—are judged dissimilar by a user. This may indicate that the currently used attributes are not sufficient to differentiate between these cases. A human could thus identify new attributes/local similarity measures to extend the similarity function.

Figure 4 gives an overview of how these inputs will be used: feedback helps to machine-learn new weights while the identification of failed differentiation will lead the human expert to re-design the general design of the global similarity function.

This example shows two different forms of human involvement: on the one hand, providing feedback on similarities of concrete cases requires only domain knowledge and can be performed by any user of a CBR system. The result of the human involvement can be seen as *data* with no internal symbolic structure.

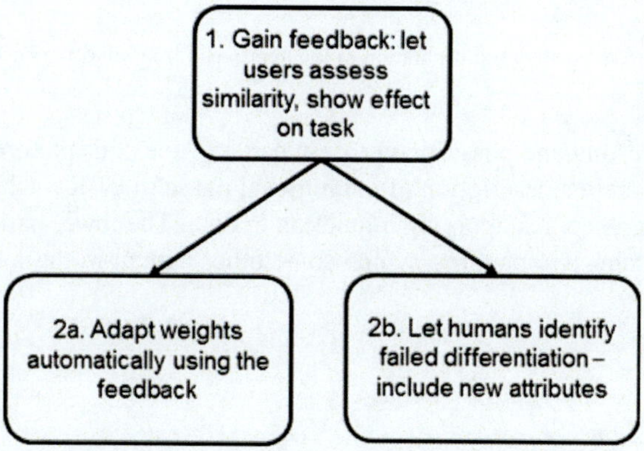


**Fig. 4** An overview of the suggested approach for similarity metric engineering

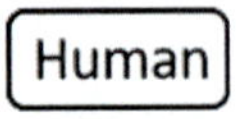


**Fig. 5** The new boxology element for humans-in-the-loop

On the other hand, re-designing the global similarity function requires both domain knowledge and engineering/mathematical skills and results in a new symbolic representation of the domain knowledge. This indicates that humans, like the two algorithmic components of the boxology, can generate both symbolic and subsymbolic output.

While later—if necessary—one could choose to differentiate various roles of humans in a learning process, we propose to maintain simplicity and just add a "Human" box to the boxology in a first step, as shown in Fig. 5.

With this element, the approach described in [17] can be modelled as depicted in Fig. 6. Note that, for a better readability of the figure, an explanation has been added to each box (which would not normally be displayed if we simply draw boxology patterns).

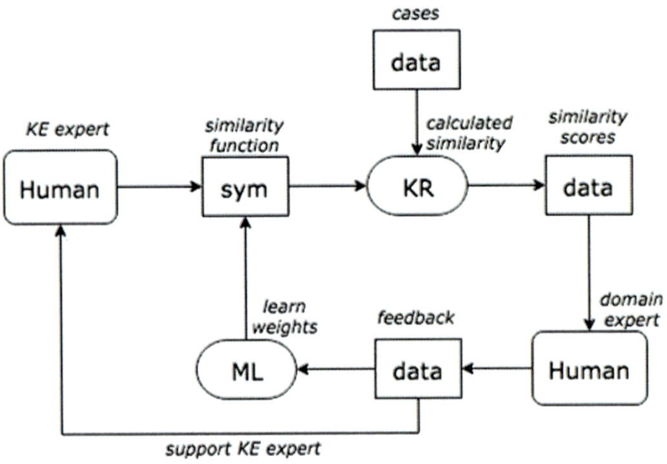


**Fig. 6** An extended boxology representation of the approach for similarity metric engineering

Which interesting and possibly repeating patterns does this picture contain? The upper part is merely a description of the retrieval phase in CBR—which is a simple example of inference, i.e. applying a function to data. The lower part is more interesting and contains two patterns, which some other human-in-the-loop approaches may also have:

1. The pattern (Human)-[data]-(ML)-[sym][4] describes, when considered in its abstract form, feedback-based learning, where human-provided feedback serves as the basis for machine learning algorithms that learn e.g. better functions.
2. Looking at the pattern (Human)-[data]-(Human)-[sym]-(KR)-[data] from a more abstract angle reveals another common pattern: humans are supported in analyzing data, with the aim to engineer better knowledge representations. This pattern could be called "data-based knowledge engineering".

This example shows that both machines and humans can sometimes learn from the same inputs (here: human feedback on computational results), but with different output, corresponding to their complementary strengths.

Another interesting insight is the fact that "Human" boxes do not necessarily need to have an input. This is not directly visible in Fig. 6 since both "Human" boxes have an input. However, if one considers the initialization of the system, one will remember that the initial formulation of the similarity function is done by the human without any input. In a more abstract sense, this reflects situations where humans use intuition and/or experience (which have been implicitly trained on data) to create outputs. We claim, however, that KR/ML combinations with a human-in-the-loop should have a feedback link (as shown in Fig. 6) that explicitly reflects human learning (otherwise, it could be regarded as part of classical knowledge engineering).

## *3.2 Learning Process Patterns from Process Models*

A sound business process model relies on correctly representing the underlying domain and adhering to the correct process modeling conventions. This involves collaboration between two expert roles—the domain experts and the modeling experts. The problem highlighted in [12] is that the domain and modeling expertise is mutually exclusive in most cases.

To solve this problem, Laurenzi et al. [12] propose to capture the domain expertise through domain-specific business processes and represent the models in an ontology-based repository. The ontology-based representation serves to provide machine-interpretability to the models, which enables reasoning (e.g. integrity and plausibility check) and forms the basis for the design of recommendation systems like in [11]. The overall approach is divided into four phases as depicted in Fig. 7.

The first phase includes the engineering of a first set of domain-specific semantic process model patterns. In this phase, various sources such as books, simulators, process mining tools as well as observations and interviews with domain experts are taken into account. What cannot be engineered, however, is the tacit knowledge in the domain experts' minds. Therefore, the approach foresees a learning approach, which relies on an extension of an ontology-based case-based reasoning (OBCBR) [5].

[4]While explaining the boxology patterns, we use the textual notations (ML), (KR) and (Human) to represent the knowledge processing components of the boxology and [data], [sym] to represent the input/output components.

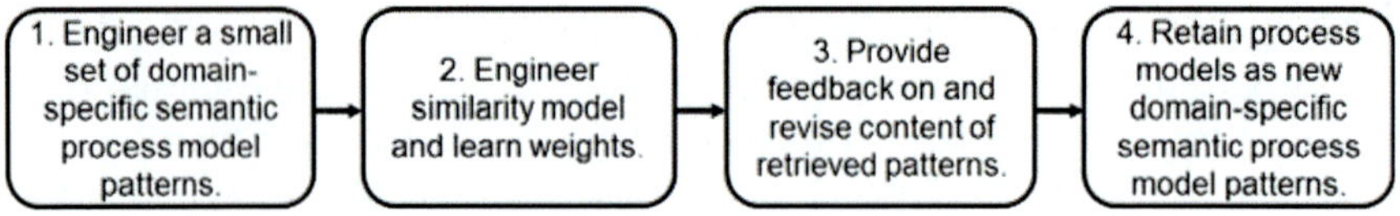


**Fig. 7** An overview of the phases involved in learning a process pattern

Phase two is similar to the approach proposed in [17]—the initial similarity model is engineered and weights are learned.

The feedback-learning approach is, however, different (phase 3). Whereas the approach in [17] allows the human to intervene directly in the similarities (e.g. by extending the similarity function with new attributes), the work in [12] foresees human intervention only on the content of the retrieved process pattern. For this, the authors propose a reinforcement learning approach. Modeling experts give a reward or penalty (i.e., positive/negative signal) for how well fragments of the retrieved process patterns match to the process being designed. In the case of a Euclidean metric (i.e. linear weighting) used for determining similarity, a positive feedback would increase weights whereas a negative feedback would decrease them. As a result, weights are altered to improve future matching operations. The long-run intention for using reinforcement learning is to reduce the modeling expert's effort in interviewing domain experts about the matching patterns. Although not ontology-based, the validity of extending CBR with reinforcement learning has been shown already in [13].

Still in phase 3, in order to keep patterns up-to-date over time the approach includes the revision of the retrieved patterns. This is done by applying (delete, insert and update) operations on classes, relations and properties that form a semantic pattern. Revision activities could include adding new tasks, re-naming the label of a gateway or removing specific tasks. An example of a revision is the replacement of manual tasks with service tasks for a higher level of automation. This type of activity refers to the "revise" step in the CBR cycle [1].

Next, phase 4 aims to increase the number of process patterns over time, the user can decide on retaining a new process model as a process pattern. By doing so, the authors in [12] aim to ensure the controlled increase of good domain-specific semantic process patterns. In this sense the semantic repository learns new semantic patterns. This step refers to the last CBR cycle phase "retention".

With the help of the newly introduced element (Human), the approach in [12] can be represented as illustrated in Fig. 8. This boxology consists of the following patterns:

1. The pattern (Human)-[data]-(ML)-[sym] serves the feedback-based learning, which foresees the CBR extension with the reinforcement learning. Similar to the pattern described in the previous section, the human-provided feedback is re-incorporated into machine learning to learn weights.
2. The pattern (Human)-[sym]-(Human)-[sym]-(KR)-[sym] serves as another abstraction of "data-based knowledge engineering" where the feedback provided

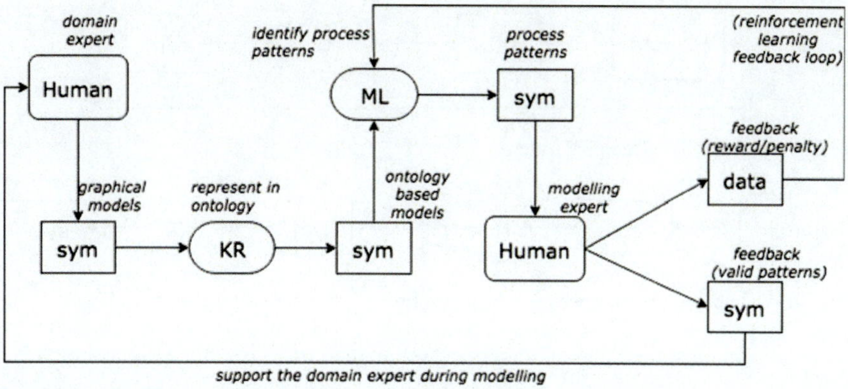


**Fig. 8** A boxology representation of the approach for learning a process pattern

by a human supports another human to accurately capture the domain, further strengthening the knowledge engineering efforts. In this particular use case, the modelling expert is enabled to (1) give feedback (i.e. revise) on the retrieved pattern and (2) decide whether the a process model can be retained as a new pattern.

## 3.3 *Library for Human-in-the-Loop Patterns*

From the discussion of the two boxologies, one can identify some commonalities about how humans influence machine learning and knowledge representation. The role of the human actor in human-in-the-loop patterns can be viewed from two perspectives: (a) feedback-provider: where the human is usually an expert e.g. expert in business domain, process modeling, knowledge engineering and (b) feedback-consumer: where the feedback provided by another human expert supports and enhances the task being performed e.g. an analytical task or in general, any knowledge-intensive task.

Based on this observation, we introduce the following abstract patterns involving human-in-the-loop.

### Feedback-Based Learning Pattern

This pattern is the generalization of scenarios where the feedback from a human actor, typically an expert, is fed back to a machine learning algorithm to incorporate the human judgment and improve the learning. These scenarios may be categorized as machine teaching [14] where human "teachers" train, build and improve machine

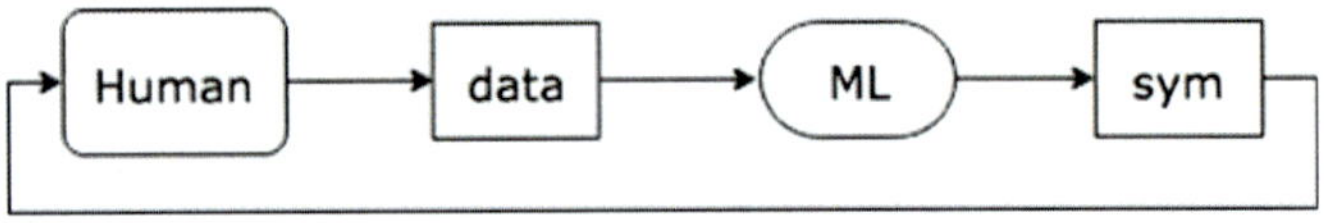


**Fig. 9** Feedback-based learning pattern

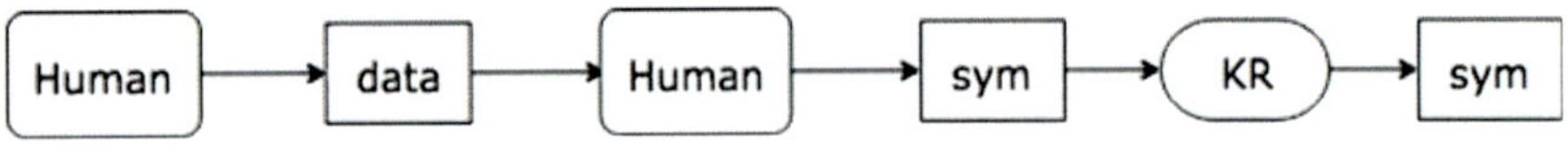


**Fig. 10** Data-based knowledge engineering pattern

learning models. Because of its cyclic nature, it represents a special case of incremental learning [9]. The graphical representation of this pattern is shown in Fig. 9. Depending on the use case, it is possible to have other versions of this abstraction where [data] and [sym] are interchanged.

### Data-Based Knowledge Engineering Pattern

The data-based knowledge engineering pattern represents the scenarios where feedback from the human experts is used by another human to support a knowledge-intensive task which in turn results in improved knowledge engineering. The graphical representation for this pattern is depicted in Fig. 10. Similar to the feedback-based learning pattern, the [data] and [sym] used as the input/output of the (Human) can be interchanged depending on the use case, but, the output of (KR) would typically be [sym].

One can observe from the boxologies in Figs. 6 and 8 that the corresponding systems consist of all three knowledge processing components, namely, (ML), (KR), and (Human). These are the examples of a typical "hybrid intelligence" system, as described in [4]. However, it is not necessary that the human-in-the-loop patterns contain all three knowledge processing components, as the patterns we abstracted can be considered atomic with either a (ML) component together with a (Human) or a (KR) component together with a (Human) (see patterns in Figs. 9 and 10). Thus, we can deduce that the involvement of (Human) in a hybrid intelligence system leads to an enhanced (ML) or (KR) component. This deduction is aligned with the observation in [4], where the role of humans usually is to undertake sense-making, training or reasoning.

The feedback-based learning pattern can also be observed in our work [8] where (ML) learns from the preferences of the user and provides whiskey recommendations. The boxology for this use case can be envisioned as a slightly modified "explainable learning systems through inspection" including humans (pattern 8 in [6]) where the explanations are the recommendations and the input to both (ML) and (KR) are the user (Human) preferences.

The role of humans in our other works [7, 10] also involve verification and training, however not exactly in the form of the patterns identified above. For instance, in [7], the accuracy of the fake news classifier would be validated by humans. Jüngling and Hofer [10] discuss a collaborative approach between (Human), (ML) and (KR) where tasks are distributed among these components and the output of each component is then verified and aggregated by another human into a richer solution.

## 4 Conclusions

In this chapter, two approaches to combining Machine Learning and Knowledge Engineering through graphical representation in a boxology were discussed. The current library of abstract patterns on hybrid systems combining ML and KR lack patterns that involve humans. Through this discussion, we identified two abstract patterns that can often be present in most systems that include humans-in-the-loop. In these systems, the human agent either performs the role of a feedback-provider or a feedback-consumer. Both roles contribute towards enhancement of machine learning and/or knowledge representation.

As a future work, we plan to confirm the general applicability of these human-in-the-loop patterns by analyzing more use cases and possibly identify new patterns.

## References

1. Aamodt, A., Plaza, E.: Case-based reasoning: foundational issues, methodological variations, and system approaches. AI Communi. **7**(1), 39–59 (1994)
2. Chen, X.W., Lin, X.: Big data deep learning: challenges and perspectives. IEEE Access **2**, 514–525 (2014)
3. Cordier, A., Fuchs, B., Mille, A.: Engineering and learning of adaptation knowledge in case-based reasoning. In: International Conference on Knowledge Engineering and Knowledge Management, pp. 303–317. Springer, Berlin, Heidelberg (2006)
4. Dellermann, D., Calma, A., Lipusch, N., Weber, T., Weigel, S., Ebel, P.: The future of human-AI collaboration: a taxonomy of design knowledge for hybrid intelligence systems. In: Bui, T. (ed.) HICSS, pp. 1–10. ScholarSpace/AIS Electronic Library (AISeL) (2019)
5. Emmenegger, S., Hinkelmann, K., Laurenzi, E., Martin, A., Thönssen, B., Witschel, H.F., Zhang, C.: An ontology-based and case-based reasoning supported workplace learning approach. In: Communications in Computer and Information Science, pp. 333–354. Springer, Cham (2017)
6. van Harmelen, F., ten Teije, A.: A boxology of design patterns for hybrid learning and reasoning systems. J. Web Eng. **18**(1–3), 97–124 (2019). https://doi.org/10.13052/jwe1540-9589.18133
7. Hinkelmann, K., Ahmed, S., Corradini, F.: Combining machine learning with knowledge engineering to detect fake news in social networks-a survey. In: AAAI Spring Symposium: Combining Machine Learning with Knowledge Engineering. Palo Alto, California, USA (2019)
8. Hinkelmann, K., Blaser, M., Faust, O., Horst, A., Mehli, C.: Virtual bartender: a dialog system combining data-driven and knowledge-based recommendation. In: AAAI Spring Symposium: Combining Machine Learning with Knowledge Engineering (2019)

9. Joshi, P., Kulkarni, P.: Incremental learning: areas and methods-a survey. Int. J. Data Min. Know. Manage. Process **2**(5), 43 (2012)
10. Jüngling, S., Hofer, A.: Leverage white-collar workers with AI. In: AAAI Spring Symposium: Combining Machine Learning with Knowledge Engineering (2019)
11. Kritikos, K., Laurenzi, E., Hinkelmann, K.: Towards business-to-IT alignment in the cloud. In: Communications in Computer and Information Science, pp. 35–52. Springer, Cham (2018). https://doi.org/10.1007/978-3-319-79090-9_3. URL http://link.springer.com/10.1007/978-3-319-79090-9_3
12. Laurenzi, E., Hinkelmann, K., Jüngling, S., Montecchiari, D., Pande, C., Martin, A.: Towards an assistive and pattern learning-driven process modeling approach. In: AAAI Spring Symposium: Combining Machine Learning with Knowledge Engineering. Palo Alto, California, USA (2019)
13. Paulson, P., Juell, P.: Reinforcement learning in case-based systems. IEEE Poten. **23**(1), 31–33 (2004)
14. Simard, P.Y., Amershi, S., Chickering, D.M., Pelton, A.E., Ghorashi, S., Meek, C., Ramos, G., Suh, J., Verwey, J., Wang, M., et al.: Machine teaching: a new paradigm for building machine learning systems. arXiv:1707.06742 (2017)
15. Stahl, A.: Learning feature weights from case order feedback. In: International Conference on Case-Based Reasoning, pp. 502–516. Springer (2001)
16. Stahl, A., Roth-Berghofer, T.R.: Rapid prototyping of cbr applications with the open source tool mycbr. In: European Conference on Case-Based Reasoning, pp. 615–629. Springer (2008)
17. Witschel, H.F., Martin, A.: Learning and engineering similarity functions for business recommenders. In: AAAI Spring Symposium: Combining Machine Learning with Knowledge Engineering. Palo Alto, California, USA (2019)

# Modeling the Instructional Design of a Language Training for Professional Purposes, Using Augmented Reality

**Terry Inglese and Safak Korkut**

**Abstract** This chapter presents the instructional design of a language-training model for professional and vocational purposes on behalf of the Swiss railway industry, specifically designed for German-speaking train drivers and train operators, who work for the Schweizerische Südostbahn (SOB). In fact, around 50 train drivers and train operators need to learn Italian and be able to communicate clearly and confidently in this language by 2021. Thanks to the opening of the Gotthard Base Tunnel in 2016, some Swiss railway companies are expanding their business portfolios also in the Italian speaking region of Switzerland. Augmented Reality (AR), specifically the Blippar app, is used here as an additional and motivating guide to learning technical terms and nouns, verbs, and dialogue structures, in short: essential railway communication features between train drivers and train operators. The final goal of the chapter is to describe how railway professional trainees, learning a new language, are actively designing their own language learning contents, using AR.

**Keywords** Augmented reality (AR) · Situation awareness (SA) · Multimedia learning theory · DART co-creation model · Bloom's learning objectives taxonomy · Language training for professional purposes

## 1 Introduction

For some Swiss railway companies, the opening of the Gotthard Base Tunnel in 2016 was an opportunity to expand their own business portfolios, also reaching the Swiss Italian-speaking destinations (from Airolo to Bellinzona). A great business opportunity indeed, but also a new challenge, because German-speaking train drivers and train operators (also defined as dispatchers) must be able to communicate confidently

T. Inglese (✉) · S. Korkut
Institute for Information Systems, School of Business, FHNW University of Applied Sciences and Arts Northwestern Switzerland, Peter Merian-Strasse 86, 4002 Basel, Switzerland
e-mail: terry.inglese@fhnw.ch

S. Korkut
e-mail: safak.korkut@fhnw.ch



R. Dornberger (ed.), *New Trends in Business Information Systems and Technology*, Studies in Systems, Decision and Control 294,
https://doi.org/10.1007/978-3-030-48332-6_14

also in the Italian language. For these railways companies, this means investing in their employees' linguistic training and offering a specifically designed language training curriculum for professional purposes for numerous German-speaking train drivers and train operators, who in addition to their already dense training programs, are obliged to learn a new language accordingly.

The authors of this chapter developed a didactical model, involving the use of AR (Augmented Reality), which actively includes the trainees, so that they become motivated participants also in the design of the instructional contents to be learned. The training is designed specifically for German-speaking railway professionals who work for the Schweizerische Südostbahn (SOB). In fact, around 50 SOB train drivers and train operators are required to learn Italian and be able to communicate clearly and confidently in this language by 2021.

First, the chapter briefly describes the SOB training case study, together with the learning needs of the train drivers and train operators. Second, the instructional design model, consisting of the DART (an acronym for Dialogue, Access, Risk, Transparency) co-creation model, the Mayer's Multimedia Learning Theory, the Situation Awareness (SA) construct and the Bloom's Learning Objectives Taxonomy will be explained. Third, as currently being in a test-phase, a concrete AR training sequence, designed by the train drivers themselves, will be described and specified. The limitations of the case study will also be highlighted.

## 2 The SOB Case Study

Driving a train is a complex task, especially in Switzerland, because on a normal working day, a train driver travels through the different Swiss language regions. In addition, train drivers and train operators are confronted with several special trainings and examinations during their career, such as regular assessments and tests to update their knowledge and the skills of their workflow and train management, as important professional goals in a continually changing environment. The railway drivers and operators must also update their emergency procedures, including firefighting, evacuation methods, and communication for safety emergencies. These *cognitive tasks* need to be communicated clearly and competently, and in multiple languages. In addition, with the introduction of rapidly changing communication and control systems with increasingly complex procedures, train drivers and train operators are equally forced to cope with high skill requirements with little autonomy [1, 2]. Moreover, recent studies [3–8] show that the lack of personal contact between train operators and train drivers is perceived as a negative condition. Furthermore, such circumstances might become the potential cause of miscommunications and/or misunderstandings, due to a lack of mutual understanding and the lack of a (mandatory) read-back of protocol-requiring commands, guaranteeing safety procedures, not only in the context of aviation [9], but also in rail transport. As a matter of fact, according to recent studies in professional language training in aviation [9], key features of effective communication depend correspondingly on *assumed expertise* and *professional behaviors*. When it falls short of expectations, communication is expanded and prolonged to fill these gaps. Nevertheless, these specific communication aspects are not adequately taken into account in language training for professional purposesin the Swiss railway sector [3–5, 8, 10].

In addition, recent work in applied linguistics for professional purposes [9] underlines the importance of redesigning the constructs of *language for specific purposes* (LSP). As a matter of fact, ensuring that the correct communication unfolds in railway safety contexts is a compelling didactic and instructional topic, which needs to be designed, taught and learnt in professional language-training curricula. Therefore, language instructors are invited to work together with professional railway experts [9] to plan meaningful and work-based training curricula, actively involving in the design of instructional learning sequences and materials *also*—and this aspect represents the added value of the present case study—the learners themselves, such as in the SOB case study.

An exploratory analysis conducted by the authors of the chapter reported some gaps in the current didactic offer of these specific language training curricula, and the following research question was formulated: How should a new didactic model be developed and implemented to meet the specific communication needs of train drivers and train operators, also taking into account the potential lack of motivation to learn a new language for these adult learners who might see the training as an additional hardship? Railway employees could be less committed to the language training, because they might prioritize their professional assessments and technical examinations over learning a new language. Therefore, the motivation to complete a new language training could be lower than expected.

## 2.1 Many Languages, Sufficient Language Skills, but at Which Language Level?

The Swiss Federal Office of Transport (FOT) determines which binding language regions correspond to the responsible operation centers [10]. In fact, the so-defined *Fahrdienstvorschriften* (FDV—the Swiss Federal Office of Transport's Driving Regulations) state that: "There are procedures in place to ensure that (especially with regard to the management of dangerous goods) the infrastructure manager can easily and promptly notify the competent person who is proficiently skilled and *has sufficient language skills*." [10]. This means that train drivers, travelling between Lausanne's and Pollegio's operation centers, must have the "*necessary language skills*" in French, German, and Italian to cross these diverse Swiss linguistic regions. An essential question that arises could be: Which level corresponds to the *necessary language skill*? In fact, between 1989 and 1996, the Common European Framework of Reference for Languages (CEF or CEFR) standardized the new language skills to six levels (A1, A2, B1, B2, C1, and C2). Level A1 refers to the basic ability to communicate and exchange information in a simple way, and A2 is the ability to deal with simple, straightforward information and begin to express oneself in familiar contexts. B1 is the ability to express oneself in a limited way in familiar situations and to deal in a general way with non-routine information and B2 is the capacity to achieve most goals and express oneself on a number of topics. Finally, C1 and C2 denote the fluent conversation level. More specifically, the required language skill for train drivers and train operators, specified in the EU (European Union) regulations, is level B2.

Within the Swiss railway operations, the responsible infrastructure managers (ISB) are the ones who regulate the use of languages for operational communication and process purposes. It is the responsibility of the railway companies themselves to determine the language skill levels of train drivers who cross the multilingual geographical areas. While the Swiss infrastructure managers (ISB) regulate the linguistic interfaces in their operating regulations, the railway companies are responsible for providing the so-called *sufficient* language training to their own professional personnel. Nevertheless, it is not yet clear in the Swiss railway network which language level could correspond to the "*sufficient language skills*", so that in the event of possible accidents there could be uncertainty due to miscommunication [3–5, 8].

## 2.2 The SOB Language Learning Framework and Andragogy

The SOB (the Schweizerische Südostbahn) currently offers intensive Italian language training (level A1-A2) for train drivers and train operators who are required to be able to communicate in Italian in early 2021. As part of this training, the language instructor, in collaboration with the trainees themselves, will design part of the instructional content actively using AR technologies to provide trainees with various modalities for stimulating the learning experience and improving the mastery of relevant safety-based dialogic structures in the Italian language. The implications for the design of a solid language-training program are at the heart of SOB: learning should be based on a real diagnosis of the adult learners' needs, based on competencies and skills embedded in realistic work-based scenarios.

The final goal for the SOB train drivers and train operators will be reaching the ability to speak Italian clearly and confidently, especially when confronted with unexpected situations, where reaction time and fast decision-making are requested. Consequently, and based on preparation and research carried out by the authors of this chapter, it is essential to develop a language training curriculum for professional purposes, based on the following three pillars: (*a*) *defining the specific phraseologies* for the most relevant routine situations, safety issues, and emergency communication with special attention to the avoidance of misunderstandings; (*b*) *focusing on a quasi-realistic training environment* that mirrors genuine work-related scenarios and events; (*c*) *designing specific training modules*, according to the requirements defined by the authorities and the railway companies to allow effortless and proficient communication in the language training. The authors of this chapter base the three communication pillars in adult learning andragogy principles.

Houle [11] was one of the first scholar who started for the first time a stream of studies putting *adult learners* in the focus of educational and psychological research, understanding their motivation to continue learning. He found out that "continuing learners" were: (a) *goal-oriented*, using education to accomplish their objectives; (b) *activity-oriented*, finding meaning in learning something new and were finally (c) *learning-oriented*, seeking knowledge for its own sake. Furthermore, Knowles [12] proposed a profile of an adult learner and defined *andragogy* as "the art and science of helping adults learn, in contrast to pedagogy as the art and science of teaching children" [12] (p. 4344; [13]). He summarized four assumptions in andragogy: first, *the concept of the adult learner*, who moves from dependency toward increasing self-directedness, but at different rates and in different moments in their lives; therefore, instructors have the responsibility to encourage the self-directed drive, although adult learners may be dependent in particular situations. The second assumption is *the role of the learners' experiences*. As adult learners grow, they develop a precious reservoir of work and life experiences, which can become part of the resources for designing the learning contents, both for themselves and for other adult learners. In addition, these learners can attach more meaning to learning gained from past and present everyday work-based experiences. The third assumption is *the readiness to learn*: adult learners are willing to learn something new, if they can project and implement new learning in real tasks and problems that need to be solved. Finally, the forth assumption is the *orientation to learning*: adult learners consider education and learning opportunities to develop improved skills, and they want to apply their new skills immediately in their professional life. The focus of the training should be on competence-oriented development and on performance-based development.

Based on the personal conversation and informal data of the six current participants in this case study (the average age is 55), for the SOB train drivers and train operators, learning a new language could be a stressful experience due to the additional pressure they might feel. To meet this challenge, a new way to motivate these trainees has been instructionally designed, with the goal to understand how Augmented Reality (AR) could promote learning and motivation. A specific instructional model was designed and will be described in the next section.

## 3 The Instructional Model Explained

For the SOB railway professionals, starting from an A1 level, the following four-layered instructional model was conceptualized, taking into account: (a) the Situation Awareness (SA) [14], (b) the Mayer's Multimedia Learning Theory [15–19], (c) the DART—*Dialogue, Access, Risk, Transparency*—Co-Creation model [20–23] and finally (d) the Bloom's taxonomy's six learning objectives [24–27].

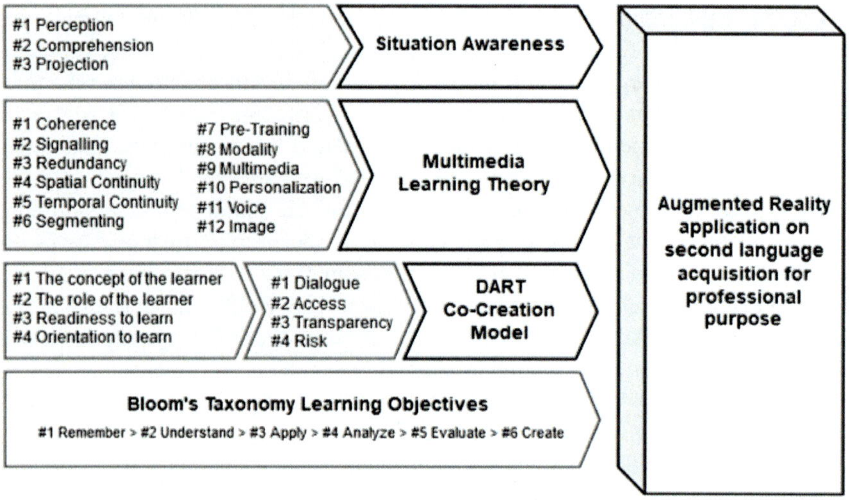


**Fig. 1** Instructional model on designing the language training for professional purposes (own illustration of the authors)

The Situation Awareness (SA) model [14] is relevant, because it is designed for complex and dynamic systems, such as the railway sector, where decisions often have to be made under time pressure and based on environmental events that could occur unexpectedly. Therefore, the cognitive comprehension of the situation and the way in which train drivers need to react to future driving situations are essential preconditions for the development of the instructional model. In addition, the Mayer's Multimedia Learning Theory [15–19] was selected on the basis of strong scientific evidence-based principles in the design of instruction sequences using multimedia. Furthermore, taking into account the six learning objectives of Bloom's taxonomy [24–27] was a guarantee that all the layers of the learning processes and objectives would be included. Finally, the DART co-creation model [20–23] is an approach to actively involve professional learners in the design of the instructional content through a bottom-up teaching and learning method.

In Fig. 1, the authors of the chapter graphically summarize the four-layered model in an illustration. In the next section, each layer of the instructional model is first described and later explained by a concrete example.

### *3.1 The Situation Awareness Model*

Situation Awareness (SA) is defined as the awareness of what is happening in and around the environment, along with the understanding of how users use this information. The concept of SA is commonly applied in complex and dynamic operative situations, such as air traffic control, military training or surgery training. In such cases, the ability to process the information plays a key role in raising awareness, perceiving and comprehending the situation, and making the right decision. Endsley [14] (p. 14) refers to SA in three levels where each level is built on the previous level:

1. Perception of elements in the current situation
2. Comprehension of the current situation
3. Projection of future status.

The first level is *perception*, related to operators perceiving the incoming information. This means that the necessary information is available and attention is focused on the important information. Without the right information, operators have no chance to understand and control a system. The second level is *comprehension*, i.e. the ability to understand the perceived information, interpret the information on one's goals and to remember the relevant parts of the information. The third level is *projection*, consequently when the important information is perceived and understood, it is possible to understand the dynamics of a system, project and anticipate future decision-making in order to influence and control the state of a system. After having described the three levels of SA, Endsley's model [28] connects SA to *decisions*, which will lead to the *performance of action*. These three steps are also influenced by task and environmental factors, such as workload, stressors, system design, and complexity, as well as individual factors such as goals, preconceptions, knowledge, experience, training, and abilities (Fig. 2).

The SA concept started to develop fully at the end of the 1980s, especially in the field of aviation, safe control, safety training and investigations ofaircraft accidents, and not surprisingly with the increase of automation in flight control [29]

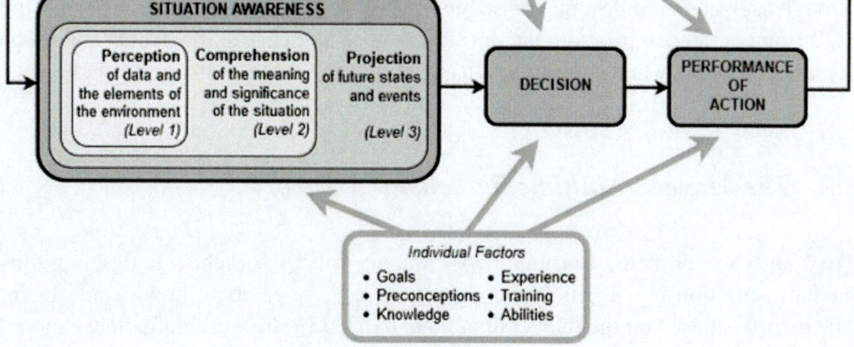


**Fig. 2** Endsley's model of situation awareness in dynamic decision-making systems [28] (p. 35)

(p. 190). Endsley [28] defined it for dynamic systems because SA plays a critical role in decision making, especially in complex and dynamic systems, followed by the execution of actions. "A person's SA is restricted to limited attention and working memory capacity… long-term memory stores, most likely in the form of schemata and mental models, can largely circumvent these limits by providing for the integration and comprehension of information and the projection of future events (the highest levels of SA), even on the basis of incomplete information and under uncertainty. SA is largely affected by a person's goals and expectations, which will influence how attention is directed, how information is perceived, and how it is interpreted. This top-down processing will operate in tandem with bottom-up processing in which salient cues will activate appropriate goals and models" [28] (p. 49).

Endsley [30] studied the phenomenon of human error and SA. According to her findings, 88% of human errors were found to be due to problems with Situation Awareness, because the situation the persons were in was misinterpreted. For Endsley, the best way to support human performance is to develop high levels of Situation Awareness. Moreover, Kaber and Endsley [31] defined that SA can be applied to contexts where there are multiple goals that must be pursued by operators simultaneously, such as multiple tasks that have different relevance to goals competing for the operators attention; operator performance under time stress, and negative consequences associated with poor performance (p. 190).

Problems occur when there are failures to detect critical cues about the state of the system, failures to interpret the meaning of perceived information, failures to understand individual task responsibilities and the responsibilities of the others, and errors in communication with other operators in the team and with other teams. In fact, this is defined as the loss of Situation Awareness. Moreover, Endsley [14] summarized that first SA "is *goal-oriented*…the elements of the environment that people need to be aware of are based on the goals associated with that job. A goal-directed task analysis […] provide[s] the basis for understanding how SA maps to goals and for creating goal-driven designs." (p. 49) Second, SA directly supports *the cognitive processes of the operator*, processes which will be represented by the system interfaces and are based on a high level of SA. For the third and final aspect, the user must have control over the expected and unexpected environments' SA through *user-centered* design. Consequently, user-centred learning scenarios for the SOB trainees need to promote the development of language skills, which should be applied consciously and continously, as SA Endsley's approach suggests [14].

### *3.2 The Mayer's Multimedia Learning Theory*

The Mayer's multimedia learning theory and his principles on how to design multimedia instruction [15–19] is based on cognitive theories about how people learn. The theory is based on the dual coding theory [32, 33], the working memory model [34], and the cognitive load theory [35, 36]. Mayer defined the twelve principles in the design of multimedia instructional content. His theory represents a benefit also in designing Augmented Reality applications for learning purposes [17]. The twelve principles are summarized in Table 1.

It is crucial to apply Mayer's *science of instruction* when designing instructional content, which needs to be in line with cognitive psychology and the learning processes. These principles guide the language instructor, as well as the trainees, in the planning and the production of the multimedia AR training contents, defined by the authors as *AR vignettes*, which will be explained later.

**Table 1** Table summarizing the Mayer's 12 multimedia principles [15–19]

| Principle | How it works |
|---|---|
| 1. The *Coherence Principle* | when extraneous words, pictures, and sounds are excluded rather than included subjects learn better |
| 2. The *Signaling Principle* | when cues that highlight the organization of the material to be learned are added subjects learn better |
| 3. The *Redundancy Principle* | subjects learn better from graphics and narration, rather than from graphics, narration and on-screen text |
| 4. The *Spatial Contiguity Principle* | when corresponding words and pictures are presented close to each other rather than far away on the page or screen, subjects learn better |
| 5. The *Temporal Contiguity Principle* | subjects learn better when corresponding words and pictures are presented simultaneously and not sequentially |
| 6. The *Segmenting Principle* | subjects learn better when a multimedia lesson is presented in user-paced segments rather than as a continuous unit |
| 7. The *Pre-Training Principle* | subjects learn better from a multimedia lesson when they know the names and characteristics of the main concepts |
| 8. The *Modality Principle* | subjects learn better from graphics and narrations than from animation and on-screen text |
| 9. The *Multimedia Principle* | subjects learn better from words and pictures than from only words |
| 10. The *Personalization Principle* | subjects learn better when in multimedia presentations words are presented in conversational style rather than formal style |
| 11. The *Voice Principle* | subjects learn better when the narrative in multimedia lessons is spoken in a friendly human voice instead of a machine voice |
| 12. The *Image Principle* | subjects do not necessarily learn better from a multimedia lesson when the speaker's image is added to the screen |

## 3.3 The DART Model

According to Prahalad and Ramaswamy [20], the customer should be actively involved in the creation processes in order to create real added value in services and/or products. Additionally, the consumer-to-business-to-consumer *co-creation* model (C2B2C) will help companies unlock competitive advantages, and employees can also establish this competitive advantage. The employees' engagement is as important as the customers' and both are considered to be co-creators of the final product or service by offering a major improvement to companies seeking advantages in overly saturated markets. The term co-creation signifies [37] a "joint creation of value by the learner and the instructional designers team; it allows the learner to co-construct the learning experience to suit his/her context and learning needs; joint problem definition and problem-solving; creating a learning experience and environment in which learners can have an active dialogue and co-construct personalized and different learning experiences, through the continuous dialogue and co-constructing personalized learning experiences and innovating the learning environments for new co-creation experiences." (p. 8) (see also [20–23]).

Prahalad and Ramaswamy suggest the DART model (an acronym for Dialogue, Access, Risk, Transparency) with its four co-creation building blocks. Dialogue implies interactivity, deep engagement, ability and willingness to act between learners and instructors; Access and Transparency relate to information in teaching and learning processes, and the Risk assessment is a clear assessment by the learners and instructors of the benefits of the risks when learners are the decision-makers of their own learning [37] (p. 9). The DART model is the framework, which aims to bring together learners and instructors as designers of training models, involving the learners' bottom-up design and the use of new technologies such as AR. Recently, Dollinger, Lodge, and Coates [21] adapted the co-creation model for higher education purposes. They summarized *co-creation* as a combination of *co-production* (knowledge sharing, equity, and interaction) and *value-in-use* (experience, personalization, and relationship). In this way, co-creation might promote results in *innovation, knowledge* and *relation* [21] (p. 224), a framework worth adapting when designing and co-designing an instructional model for teaching and learning a new language for professional purposes with adult learners such as train drivers and train operators.

## 3.4 The Bloom's Learning Objective Taxonomy

The Bloom's six learning objectives taxonomy [24–27] is a classic construct in designing instructional activities. It implies six cognitive processes: 1. remembering; 2. understanding; 3. applying; 4. analyzing; 5. evaluating/synthesizing and 6. creating. The first two are considered the lower cognitive processes and the last four are considered the higher cognitive processes. In terms of learning a new language, *remembering* and *understanding* technical terms and verbs are two basic cognitive processes, which set the foundations for *applying* the linguistic elements for e.g., creating small dialogues. With the higher cognitive processes of *analyzing* and *evaluating/synthesizing* the learners must make use of what they have learned and be able to adapt it to new communication scenarios. The last higher cognitive process, *creating*, reflects the ability to generate something new byapplying the previous five cognitive processes, and designing a more complex dialogue.

The aim of the SOB case study is to create learning conditions under which train drivers and train operators are able to competently and safely create their own learning content, such as dialogues on work-based situations by using specific railway terminologies and verbs through dialogues. To achieve this goal, the SOB case study also entails the use of Augmented Reality (AR) by designing a series of so-called *AR vignettes*.

## 4 Augmented Reality Technology

AR technology uses a device to bring digital holographic information to the users' real environment, i.e. a head-mount display, such as Microsoft Hololens or a modern smartphone, such as iPhone X. The projection of virtual elements to the real environment enhances user perception and opens up new possibilities to interact with these elements between virtuality and reality, and in real-time [38]. The most broadly accepted definition of AR comes from Azuma et al. [39] and specifies AR as a system that "supplements the real world with virtual (computer-generated) objects that appear to coexist in the same space as the real world".

Today's AR technology is resourceful to overlay text, image, audio, video, and three-dimensional elements in the real environment. It is able to anchor these elements to real objects so that when the user moves in the real space the virtual elements remain in the anchored position. This provides flexibility to observe digital artifacts from multiple angles. The virtual information is integrated into the users' environment with few or no glitches. AR provides the ability to consolidate and visualize information for users so that they can be more productive, more effective, or make better decisions [40]. The ability to overlay digital information in the immediate environment makes AR an ideal tool to provide information in unknown or unexpected situations. The technology has a positive impact on how users work with digital information on an additional layer and how they access data [40]. This provides a door to immersive learning experiences with the use of AR technology. Moreover, modern education scenarios consider multimedia material indispensable for effective learning outcomes [41]. The advantage of physical objects, however, is that they can be touched and allow natural interaction [42], which is not possible with digital objects. As the printed book is still very important for language learning, AR offers the potential to overlay additional useful information [43].

In the field of education, where most research evidence for book augmentation is found, the combination of a printed book and AR shows great learning potential. The instructive added value of using digital data on books can be further developed to overcome the analogue limitation of space, and to ensure the topicality of information in printed books. There are several examples in education that use AR solutions in combination with printed books. Studies have shown that AR has the potential to digitally enhance books in the field of education, to achieve better interaction or learning effects [42, 44, 45].

However, one of the major drawbacks of AR technology is the novelty effect [46]. Researchers have addressed this tendency when most of the users try it for the first time [47]. The novelty effect leads to improved performance with the interest in using new technology. As the interest in the new technology decreases, the user engagement might not be as high as it was at the beginning because users have become familiar with the technology [48–50]. Therefore, the AR implementation strategy should take this tendency into account and balance the workload and the commitment to designing the own AR learning content, through a bottom-up approach which is explained in the next section with a concrete instructional example.

## 5 An Augmented Reality *Vignette* Example

The SOB railway professionals will be actively engaged in designing their own *AR learning vignettes.* The authors of the chapter call a short instructional AR sequence a *vignette,* using an app called Blippar. In fact, Blippar (www.blippar.com) is a Software as a Service (SaaS) company founded in 2011 that specializes in mobile AR and provides a platform for enhancing images with AR without prior programming knowledge with the aim to gradually standardize AR formats. The SOB learners will use the Blippar app together with the multimedia learning theory to help shape their own instructional content, by selecting and designing their own learning materials based on their rich and diversified work experience. In collaboration with the language instructor who will apply the DART model, the SOB learners select work-based Situation Awareness scenarios (SA) and design their own set of learning materials.

Train drivers and train operators need to be ready to react to numerous unexpected situations (in reference to SA) and be able to communicate in a foreign language with confidence and clarity. Sixteen common SA-related situations while driving a train were identified. These will be visually designed and represented using the *AR learning vignettes.* The sixteen situations include: 1. calling and starting a conversation; 2. avoiding misunderstandings; 3. communicating shunting movement; 4. communicating being "ready to drive"; 5. standing and waiting in front of a signal; 6. communicating a signal fault or defect; 7. communicating a brake failure or defect; 8. being ready to write the technical forms; 9. reporting irregularities; 10. communicating a locomotive defect; 11. being able to communicate an alarm ZKE (train control device); 12. communicating in case of accidents; 13. what to do in case of release of dangerous substances; 14. communicating clear announcements; 15. spotting events on the platform; and finally 16. communicating emergency calls. Each of these situations is outlined by the perception of the Situation Awareness (SA), by

**Fig. 3** Photo taken by a train driver

the comprehension of it, by the projection in making the right decision and finally in communicating in a clearly and securly in Italian.

Based on the Bloom's taxonomy [24–27], the instructional design of the *AR vignettes* is applied in a class of six SOB train drivers. All six trainees have an A1 level (beginners). The training is designed also to reach an A2 level, specially designed for mastering technical terms, verbs, and being able to communicate through short but clear dialogues, where safety expressions are a requirement.

The training involves class participation during 12 lessons of two hours each, together with an additional three hours of self-study. At the end of the 12 lessons, the trainees will perform a one-hour language exam within a simulator. Their language performance will be voice-recorded and analyzed, according to the following criteria of pronunciation, structure, vocabulary, fluency, comprehension, and interactions [9]. Additionally, reaction time and decision-making will also be examined.

When creating the *AR vignettes*, each trainee selects a relevant situation and the visual representation of thier own work environment which corresponds to an SA-related task, and provides self-made photos. For example, Fig. 3 represents a derailed train. This is a picture taken by a train driver. Using the DART model, the instructor adapts the instructional contents *dialoguing* with the trainees. The instructor is given *Access* to new instructional content, according to the professional background of the learners, opening up to the *Risk* of using new instructional content and finding meaningful ways (*Transparency*) to create the instructional contents and the learning sequences together with the learners.

Each learner then annotates technical terms and specific verbs in Italian on a transparent sheet over the photos, which they previously produced (Fig. 4).

Figure 4 shows how technical terms, such as *il segnale di rallentamento, la locomotiva, la linea di contatto, il pilone, la gru* (the deceleration signal, the locomotive, the contact line, the pylon, the crane) are annotated in the picture by the train driver trainee.

**Fig. 4** Photo taken by a train driver, enhanced by specific terms chosen by the train driver, implemented in the *AR vignette*, using the Blippar app

In a second step, the instructor allows the trainees to read these words out-loud to an audio-recording the correct pronunciation. These audio files will be part of the *AR vignette*. Using the Blippar mobile AR, the instructor will combine words and verbs with audio files to create the multimodal *AR vignettes* ready to be shared in class and at work. In a further step, the trainees will use the same photos to create complete sentences and embed them in short and structured dialogues, first in a written format and in a second step as audio files. Through this co-created and co-learned approach, the number of words, verbs, and dialogues is multiplied by the number of trainees and by the number of *AR-learning vignettes* produced thatreach a proportional size that cannot otherwise be achieved in traditional language-learning scenarios.

Through these instructional design processes, all six Bloom's learning objectives will be experienced using AR: trainees *remember* and *understand* new words and verbs; they *apply* these through reading, writing, pronouncing, and designing their own collection of *AR vignettes*. They *analyze* new work-based situations in which they *evaluate* the use of learned words, verbs, and sentence structure, and finally *create* new dialogues.

The instructor ensures that the Mayer's multimedia learning principles [17] are applied: the *Multimedia Principle*: subjects learn better from words and pictures than from words, followed by the *Coherence Principle*: subjects will learn better when extraneous words, pictures, and sounds are excluded rather than included. In this case, learners use only their specific technical words and verbs and create specific dialogues based on selected Situation Awareness events. Additionally, the *Signaling Principle* will guide the instructional design: when they add cues (visual and audio) to the essential material to be learned, subjects learn better. In fact, this is related to the *Redundancy Principle and the Modality Principle*, because subjects learn better from graphics and narration. The *Spatial and Temporal Contiguity Principles* will also promote learning, because when corresponding words and pictures are presented near to each other and simultaneously, rather than far away from each other on the screen, the subjects learn better. The *Segmenting Principle*: the subjects learn better when a multimedia lesson is presented in user-paced segments, rather than as a continuous unit. The *Pre-Training Principle* states that subjects learn better from a multimedia lesson if they know the names and characteristics of the main concepts. The P*ersonalization Principle* will also be taken into account because words are presented in conversational style rather than in formal style. In this process, the V*oice Principle* will sustain learning, because if the narrative is spoken with a friendly human voice and not with a machine voice, the subjects learn better.

## 6 Conclusion

In this chapter, the SOB case study was first described, together with the training needs of the train drivers and train operators. Secondly, the four-layered instructional design model, composed of the Situation Awareness (SA) construct, the Mayer's Multimedia Learning Theory, the DART co-creation model, and finally the Bloom's Learning Objectives Taxonomy, was explained. Thirdly, as they are currently in a test-phase, the authors described a concrete AR training sequence, which they defined as *AR-learning vignette*, designed by an SOB train driver. In fact, with today's AR technology, overlaying information on photos that enhances the information to be learned, (in this case technical terms and nouns, verbs and dialogues created by the learner) is effortlessly created on individual mobile devices. Moreover, AR technology has the capability to add motivation to the learning experience, because adult learners become active instructional designers of their own learning content. A follow-up project could consider adding a Virtual Reality (VR) component to the training, which is currently in the brainstorming phase.

One relevant limitation is that no official data sets have yet been gathered, as the instructional model and the teaching is currently in a test-phase. The *AR-learning vignettes* are still in the process of being produced and tested. Nevertheless, one lesson learned which could be shared is that the six train drivers are actively engaged and very motivated in the co-creation of their own learning *AR vignettes*. In informal exchanges, all trainees strongly advocated a curriculum tailored to their learning needs and based on real work-based scenarios and the use of responsive technologies.

A follow-up publication is foreseen containing the quantitative and qualitative data and the results of the official language performance of an oral exam conducted in a simulator. The current instructional model is not limited to learning a new language for professional purposes. Consequently, the authors suggest testing the *AR-learning vignettes* embedded in the four-layer instructional model in similar teaching-learning professional scenarios.

## References


1. Zoer, I., Sluiter, J.K., Frings-Dresen, M.H.W.: Psychological work characteristics, psychological workload and associated psychological and cognitive requirements of train drivers. Ergonomics **57**(10), 1473–1487 (2014). https://doi.org/10.1080/00140139.2014.938130
2. Tichon, J.G.: The use of expert knowledge in the development of simulations for train driver training. Cogn. Technol. Work **9**(4), 177–187 (2007). https://doi.org/10.1007/s10111-006-0048-6
3. Suter, J.: Gestaltung eines Eisenbahn-Systemmodells für die Untersuchung von komplexen soziotechnischen Problemstellungen, Doctoral Dissertation, Braunschweig. Technische Universität Braunschweig, Germany (2014)
4. Stoller, N.: Situation Awareness von Lokführenden während sicherheitskritischer Ereignisse im Bahnverkehr. In: Bachelor Thesis in Applied Psychology, FHNW—Fachhochschule Nordwestschweiz, Olten, CH (2013)
5. Schwager, M.: Risiken der Automatisierung für die Zusammenarbeit von Lokführer und Zugverkehrsleiter Lösungsansätze für die Entwicklung eines gemeinsamen Verständnisses. Bachelor Thesis in Applied Psychology, FHNW—Fachhochschule Nordwestschweiz, Olten, CH (2013)
6. Suter, J., Inglese, T.: Einfluss von Sprachkenntnissen auf die Sicherheit im Bahnbetrieb: Simulationsbasierte Untersuchung und Schulung von Lokführern und Fahrdienstleitern. In: Tagungsband. 7. Interdisziplinärer Workshop. Kognitive Systems: Mensch, Teams, Systeme und Automaten. pp. 59–68. Braunschweig (2018)
7. Inglese, T., Suter, J.: Simulationsbasierte Sprachschulung für Lokführer. Konzept. Goldiwil (Thun). Internal Project Document. (2017)
8. Suter, J., Stoller, N.: Use of simulators to investigate complex issues relating to human factors. In: FORMS/FORMAT 2014 10th Symposium on Formal Methods for Automation and Safety in Railway and Automotive Systems. pp. 86–96. Braunschweig (2014)
9. Kim, H.: What constitutes professional communication in aviation: is language proficiency enough for testing purposes? Lang. Test. **35**(4), 403–426 (2018)
10. Bundesamt für Verkehr BAV Fahrdienstvorschriften [FDV-federal office of transport driving regulations—only in German, French and Italian]. https://www.bav.admin.ch/bav/de/home/rechtliches/rechtsgrundlagen-vorschriften/fdv.html. Accessed 18 Jan 2020
11. Houle, C.O.: The inquiring mind. Madison, WI. University Wisconsin Press. Hoyt, DP, Rhatigan, JJ (1968). Prof. Prep. Jr. Sr. Coll. student Pers. Adm. Pers. Guid. J. **47**, 263–270 (1961)
12. Knowles, M.S.: The Modern Practice of Adult Education. Cambridge Adult Education. Prentice Hall Regents, Englewood Cliffs, NJ (1980)
13. Knowles, M.S., Holton III, E., Swanson, R.: The Adult Learner, 6th edn. Elsevier, Burlington, MA (2005)
14. Endsley, M.R.: Designing for situation awareness: an approach to user-centered design. CRC Press (2016)
15. Mayer, R.E.: Multimedia learning. In: Psychology of learning and motivation, vol. 41, pp. 85–139. Academic Press (2002)
16. Mayer, R.E.: Advances in applying the science of learning and instruction to education. Psychol. Sci. Public Interest J. Am. Psychol. Soc. **9**(3), i–ii (2008)
17. Mayer, R.E. (ed.): The Cambridge Handbook of Multimedia Learning. Cambridge University Press (2005)
18. Clark, R.C., Mayer, R.E.: E-learning and the science of instruction: proven guidelines for consumers and designers of multimedia learning. Wiley (2016)
19. Mayer, R.E., Alexander, P.A.: Handbook of Research on Learning and Instruction. Taylor and Francis, Routledge, New York (2016)
20. Prahalad, C.K., Ramaswamy, V.: The Future of Competition: Co-creating Unique Value with Customers. Massachusetts, USA, Harvard Business Press, Brighton (2004)

21. Dollinger, M., Lodge, J., Coates, H.: Co-creation in higher education: towards a conceptual model. J. Mark. High. Educ. **28**(2), 210–231 (2018)
22. Ramaswamy, V.: Co-creation of value—towards an expanded paradigm of value creation. Mark. Rev. St. Gallen. **26**(6), 11–17 (2009)
23. Ramaswamy, V., Ozcan, K.: What is co-creation? An interactional creation framework and its implications for value creation. J. Bus. Res. **84**, 196–205 (2018)
24. Bloom, B. Taxonomy of educational objectives. In: Cognitive Domain, vol. 1. McKay, New York (1956)
25. Bloom, B., Krathwohl, D., Masia, B. Bloom Taxonomy of Educational Objectives. Allyn and Bacon. Pearson Education (1984)
26. Krathwohl, D.R.: A revision of Bloom's taxonomy: an overview. Theor. Into Pract. **41**(4), 212–218 (2002)
27. Krathwohl, D.R., Anderson, L.: A taxonomy for learning, teaching, and assessing: a revision of Bloom's taxonomy of educational objectives. Longman (2009)
28. Endsley, M.R.: Toward a theory of situation awareness in dynamic systems. Hum. Factors **37**, 85–104 (1995)
29. Stanton, N., Chambers, P.R., Piggott, J.: Situation awareness and safety. Saf. Sci. **39**(3), 189–204 (2001). https://doi.org/10.1016/S0925-7535(01)00010-8
30. Endsley, M.R.: Measurement of situation awareness in dynamic systems. Hum. Factors **37**(1), 65–84 (1995)
31. Kaber, D.B., Endsley, M.R.: Out-of-the-loop performance problems and the use of intermediate levels of automation for improved control system functioning and safety. Process Saf. Prog. **16**(3), 126–131 (1997)
32. Paivio, A.: Imagery and Verbal Processes. Holt, Rinheart & Winston, New York, NY (1971)
33. Paivio, A.: Mental representations: a dual coding approach. Oxford: Oxford University Press (1986, 1990)
34. Baddeley, A.D.: Working memory. Oxford University Press, Oxford (1986)
35. Sweller, J.: Cognitive load theory, learning difficulty, and instructional design. Learn. Instr. **4**(4), 295–312 (1994). https://doi.org/10.1016/0959-4752(94)90003-5
36. Sweller, J.: Instructional design in technical areas. Victoria: ACER Press (1999)
37. Prahalad, C.K., Ramaswamy, V.: Co-creation experiences: the next practice in value creation. J. Interact. Mark. **18**(3), 5–14 (2004)
38. Carmigniani, J., Furht, B.: Augmented reality: an overview. In: Handbook of Augmented Reality. pp. 3–46. Springer, New York, NY (2011)
39. Azuma, R., Baillot, Y., Behringer, R., Feiner, S., Julier, S., MacIntyre, B.: Recent advances in augmented reality. IEEE Comput. Graph. Appl. **21**(6), 34–47 (2001)
40. Porter, M., Heppelmann, J.E.: Why every organization needs an augmented reality strategy. Harvard Bus. Rev. **95**(6), 46–57 (2017)
41. Mayer, R.E. (ed.): The Cambridge Handbook of Multimedia Learning. Cambridge University Press, Cambridge (2005)
42. Dachselt, R., Al-Saiegh, S.: Interacting with printed books using digital pens and smart mobile projection. In: Proceeding of the Workshop on Mobile and Personal Projection (MP2)@ACM CHI, vol. 11 (2011)
43. Shatte, A., Holdsworth, J., Ickjai L.: Hand-held mobile augmented reality for collaborative problem solving: a case study with sorting. In: 47th Hawaii Conference on System Sciences, pp. 91–99. IEEE (2014)
44. Billinghurst, M., Kato, H., Poupyrev, I.: The magicbook—moving seamlessly between reality and virtuality. IEEE Comput. Graph. Appl. **21**(3), 6–8 (2001). https://doi.org/10.1109/38.920621
45. Grasset, R., Dunser, A., Billinghurst, M.: The design of a mixed-reality book: Is it still a real book? In: 2008 7th IEEE/ACM International Symposium on Mixed and Augmented Reality, pp. 99–102. IEEE Computer Society (2008)
46. Akçayır, M., Akçayır, G.: Advantages and challenges associated with augmented reality for education: a systematic review of the literature. Educ. Res. Rev. **20**, 1–11 (2017). https://doi.org/10.1016/J.EDUREV.2016.11.002

47. El Sayed, N.A.M., Zayed, H.H., Sharawy, M.I.: ARSC: augmented reality student card an augmented reality solution for the education field. Comput. Educ. **56**(4), 1045–1061 (2011)
48. Gavish, N., Gutiérrez, T., Webel, S., Rodríguez, J., Peveri, M., Bockholt, U., Tecchia, F.: Evaluating virtual reality and augmented reality training for industrial maintenance and assembly tasks. Interact. Learn. Environ. **23**(6), 778–798 (2015). https://doi.org/10.1080/10494820.2013.815221
49. Di Serio, Á., Ibáñez, M.B., Kloos, C.D.: Impact of an augmented reality system on students' motivation for a visual art course. Comput. Educ. **68**, 586–596 (2013). https://doi.org/10.1016/J.COMPEDU.2012.03.002
50. Hsiao, K.-F., Chen, N.-S., Huang, S.-Y.: Learning while exercising for science education in augmented reality among adolescents. Interact. Learn. Environ. **20**(4), 331–349 (2012)

# Frictionless Commerce and Seamless Payment

**Michael H. Quade**

**Abstract** Improving the user experience and optimizing conversion—these challenges have been at the top of e-commerce managers' task lists for years. In order to improve the customer experience when shopping online, time-consuming and repetitive entries must be reduced or even eliminated before the purchase is completed. For example, the less data the customer must enter during the check-out process, the faster the purchase can be completed. If customers use smartphones for online shopping, cumbersome data entry can quickly lead to termination. Not only should the purchase be completed quickly and smoothly, it should also be secure and legally binding for the parties involved. The concepts "frictionless commerce" and "seamless payment" support this. Based on the differences from conventional concepts, this chapter explains the two concepts and illustrates under which aspects they can be used together. In addition, practical examples will show how these concepts are already being applied.

**Keywords** E-Commerce · Frictionless commerce · Check-out · Seamless payment · Mobile apps

## 1 Introduction

The term "frictionless commerce" was first used in 2000 by Brynjolfsson and Smith [1]. At that time, the term referred primarily to the transparency of supply and prices on the market resulting from the emergence of e-commerce. It was observed that e-commerce makes the market more "frictionless", that the search effort for consumers e.g. in stores has been reduced by the electronic availability of information, regardless of time and place. Today, frictionless commerce also means integrating digital purchasing processes as easily and seamlessly as possible into the everyday life and the natural environment of consumers [2]. The goal of frictionless

M. H. Quade (✉)
School of Business, Institute for Information Systems, FHNW University of Applied Sciences and Arts Northwestern Switzerland, Peter Merian-Strasse 86, 4002 Basel, Switzerland
e-mail: michael.quade@fhnw.ch

 
R. Dornberger (ed.), *New Trends in Business Information Systems and Technology*, Studies in Systems, Decision and Control 294,
https://doi.org/10.1007/978-3-030-48332-6_15

commerce, also known as "contextual commerce" [3], is that each customer can buy a product or service with as few clicks or other steps as possible, whenever and wherever they want. A purchase process can be described as completely frictionless if a consumer only has to express the will that they want to have a particular product or want to use a certain service. The purchase process is executed immediately and only completed by the expression of will.

In order to be able to place an order with only one interaction such as a click, a customer must conclude a master agreement with the supplier. The master agreement regulates everything that needs to be specified during a normal check-out. This includes the delivery address, the method of delivery, and the payment modalities. The customer must provide a credit card, bank account, or other means of payment that may be charged by the supplier. In particular, the customer must agree that their means of payment may be charged by the supplier without any further action on their part. The master agreement could be a hurdle for frictionless commerce and has the effect that the procedure is not advantageous for one-off or rarely repeated purchase transactions.

However, some companies have already implemented the concept of completely frictionless commerce. Very early on, Amazon opted for frictionless commerce with its 1-Click ordering concept, when it was still primarily a bookseller [4]. Amazon customers could register for 1-Click quickly and easily. Amazon also patented the invention of the "one click" in September 1999 [5]. One of the largest licensees became Apple, which used the concept for its iTunes Store [6]. Amazon's patent expired in September 2017 after 18 years, and other vendors can now freely use the concept. Similar one-click-purchase buttons now exist on different platforms, even on sites that are not online shops e.g. the Action Button on Instagram or Pinterest [7].

Frictionless commerce and seamless payment differ from conventional concepts in several aspects. The aim of this chapter is to present these aspects in a structured way and to illustrate them with examples.

The chapter is structured as follows: After a brief introduction to frictionless commerce beyond the web browser, this article will focus on seamless payment, a concept to further improve frictionless commerce. After clarifying the terms and aspects, several current examples will be presented and discussed. The examples focus mainly on the use of frictionless commerce and seamless payment with smartphones. The chapter ends with the conclusion.

## 2 Frictionless Commerce Beyond the Web Browser

Amazon also transferred the concept of 1-Click to physical order buttons: the "Dash Button" [8]. The aim of the physical buttons is to trigger an order process outside an online shop, but to transmit the order to an online shop without friction. Brack.ch, a Swiss online retailer offers a similar physical button: the "Brack.ch Button" [9]. The physical buttons of Amazon and Brack.ch function mostly the same. While the Dash button already contains an article from the Amazon range, the Brack.ch Button can contain up to four articles, which are ordered automatically when the corresponding button is pressed.

Amazon often automatically charges the credit card that is stored with the Amazon account; with Brack.ch, payment is currently only possible with the payment slip included in the invoice. Amazon's frictionless commerce concept therefore goes beyond that of Brack.ch. Payment at Amazon is seamlessly integrated and without customer interaction during order processing. At Brack.ch, the money transfer with the payment slip is an additional step. Customers must enter the payment slip in their e-banking or submit it to their financial institution. It will probably take some time before the physical order buttons are established as a common concept in e-commerce and are offered by other retailers. The legal situation in Europe regarding these physical buttons is not yet fully clarified and there are some legal questions to answer [10]. The main problem with the currently existing physical order buttons is that customers only find out afterwards at what price they have bought a product.

The implementation of frictionless commerce with mobile apps can also be counted as "beyond the web browser". Frictionless commerce has been implemented on smartphones with apps such as "Google Play" or "Apple App Store", where mobile apps can be purchased with few touches. The concept is also frequently used in the area of transport services [11]. Some examples will be provided later in this chapter.

The frictionless commerce concept can be applied even more extensively than with a physical order button or a mobile app. Yet, this can only be done with massive technical effort, which remains practically invisible to the customer. An example for such an implementation is Amazon Go [12]. It is the first self-service brick and mortar shop in the retail trade without a cash register. Amazon Go is a mixture of off- and online trading. The customers must identify themselves with their smartphone and the Amazon Go App to enter the shop. In the shop, they can take what they want to buy from the shelves. Customers can simply complete the purchase by leaving the shop with the goods. The concept behind Amazon Go will be further explained later in this chapter.

## 3 Frictionless Commerce Versus Conventional Commerce

The frictionless commerce approach is suitable for purchases of all products and services. Especially the regularly repeated demand for a product or service favors the use of frictionless commerce—e.g. if customers already know the price of a product or service from previous purchase transactions. From the customer's point of view, providing or confirming payment information is an additional effort that hinders a quick purchase. If the price of the product or service is relatively low, customers are more likely to accept automated purchasing processes. This is mainly due to the lower risk of loss of the customer.

Table 1 shows a comparison between purchase situations that are still more suited to conventional purchasing methods and those that are suitable for frictionless commerce and thus for seamless payment.

**Table 1** Purchase situations with suitability for conventional purchasing procedures and for frictionless commerce

| Purchase situations primarily suitable for conventional purchasing procedures | Purchase situations suitable for frictionless commerce |
|---|---|
| Single or seldom repeated purchase of goods or services | Regular or more frequently repeated purchases. (public transport ticket, car sharing, convenience food etc.) |
| Any purchase amounts | Low purchase amounts |
| More stationary purchase process | More mobile purchase process |
| More related to physical products | More related to services |
| Manual purchasing processes | Fully or partially automated purchasing processes |

Depending on the purchase situation or type of product or service, frictionless commerce is a sensible option. Frictionless commerce is less useful for single or less frequent purchase transactions, because, as mentioned in the introduction, the hurdle of the master agreement exists. Customers are not necessarily willing to complete a master agreement in case of a one-time purchase. A frequent buyer, on the other hand, benefits because of fewer steps in the check-out.

Frictionless commerce is also suitable in conjunction with mobile purchase situations where the rather limited input and output interface of a smartphone must be used to complete a purchase. With frictionless commerce, there is no need for the buyer to enter extensive information on the small screens.

When a mobile app is used for frictionless commerce, additional aspects can be considered to make a purchase more convenient for the customer: For example, useful additional information about the product or service, ratings from other customers nearby, and special offers for the next purchase such as vouchers or collecting points in customer loyalty programs [13, 14].

Frictionless commerce is easier to implement with services than with physical products. A service can be presented digitally, e.g. a ticket displayed on the screen of the smartphone. In the examples later in this chapter, frictionless commerce solutions therefore predominate for services. The examples with physical products show the immense effort required to securely record purchases.

If an entire purchasing process is to be fully automated, the beginning and end of the process must also be automated. The purchase intention of the customer must be automatically detected, and the customer must also be identified in an appropriate way. Persons can be identified via biometric features for payment [15, 16]. These can be recorded via cameras, e.g. when entering and leaving a shop, and compared with the data of the registered customers. Purchasing processes based on biometrics have been discussed and tested for years, but currently such systems are mainly used for the detection of unwanted persons such as shoplifters [17, 18]. Alternatively, identification can also be made based on a personal device such as a smartphone. For identification purposes, the smartphone exchanges a small amount of data at an entrance point, e.g. via a QR code or NFC transmission [19, 20].

## 4 Seamless Payment

The seamless payment concept differs from the conventional payment concept. Conventional payment offers more flexibility in the choice of means of payment and control over the steps in processing than seamless payment. Conventional payment requires more interaction and presence in the process, whereas seamless payment primarily represents a simplification of the entire payment procedure. The simplification allows the customer to complete a purchase procedure with fewer steps and thus faster, but with less flexibility.

Fewer steps also mean fewer hurdles to completion and more convenience for customers. Therefore, the benefit of seamless payment for customers lies primarily in simpler and faster payment processes. On the seller's side, the benefit is an increase in the attractiveness of their offering through low-threshold handling as well as a lower bounce rate for customers in the check-out process. A comparison of the different aspects of conventional payment and seamless payment is shown in Table 2.

Based on these considerations, the following definition forms the basis for further explanations of the seamless payment concept: *Seamless payment is a payment that is automatically triggered by the provider and integrated into a purchase process. With every purchase of goods or services, a purchase contract is concluded. The declaration of intent can be made by an activity of the buyers themselves or by a personal identification feature, e.g. a personal device. Payment of the goods or*

**Table 2** Differences in the payment process between conventional payment methods and seamless payment

| Conventional payment | Seamless payment |
|---|---|
| Separate payment sequentially separated from product selection or service purchase | Payment directly linked to the purchase decision for a product or service purchase |
| Manual triggering of the payment by the buyer | Automatic triggering of the payment by the provider |
| In the purchase process, the payment method is selected by the customer | In the purchase process, the stored payment method is automatically proposed and selected |
| Payment requires presence at the point of sale or the check-out of the online shop | Payment requires no presence at the point of sale or the check-out of the online shop |
| Confirmation of each individual payment with PIN, password, or biometric sensors (fingerprint, etc.) | No confirmation of a single payment |
| Spontaneous payment possible (with necessity of recording a means of payment) | Spontaneous payment not possible (due to the master agreement, including regulation of procedures for user identification and payment processing) |
| Detailed payment information is displayed before execution | Detailed payment information is not necessarily displayed before the payment is executed |
| Recurring credit check by the supplier | One-time credit check by the supplier |

*services is made without any interaction of the buyer. The prerequisite for this is that seamless payment has been agreed upon between buyer and seller in a master agreement. The master agreement includes the deposit of personal data, in particular the means of payment.*

For seamless payment, customers must specify a valid means of payment at least once. As a rule, this is done during the registration process for the use of regular and recurring simplified purchasing processes for products and services. The supplier must ensure with appropriate terms and conditions that they have the rights to charge the means of payment specified by the customer.

Customers can be informed about the amount of the upcoming payment, but their intervention in the process is not possible. In the event of complaints, customers can only submit them to the supplier after payment has been made.

The supplier determines the time at which the means of payment is debited. This can be before, during, or after the receipt of the service. For example, when purchasing tickets for public transport with the app "TNW Tickets" of the company Baselland Transport BLT [21], the means of payment is charged before the journey. In the app, the desired ticket must be "purchased" in the same way as at a vending machine. For the FAIRTIQ [22] or Lezzgo apps for public transport [23], the ticket must be "checked-in" at the beginning of the journey and "checked-out" at the end. The ticket price will be calculated retrospectively based on all journeys made in one day and debited the following day via the deposited means of payment.

## 5 Implementations of Frictionless Commerce

In the following, a few current implementations of frictionless commerce and seamless payment are described to illustrate the concepts. As argued here, frictionless commerce is more suitable for use in mobile situations and for services. The implementations come from the public transport, taxi industry, and retail sectors. All presented implementations of the concepts are based on the use of a mobile app with a smartphone.

### *5.1 FAIRTIQ*

FAIRTIQ marks the beginning of a new era of public transport use in Switzerland [22]. Previously, a ticket for the planned journey had to be purchased before departure. Within tariff networks, a zone ticket is required, and at national level, a ticket is issued for the destination. Tickets for public transport journeys can now be purchased via the FAIRTIQ mobile app. The destination does not have to be set before the beginning of the journey and the price of the ticket is not predetermined. The ticket will be invoiced retrospectively based on the actual distance travelled.

When the mobile app is opened for the first time, a short video introduction is offered to explain the app. In the next step, the app requires access to location data and mobile communications. Once granted, customers can register in three steps: The mobile number, a means of payment and, in accordance with the regulations of the Swiss Public Transport Association for tickets purchased online, name and date of birth must be recorded. The name is only a security element to prevent fraud by unauthorized copying of a ticket. In addition, customers can indicate whether they have a "half-fare" card or a tariff network card. These are automatically taken into account when ticket prices are calculated.

The journey can now be started by simply dragging the slider in the app. With this check-in, the nearest public transport stop is automatically saved as the starting point of the journey (Fig. 1 left side). During the journey, the digital ticket can be displayed in the app and presented during checks (Fig. 1 right side). At the destination, the slider in the app is retracted again. This check-out ends the journey and saves the nearest public transport stop as the destination. A provisional price is displayed after the end of the transport is accepted by the app. The ticket price remains provisional until the end of the day: the system is designed so that it always calculates the cheapest ticket price for a day. If a customer makes several trips during the day with FAIRTIQ, which exceed the price of a day ticket in the sum of the individual prices, the customer will be charged the price of a day ticket. During the trip or if the customers are checked-in, the app continuously transmits the location data to FAIRTIQ. The back-end system continuously analyzes this data. If a customer forgets to check-out after disembarking from the means of transport, the system sends a message via app that the end of the

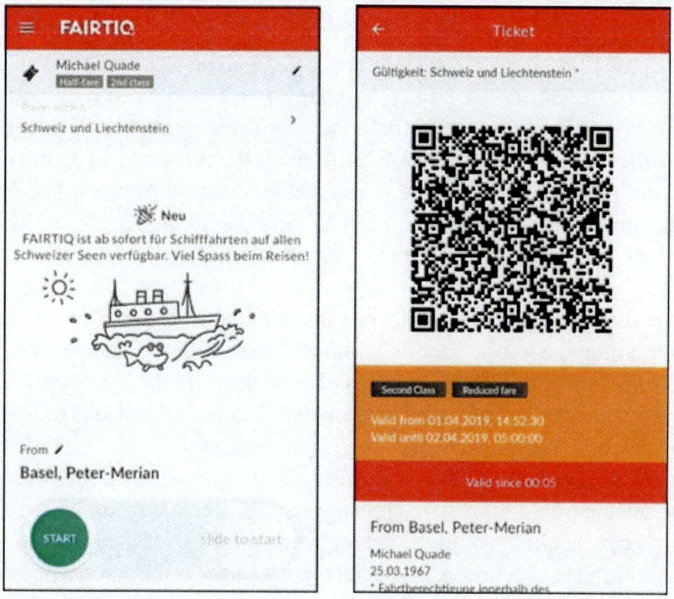


**Fig. 1** Left side: start screen with advertisement and slider, right side: ticket screen

journey has now been recognized and that the customer should check-out. If the customer does not check-out manually within an hour, this is done automatically. In order to improve the recognition of the different situations of the customers, the received data are analyzed with machine learning algorithms.

The FAIRTIQ system massively simplifies the use of public transport. The customer no longer has to deal with the tariff zones and borders of the transport associations in Switzerland. FAIRTIQ is a relief, especially if the traveler does not know the local public transport system. All Swiss tariffs are stored in the back-end system, and the cheapest ticket is always calculated automatically. The possibilities of the system have not yet been exhausted. FAIRTIQ is currently designed for one day. In principle, however, it is also conceivable to extend it to one month or one year. Depending on how much a customer is traveling, a monthly subscription or a general subscription will be billed. Trips can also be billed using vouchers or promotion codes. For example, an advertising organization can offer a discount to a specific location and back with a specially designed voucher, or it can cover the expenses in full, e.g. in connection with an event. In the same way, companies could cover travel expenses between several locations for their employees and eliminate the need for costly expense reports. In future, it should be possible to replace the still explicit check-in and check-out procedures with automated recording of boarding and disembarking.

## 5.2 *go!—Taxi Is so Easy*

The mobile app "go!—Taxi is so easy." from yourmile AG is also located in the transport sector [24]. With the mobile app go!, taxi rides can be booked and paid. The vehicles of the Zurich-based taxi company $7 \times 7$ associated with yourmile and numerous other associated taxi enterprises are mediated throughout Switzerland. The app is the official industry solution of TaxiSuisse, the professional group of the taxi industry within the Swiss Commercial Vehicle Association ASTAG. The integration of the companies is carried out either via an interface to the customary dispatching systems of the taxi centers or, for the drivers of smaller taxi companies, via the app "go! Driver".

After the mobile app is opened for the first time, the customer is asked to create a profile with name, address, mobile number, e-mail address, and password, and to register by confirming the general terms and conditions. In the next step, one or more means of payment can be entered. This step can also be skipped and postponed until the first booking is completed.

A ride can be booked in three steps (Fig. 2): In the first step, the destination is entered or selected on a map. The app suggests the current location as the starting point, which can be changed. In the second step, the class of the means of transport and the taxi company are chosen. A special feature of go! is that a fixed price is always displayed and calculated based on the measured distance and the stored tariffs of the selected class and taxi company. The displayed price is billed after the trip, even if

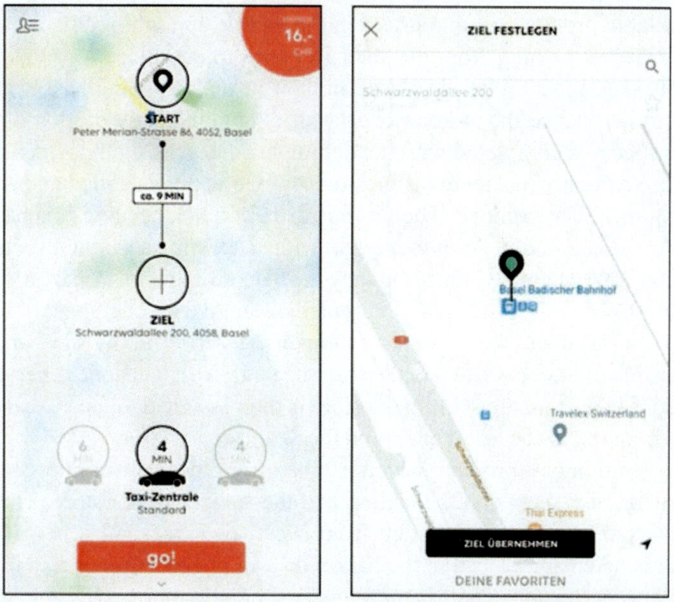


**Fig. 2** Left side: start screen with booking steps, right side: destination selection screen

detours had to be made or the trip took longer. Finally, the booking must be confirmed. Once the taxi company has accepted and confirmed the booking, the passenger is informed of the make and model, as well as the license plate number of the vehicle via the app. In addition, the passenger sees how long it will take for the driver to arrive. During the journey, the passenger is offered the opportunity to tip the driver via the app. As soon as the driver confirms the end of the journey, the deposited means of payment is debited. The passenger can now rate the trip via the app as positive or negative. In the case of a negative rating, the customer can enter a comment.

By depositing several means of payment, go! allows to distinguish between private and business trips. The app can also be loaded with voucher codes. For example, a company can take over taxi rides for its guests.

## 5.3 *Amazon Go*

With "Amazon Go", Amazon enters the world of supermarkets and introduces a frictionless commerce solution in a stationary store [12]. The concept, which has been available to the public in a pilot store in Seattle since January 2018, has great potential for transformation in the distribution of consumer goods and for stationary retail. Amazon Go is a store mainly for convenience products.

Amazon Go is also a solution, which combines a stationary shop with an online shop. This is shown in several ways: Customers can find out about the product range, available promotions, and prices in the mobile app or directly in the store. If customers take something from the shelf in the store and place it in their physical shopping basket, the item is also placed in their virtual shopping basket in the app. When customers leave the store, a check-out is automatically carried out and the purchase process is completed with the debiting of the deposited payment method.

The store can only be entered if the Amazon Go app is installed and configured on the customer's smartphone. The setup is usually quick because it is based on an ordinary Amazon account. After logging in with an existing account, the customer is asked to confirm the payment method they have deposited with Amazon, or to enter a new one. After the registration is completed, an Aztec code appears in the app (Fig. 3). With this code, the customer can open the barrier to the store and enter. It is also possible to have several people enter the store with one code. Everything that these persons take from the shelf in the store is then assigned to the virtual shopping cart corresponding to the account behind the Aztec code.

The registration of a product, which is taken from the shelf or returned, is made with hundreds of cameras on the ceiling and the shelves of the store. The cameras capture every movement in the store from every angle, several times over. Every article that is taken and placed in the basket or own pocket is immediately captured and displayed in the app. Theft is therefore practically impossible. The articles are not specially marked. These are the same product packaging and labels as in other stores. The customer leaves the store again via the barrier, which is automatically registered as a check-out and the app indicates that the purchase was completed.

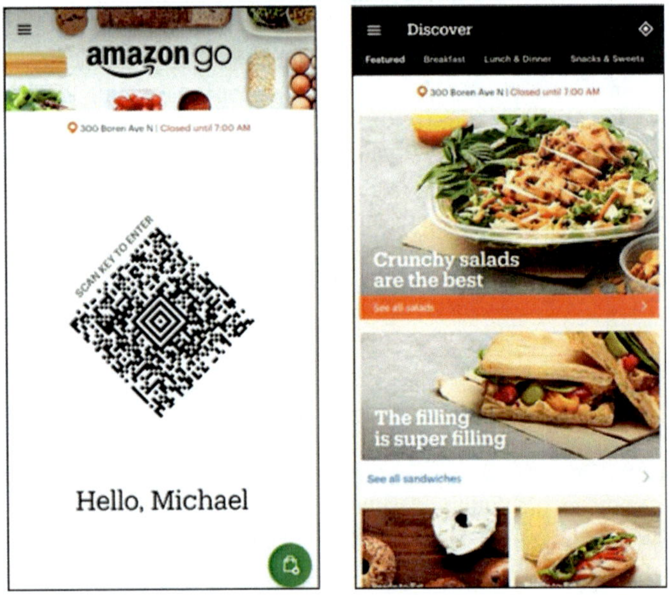


**Fig. 3** Left side: start screen with Aztec code, right side: featured offers of a specific store

Approximately 15 min later, the app also displays the receipt for the purchase, and Amazon debits the selected means of payment.

Amazon developed and tested the solution for five years before it was released to the customers. The underlying technology is based on machine learning and is the same as for Amazon Echo with the Alexa assistance system, which is controlled via voice [25]. Unlike Amazon Echo, Amazon Go processes image data. The system can track individuals through the store without capturing biometric information such as facial features.

For customers, Amazon Go can be regarded as complete frictionless commerce. The additional action of a buyer, which is explicitly necessary for the billing, is only the check-in with the Aztec code at the barrier. Amazon Go, however, is not the only project for testing a cashless business. In the summer of 2017, Alibaba opened a Tao Cafe in the Chinese city of Hangzhou [26], which includes a café and a cashless supermarket. There, customers are biometrically recorded via facial recognition. The Tao Café also uses optical sensors to detect purchases. Further technical details are unknown.

## 6 Conclusion

In this chapter, the aspects of frictionless commerce and seamless payment were examined and the differences to conventional concepts pointed out. The new concepts should not only make online shopping easier, but also safer. However, the individual differences with conventional purchasing and payment concepts must be taken into account by the provider. Customers must be able to rely on the provider not to use the authorization assigned to them, such as automatically triggering an order or automatically debiting a credit card. A balance must be found between convenience and safety.

Frictionless commerce and seamless payment solutions are based on technologies that were not available ten years ago. Further developments are therefore to be expected. In the examples, customers have control over the triggering of a purchase transaction. They must either identify themselves with an app (Amazon GO) or start the purchase process with an explicit function in the app (FAIRTIQ and go!). Future solutions could automate these activities. The triggering occurs solely through the data measured by sensors in smartphones, such as location, speed, and direction, and the comparison with data from e.g. a bus in public transport.

Future studies could shed light on data protection aspects and these new concepts. To what extent are customers willing to have data captured automatically, and where are there limits? Regarding the procedures and amounts used, this chapter assumes that customers agree to the use of seamless payment, especially for repeated purchases and rather low amounts. It remains to be seen whether these assumptions will be confirmed in the future and whether frictionless commerce and seamless payment will become widely accepted.

## References

1. Brynjolfsson, E., Smith, M.D.: Frictionless commerce? A comparison of internet and conventional retailers. Manag. Sci. **46**, 563–585 (2000)
2. Rouse, M.: What is frictionless commerce? https://whatis.techtarget.com/definition/frictionless-commerce
3. Gibson, J.F.: Contextual commerce: the next wave of content and commerce integration. http://www.econtentmag.com/Articles/Editorial/Commentary/Contextual-Commerce%c2%a0The-Next-Wave-of-Content-and-Commerce-Integration-114160.htm
4. Filson, D.: The impact of E-commerce strategies on firm value: lessons from Amazon.com and its early competitors. J. Bus. 77, S135–S154 (2004)
5. Hartman, P., Bezos, J.P., Kaphan, S., Spiegel, J.: Method and system for placing a purchase order via a communications network. https://worldwide.espacenet.com/publicationDetails/biblio?FT=D&date=19990928&DB=&locale=en_EP&CC=US&NR=5960411A&KC=A&ND=2 (1999)
6. Apple Computer, Inc: Apple licenses Amazon.com 1-click patent and trademark [Press Release]. https://www.apple.com/newsroom/2000/09/18Apple-Licenses-Amazon-com-1-Click-Patent-and-Trademark/
7. Carey-Simos, G.: Instagram starts selectively rolling out "Action Buttons" for business profiles. https://wersm.com/instagram-starts-rolling-out-action-buttons-for-business-profiles/ (2018)
8. Günder, A.: Wie nützlich ist der Amazon Dash Button? https://www.homeandsmart.de/amazon-dash-button-bestellung-per-knopfdruck
9. Gröflin, S.: Brack hat jetzt einen Bestellknopf. https://www.pctipp.ch/news/hardware/artikel/brack-hat-jetzt-einen-bestellknopf-86541/?forcedesktop=1&cHash=46f75a494c604746dcefa403c7f1f36e
10. Wilkens, A.: Amazon dash button: Gericht verbietet WLAN-Bestellknöpfe. https://www.heise.de/newsticker/meldung/Amazon-Dash-Button-Gericht-verbietet-WLAN-Bestellknoepfe-4271107.html
11. Dinning, M., Weisenberger, T.: Multimodal transportation payments convergence—key to mobility. In: Disrupting Mobility, pp. 121–133. Springer (2017)
12. Wingfield, N.: Inside Amazon Go, a store of the future. https://www.nytimes.com/2018/01/21/technology/inside-amazon-go-a-store-of-the-future.html (2018)
13. Cameron, D., Gregory, C., Battaglia, D.: Nielsen personalizes the mobile shopping app: if you build the technology, they will come. J. Advert. Res. **52**, 333–338 (2012)
14. Groß, M.: Mobile shopping: a classification framework and literature review. Int. J. Retail Distrib. Manag. **43**, 221–241 (2015). https://doi.org/10.1108/IJRDM-06-2013-0119
15. Garg, R.K., Garg, N.K.: Developing secured biometric payments model using tokenization. In: 2015 international conference on soft computing techniques and implementations (ICSCTI), pp. 110–112. IEEE (2015)
16. Smowton, C., Lorch, J.R., Molnar, D., Saroiu, S., Wolman, A.: Seamless customer identification. Tech. Rep. MSR-TR-2013-31, Microsoft Research (2013)
17. Jones, R.: Digital rule: punishment, control and technology. Punishm. Soc. **2**, 5–22 (2000)
18. Mansfield-Devine, S.: Biometrics in retail. Biom. Technol. Today 5–8 (2013)
19. Finžgar, L., Trebar, M.: Use of NFC and QR code identification in an electronic ticket system for public transport. In: SoftCOM 2011, 19th International Conference on Software, Telecommunications and Computer Networks, pp. 1–6 (2011)
20. Urien, P., Piramuthu, S.: Framework and authentication protocols for smartphone, NFC, and RFID in retail transactions. In: 2013 IEEE Eighth International Conference on Intelligent Sensors, Sensor Networks and Information Processing, pp. 77–82 (2013)
21. BLT Baselland Transport AG: App "Tickets", https://www.blt.ch/reisen/tickets/app-tickets/
22. FAIRTIQ: Home|FAIRTIQ. https://fairtiq.com/de-ch/
23. lezzgo c/o BLS AG: lezzgo—Einsteigen, fahren, fahren, fahren… bezahlen! https://www.lezzgo.ch/

24. yourmile AG: go! App—The taxi app—go! Mobility. https://go-app.ch/en/
25. Herbrich, R.: Künstliche Intelligenz bei Amazon Spitzentechnologie im Dienste des Kunden. In: Buxmann, P. und Schmidt, H. (hrsg.) Künstliche Intelligenz: Mit Algorithmen zum wirtschaftlichen Erfolg. S. 63–75. Springer Berlin Heidelberg, Berlin, Heidelberg (2019)
26. Peng, T.: Amazon Go vs Alibaba Tao Cafe: staffless shop showdown. https://medium.com/syncedreview/amazon-go-vs-alibaba-tao-cafe-staffless-shop-showdown-3f3929393d62 (2018)

# Direct to Consumer (D2C) E-Commerce: Goals and Strategies of Brand Manufacturers

**Uwe Leimstoll and Ralf Wölfle**

**Abstract** Contrary to expectations expressed in the 1990s, e-commerce has so far led to relatively little disintermediation. Even large and well-known brands continue to generate the majority of their sales through traditional distribution channels. However, many end customers expect to be able to buy directly from the brand. In addition, structural changes in the value chain have made it necessary to present the brand on the Internet in a brand-specific manner. Depending on the structure of their online activities, however, brand manufacturers are competing with their traditional indirect distribution channels, which remain important for them. This raises the question of which e-commerce strategies brands can use to meet the needs of end customers without harming their distribution partners. To answer this question, qualitative expert interviews were conducted with e-commerce managers of market-shaping Swiss companies. The results confirm that the importance of brand engagement in communication and interaction with end customers is increasing. For the implementation of online direct sales, three strategies have emerged that avoid or at least minimize conflicts with traditional sales partners: a pure online direct sales strategy, a multi-touch-point strategy, and a platform strategy. The multi-touch-point strategy can in turn be designed in four different variations.

**Keywords** Online direct sales · Direct-to-consumer e-commerce · Direct distribution channels · Disintermediation · Intermediary functions

## 1 Introduction

Brand manufacturers (also brand suppliers or just brands) have not actively pushed online direct sales for many years. Depending on the industry, they are more or

U. Leimstoll (✉) · R. Wölfle
School of Business, Institute for Information Systems, FHNW University of Applied Sciences and Arts Northwestern Switzerland, Peter Merian-Strasse 86, 4002 Basel, Switzerland
e-mail: uwe.leimstoll@fhnw.ch

R. Wölfle
e-mail: ralf.woelfle@fhnw.ch



R. Dornberger (ed.), *New Trends in Business Information Systems and Technology*, Studies in Systems, Decision and Control 294,
https://doi.org/10.1007/978-3-030-48332-6_16

less careful in selling directly to consumers (D2C). Consequently, the degree of disintermediation is still rather low in contrast to considerations made at the beginning of the e-commerce era [1]. The reasons for the cautious attitude towards online direct sales are manifold. One of the main reasons is that even manufacturers of famous brands strongly depend on their traditional trading partners because in many industries the vast majority of sales are still generated via traditional distribution chains and channels consisting of intermediaries such as general agents, importers, wholesalers, retailers, and others. Because of the significance of these partners, many brand manufacturers avoid creating conflicts, which typically occur when they build up their own direct distribution channels and thus become a competitor of their partners [2]. Other possible reasons for the reserved attitude of brand manufacturers towards D2C might be missing know-how and preconditions for realizing direct sales. As Sarkar et al. [3] pointed out, intermediaries fulfill specific functions, which partly must be taken over by the brands themselves when they sell directly to consumers.

On the other hand, brand manufacturers have several good reasons and even needs to invest in direct sales channels [4]. Consumers do not only expect an informative website with an engaging online presentation of brand products by the manufacturer. Above all, they do not understand if the brand's website does not offer the opportunity to purchase the articles online. Particularly within industries, which are affected by product piracy, customers strive to reduce the risk of purchasing fake products by ordering directly from the brand manufacturer's flagship or online store. Additionally, for consumers identifying with the brand, shopping directly from the brand is part of their brand experience.

From the point of view of brand manufacturers, complete control over distribution and prices often builds the motivation to invest in direct sales channels. These investments could also be driven by the reduction of distribution costs by bypassing the intermediaries, the increase of market coverage, the provision of a brand-specific market presence, or the direct relationship between supplier and customer [5]. The latter enables the realization of services that are better tailored to customer requirements and behavior and thus lead to stronger customer loyalty.

Finally, brand manufacturers face a trade-off between maintaining their relationship with traditional distribution partners on the one hand and meeting consumer expectations and their own distribution goals on the other hand. Thus, the topic of online direct sales implies an enormous challenge for brand manufacturers. It needs solutions that show possible ways of how to solve or reduce the described trade-off.

The goal of this paper is to find useful strategies brand manufacturers can apply to meet the needs of their end customers without harming the business of their traditional trading partners too strongly or even with offering benefits for them. The hypothesis followed in this paper is that there exist direct online sales strategies for brand manufacturers, which avoid creating conflicts with traditional distribution partners. The research questions derived from this hypothesis are the following:

(1) Which online channels belong to the area of online direct sales?
(2) What kind of online strategies are possible for brand manufacturers?

(3) Which strategies have the potential to avoid conflicts with other partners in the value chain?

This chapter focuses on the online direct sales of brand manufacturers or comparable companies that also sell via the classic distribution channels, i.e. via importers and the wholesale and retail trade. Pure direct sales, as known from big brands such as Tupperware or Thermomix, are not considered. Vertically integrated companies that realize the wholesale and retail level only through their own companies, such as H&M, Tally Weijl or Ikea are also not considered.

However, sales via digital platforms such as electronic marketplaces or social media platforms such as Facebook and Instagram are included. The brand manufacturer can implement sales via such platforms mostly independently and self-determined. It therefore seems justified to interpret this distribution channel as a direct sales channel.

The following section defines relevant terms and gives a short overview of the recent literature in the field of direct to consumer e-commerce. Section 3 describes the research approach and the procedure of collecting and analyzing data. In a next step, the transformation of the Swiss retail market will be explained in order to understand the initial situation of the brand manufacturers and their need to act (Sect. 4). Section 5 reports the motivation and goals of the brand manufacturers, followed by an explanation and discussion of possible online direct sales strategies (Sect. 6). Section 7 summarizes the results, draws conclusions, and indicates limitations and topics for further research.

## 2 Literature Review

Direct sales in general can be described as sales, which are realized via vertically integrated distribution channels. The channels or distribution chains can look different: they can consist of several value chain stages such as wholesaling and retailing (e.g. flagship stores, outlets), which are all operated or at least controlled by the brand manufacturer, or they can be more directly connected to the consumer via sales persons or online shops. However, the operation of supporting functions such as payment, transport, or advertising can be outsourced to third party service providers [6].

A narrower definition limits direct sales to personal selling activities to private end-users (consumers) outside physical selling locations as it is realized e.g. by well-known brands like Vorwerk or Tupperware [7]. In this chapter, the narrow definition is not appropriate because it covers only a very specific kind of distribution channel. Using a broader definition is necessary in order to show how online activities of brand manufacturers are established parallel to distribution activities along the traditional distribution chains.

As online activities build the focus of this chapter, the area of online direct sales channels is of particular relevance. They can be defined as online sales channels,

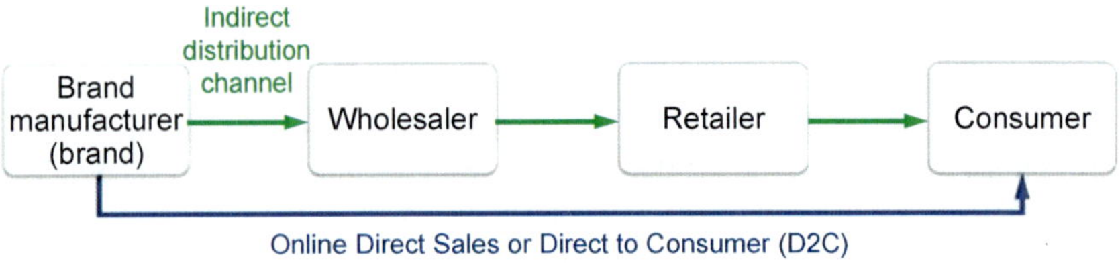


**Fig. 1** Concurrent channels in a manufacturer business model

which are operated by the brand manufacturer to sell products directly to the business or private end customer. This means that no other intermediaries, neither electronic marketplaces nor other brokers or resellers, are involved in the transaction process [6]. Li et al. [6] additionally require that the complete transaction process including agreement and ordering can be realized on the online platform.

The opposite of direct sales are indirect sales. They are created via distribution channels, which incorporate third party partners such as general agents, importers, wholesalers, and retailers. Indirect and direct distribution channels can be combined (Fig. 1). If they are combined in one regional market with the same product line in both channels, they are called concurrent channels [8]. The operation of concurrent channels gives customers the opportunity to choose the channel that best fits their needs. At the same time, it gives the brand manufacturer the chance to achieve a better market coverage [8, 9]. International brand manufacturers with rather long and diverse distribution chains often operate their own distribution centers in the target countries supporting their own and—if appropriate—third party retailers with marketing communication or logistics services. Nevertheless, concurrent sales channels usually lead to intra-brand competition and conflicts among the different channels. Sa Vinhas and Anderson [8] analyze these effects in the B2B area in detail.

Selling directly to consumers is a major challenge for brand manufacturers who have worked with traditional distribution partners in the past. Although they are familiar with business-to-business (B2B) relationships and processes, they must acquire business-to-consumer (B2C) know-how anew. Accordingly, several aspects cited in the literature illustrate the problem on the side of the brands:

- Specialized resources and functions, which are needed in the relationship with consumers (e.g. sales, logistics, communication), are often not available [10].
- Specialized B2C online-distribution know-how, e.g. country- or segment-specific knowledge regarding legislation, customs regulation, customer requirements, is often missing.
- Conflicts between the goals of presenting products and selling products arise.
- The competitive situation in a B2C context is different from a B2B situation [10].
- Retailers who are also important partners become competitors. The relationship with them must therefore be reconsidered [10].

One aspect that is controversially discussed in the literature is the question of whether sales via digital intermediary platforms can still be counted as online direct sales. Li et al. [6] define that online direct sales exclude the involvement of brokerage platforms such as marketplaces or price comparison platforms, whereas fulfillment services can, for example, be outsourced to third parties [6].

The basic definition of direct sales states that direct distribution chains are vertically integrated, and distribution can therefore be controlled by the brand manufacturer. With reference to this possibility of control, sales via brokerage platforms can also be counted as direct sales, because brand manufacturers can control the content and prices offered themselves. In addition, the contract is usually concluded directly between the supplier and the buyer and not between the intermediary and the buyer.

## 3 Research Design

The behavior of companies can be observed directly only to a limited extent. In order to gain deeper insights into the goals pursued and strategies implemented by brand manufacturers, an explorative and interpretative research approach with qualitative expert interviews is therefore suitable [11, 12]. E-commerce managers of companies that are market-shaping e-commerce providers in Switzerland are interviewed as experts. They are involved in the strategic positioning of their e-commerce offers and at the same time actively involved in the value creation processes. These views enable them to make statements about current developments in their industry and in Swiss e-commerce. Market-shaping e-commerce providers are defined here as companies that significantly influence the development of e-commerce in their industry. These include companies that are leaders in e-commerce in their industry (e-commerce leaders), that shape the industry by their size or reputation and are committed to e-commerce (industry leaders), or that promote e-commerce through innovative business models (e-commerce innovators).

The data and statements relevant to the research questions examined here were collected at the beginning of 2016 [4] as part of a large-scale study, the E-Commerce Report Switzerland. The E-Commerce Report Switzerland is an annual panel study on the development of e-commerce in Switzerland, which has been conducted since 2009 (see for example [13–15]. A total of 36 companies from different industries—mainly consumer electronics, fashion, food and beverage, lifestyle, media, travel, and marketplaces—took part in the study in 2016. Among them are twelve companies that act as brand manufacturers or brand suppliers and sell online directly to consumers (e.g. FREITAG lab., Nespresso Suisse, Swiss International Air Lines, SBB Swiss Federal Railways, Scott Sports, Victorinox).

The survey is designed as a structured expert discussion with open and closed questions, which allows a constructive and in-depth examination of the topics [12]. The questionnaire is not identical for all companies, as in some cases it also contains questions that are adapted to specific industries and companies. The study participants received the printed questionnaire at the beginning of the interview. They were only pre-informed about the general topics a few days before.

Company representatives who have already been interviewed several times in previous years are sometimes interviewed in writing. The questionnaire for the written survey is largely identical to that used for the interviews. In 2016, six new participants were added to the panel specifically for the topic "Online direct sales". A modified questionnaire with some topic-specific questions was used for the brand manufacturers. In total, the authors conducted 32 personal interviews, and four experts answered the questionnaire in writing.

The written documentation of the interviews was done by listening to the audio recordings and writing a summary transcript [16]. The transcription follows previously formulated rules to ensure a consistent approach [17]. The transcription was done question by question and followed the course of the interview.

The level of detail of the answers, the range of the statements, and the confidentiality of the individual statements made an interpretive condensation necessary. The interpretation was carried out by the authors themselves because of the expertise this requires.

## 4 E-Commerce Drives the Transformation of Distribution Chains

This section describes the current situation of the trade in general and specifically of B2C e-commerce in order to show the current developments, which determine the conditions for the market players. All experts interviewed in this study rather agree or fully agree that a transformation of the value chains is currently taking place in their industry. As the study is about B2C e-commerce, the descriptions focus on the distribution chains between suppliers and consumers.

In the eyes of the experts, structural change in Switzerland is primarily reflected in the following changes: the appearance of innovative business models driven by the Internet and mobile computing, ever-increasing customer requirements, falling margins and prices, superior foreign suppliers, and a growing range of products and services (more products and providers). In some industries, these changes occur in parallel, so that it is obvious that the situation of many suppliers is becoming increasingly acute as demand stagnates. In addition to digitalization and increasing networking, developments in foreign markets are emerging as drivers of these developments. Foreign suppliers have more sales potential due to the size of the market, motivating them to make higher investments. This in turn allows them to achieve positive economies of scale and to offer a higher service level—often at lower prices.

On the supplier side, an increasing division of work and specialization can be observed as a result of increasing digitalization and networking along the value chain. This means that countless service providers are emerging to take over individual retail functions. This in turn leads to the effect that typical functional bundles of the retail trade dissolve [18–20], more alternatives for the provision of retail functions are emerging and the classic distribution channels are losing their importance for brand manufacturers.

On the consumer side, this loss of importance is driven by the general shift of purchase-related activities to online channels. The volume of online trade in Switzerland has grown by approximately 10% annually over the past ten years. Depending on the product category, it now accounts for between 2 and 50% of private household consumption expenditure [21].

In order to compensate for the loss of importance of the traditional distribution channels and to ensure the service level defined by the brand, the importance of direct sales is increasing for brand manufacturers. Within the framework of the series of studies described here, it was observed that Swiss e-commerce retailers have been aware of increased direct sales activities by brand manufacturers since 2012, coinciding with a strong depreciation of the euro. For this reason, the topic of direct sales was taken up as the main topic of the study in 2016. In the interviews, a large majority of the respondents agreed that many brand manufacturers have been trying to control the online sales of their products themselves or at least sell parts of their product range directly to end customers since around 2012.

## 5 Motivation and Goals for Direct Sales

In the discussions with brand manufacturers, it becomes clear that the structural changes require a stronger commitment of the brand manufacturers. One manufacturer describes that in the years to come, brands will have to take on more retail functions in order to ensure the customer's brand experience. The reason for this is the declining margins, which means that retailers will no longer be able to provide personnel-intensive services. To compensate for this, brands will have to become more involved in this area, which will lead to a shift of retail functions to the brand manufacturers. Compared to retailers, brand manufacturers have the advantage that their specialization and size enable them to provide the necessary services at a higher quality and possibly also at lower cost.

Other statements reflect the fact that retailers are increasingly being restricted: They cannot physically display all brands or their complete assortment, not least because of the decline in physical sales space. Even in specialist shops, sales staff can no longer know all the details of the wide range of products, which limits their advisory skills. This is exacerbated by the changed customer behavior: Customers are often better informed than the sales staff in the retail stores due to the information options available today. In order to obtain authentic information, they increasingly turn directly to the brands for information and questions. By doing so, they expect to be able to buy the products from the manufacturer, e.g. to ensure that they receive original and up-to-date articles. This is clearly expressed in a statement by Kilian Eyholzer of Victorinox:

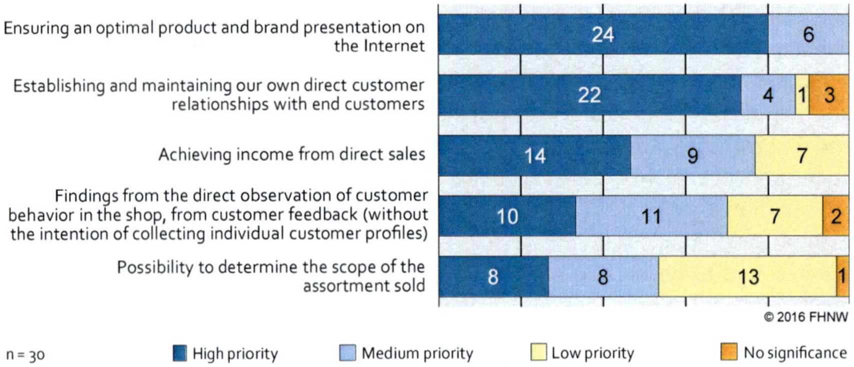


**Fig. 2** Goals for D2C sales of brand manufacturers [4]

> Customers today expect an online shop from us. In a survey, two customer concerns stood out above all others: Customers want to see the prices and be able to order directly online.

The goals that brand manufacturers want to achieve with their own online sales are multifaceted. Surprisingly, only half of the panel participants believe that generating income from direct sales is a high priority for brands—this motive is in the middle of the field compared to other motives (Fig. 2). This means that a direct distribution channel is not always intended to mainly serve as a sales channel. Less surprisingly, ensuring optimal product and brand presentation on the Internet is given the highest priority. After all, this is a core task of brand suppliers. The results also clearly show the efforts to build direct customer relationships with end customers and to derive insights from the interaction with customers.

## 6 Strategies for D2C Sales of Brand Manufacturers

This section describes and analyzes the strategic options that brand manufacturers in Switzerland have developed to position themselves in the online business. It is apparent that some strategies are designed to avoid conflicts with traditional distribution partners in order not to jeopardize the indirect distribution channel, which remains important. Three fundamentally different strategies can be observed: pure online direct sales strategy, multi-touch-point strategy, and platform strategy.

### *6.1 Pure Online Direct Sales Strategy*

A very consistent way to avoid conflicts of interest with distribution partners in the classic value chain is to position the company as a purely direct selling company. This naturally presupposes that sufficient access to consumers is possible without intermediaries and that the direct distribution channel is viable enough to achieve a critical sales volume. If direct sales are made purely online, it is also necessary to perform value-added functions that would otherwise be performed by the retailer, either by oneself or with the help of specialized service providers.

This is the path taken by the ifolor photo laboratory in Kreuzlingen, originally an indirect distribution company. Its aim is to ensure that its positioning as an online pure player is diluted as little as possible by additional channels. ifolor has mastered the digital transformation in photography and is now exploiting the potential arising from digitalization, e.g. mass production of individualized photo books and other photo products. The efficiency of digitally controlled production processes is so high that the company can produce in Switzerland—even for foreign countries. The restriction to an online direct sales channel also contributes to the high efficiency and the result of the business model. Sven Betzold from ifolor puts it like this:

> B2B customers are a completely different clientele than B2C customers. In marketing and sales, completely different instruments are required and this has an impact on the cost structure.

The company mySwissChocolate, also a provider of individualized mass products, follows similar principles and considerations. The manufacturer of chocolate bars and chocolate greeting messages started as an Internet start-up, which made it easier to enter the pure direct online business. However, the manufacturer of chocolate bars had to shift its business model to business customers in order to scale sufficiently in a small country like Switzerland.

The pure online direct sales strategy cannot only be applied to individualized products. The Swiss company Blacksocks, for example, sells standardized mass products under the brand of the same name, primarily socks and shirts, very successfully in more than one hundred countries. Blacksocks also started out as a pure online player.

## 6.2 Multi-touch-point Strategy

ifolor's transformation from purely indirect to purely direct sales is rather unusual and went hand in hand with the technological change from analogue to digital photography. For existing manufacturers with an established distribution structure via the retail trade, the focus is usually on being able to incorporate new customer touch points into their sales concept and to avoid conflicts with existing distribution partners wherever possible. A multi-touch-point strategy has emerged as the most important approach to solve this problem. It differs from a multi-channel strategy in that a touch point is not necessarily a sales channel. The primary goal is to be present in the customer journey in as many places as possible. Michael Lipburger from Jura Elektroapparate observes:

> When buying high-quality coffee machines, the pre-purchase phase and the purchase phase are often decoupled and take place on different channels.

At the heart of the direct customer contact of brand manufacturers via online channels is their own website, online product catalogue, or online shop. If a purchase opportunity exists, the question arises to what extent the manufacturers compete with their traditional trading partners by structuring their conditions. According to previous observations, only a few, typically very strong brands are entering into fierce competition with their trading partners. Examples of this can be found in the aviation industry in the sale of airline tickets; Apple is mentioned several times as another example.

In order to reduce conflicts with distribution partners in a multi-touch-point strategy, various measures can be considered. On the one hand, it can be observed that brands *divide their business activities* in order to avoid conflicts. Online direct sales are then only operated in a separate business segment. Two examples come from the publishing industry: In the case of specialist and textbooks, publishers are establishing new forms of distribution via databases and licenses, completely bypassing the book trade. In the case of e-books, they are reducing the book trade to a role of intermediary via agency contracts. This way, publishers can determine the sales price themselves and eliminate price competition at the retail level.

On the other hand, it is common practice to *provide different distribution channels for different product lines*. To do this, a product hierarchy and a channel hierarchy are defined and combined with each other depending on the positioning. Exclusive products, for example, are only offered through the company's own channels or through particularly loyal sales partners, where brand presentation and pricing policy correspond to the manufacturer's ideas. Other products are distributed more widely and in the traditional way, where differing forms of presentation and sales prices are accepted to a certain extent. Nespresso has been very successful in doing this. Patrick Th. Oken of Nespresso Suisse describes why the machines are sold through stationary retailers:

> Getting started with the Nespresso system begins with the purchase of a machine. For this, Nespresso uses the reach and the stimulation potential of the stationary trade.

A third measure can be observed in young companies. Right from the start, they establish *indirect sales and online direct sales side by side* and try to maintain a roughly uniform price level in both channels. In this constellation, the brand achieves reach through its stationary trading partners and direct customer contact via the online channel. An example of this is ON-Running, a running shoe brand that is currently in high demand.

The fourth measure, *cooperation between brand manufacturers and retailers*, is ideally designed to benefit both parties. For example, the brand manufacturer could support its dealers in presenting its brand appropriately on the dealers' online channels. The brand manufacturer could also refer to dealers' shops on its website—however, this is often not well resolved at the moment. The brand Neue Wiener Werkstätten (NWW) has implemented an exemplary cooperation with its dealers. The products are high-quality furniture, which are sold through specialist trade partners. As the online activities of the specialist trade partners were not always satisfactory from the brand's point of view, NWW now conducts the customer dialogue itself via all online touch points. For this purpose, an online shop for the furniture was set up. With the trading partners, extended distribution agreements are being made: They provide delivery, assembly, customer service, etc. for online orders in the NWW online shop and are credited the full trading margin in return. They also support and advise customers in the selection of furniture. In the end, both brand manufacturer and dealers are in continuing contact with the customers. The special thing about this example is that a weaker brand, in coordination with its existing distribution partners, implements a multi-touch-point strategy with online sales consistently and with little conflict.

## *6.3 Platform Strategy*

As already described in the introduction, self-determined distribution via digital platforms is also counted as online direct distribution here. Sales via such—partly global—platforms thus represent a further strategic option for direct sales to consumers. Examples of digital platforms with purchasing options are electronic marketplaces and, gradually, search engines, price comparison platforms, rating portals, and social networks.

Although consumers are looking for products not only on Google, but increasingly also on electronic marketplaces and other digital intermediary platforms, many manufacturers find it difficult to offer their brands on such platforms. When doing so, they have the chance to meet their customers' needs in a partly self-controlled way: They can determine part of the content and presentation themselves and give customers the security of receiving original products.

On the other hand, brands are usually not able to present their products on these platforms as well as they would like. They cannot determine, which other products are presented on the platform, and are exposed to price competition. If the branded product is offered on the same platform also by dealers and if these dealers are part of the traditional distribution chain of the brand manufacturer, other undesirable competitive situations arise. These are particularly problematic where there are relevant price differences between the manufacturer's and the dealer's price.

Due to the pros and cons, the brand suppliers represented in the study panel are divided in terms of their own involvement in digital platforms. Many brands would prefer to completely prevent the supply of their products for example on marketplaces, which is hardly possible for antitrust and other reasons. Many brands, however, have abandoned their initial reluctance towards online platforms. One participant in the study states that it is better to sell the goods via a third party on the Internet than not to sell them at all.

## 7 Conclusion

This chapter analyzes the online direct sales activities of brand manufacturers. Despite online channels and digitalization, most of them still use the traditional distribution channels via wholesale and retail. The aim is therefore to determine how brand manufacturers can structure their online distribution activities without getting into too much competition with their traditional distribution partners.

To answer this question, an explorative and interpretive research approach was chosen. For this purpose, 36 e-commerce managers from potentially market-shaping companies were interviewed. Among them, there were twelve brand manufacturers.

The results initially show that the Swiss retail trade is characterized by a far-reaching transformation process, in which, among other things, there is a shift from offline to online channels. The importance of traditional retailers for the distribution of branded products is thus diminishing. As a result, brand manufacturers have to find alternative distribution channels to compensate for the decline in sales in traditional retail, to present the brand optimally in the online world and also to provide consumers with direct contact.

This is then also expressed by the motivation that has been driving brand manufacturers to expand their online activities since 2012. It is not so much the achievement of additional revenues, but rather the establishment of direct communication relationships with consumers and the use of the resulting data.

Three basic strategies have emerged as possible strategies for direct to consumer e-commerce of brand manufacturers: the pure online direct sales strategy, the multi-touch-point strategy, and the platform strategy. The first two are designed or can be designed in such a way that conflicts with other distribution partners are largely avoided. In the case of the platform strategy, this depends on whether brand manufacturers and retailers are active with the same products on the same platform or not.

The pure online and the multi-touch-point strategy are mutually exclusive, while both can be combined with a platform strategy. Direct sales via a digital platform can therefore be used by both pure online brand manufacturers and manufacturers who maintain several contact points.

One limitation is that the analysis is based primarily on the statements of individuals who all hold similar positions in e-commerce. Thus, the perspective of the respondents is quite similar. Since the panel participants are potentially market-shaping e-commerce companies, the database from the interviews is not representative. Finally, only twelve companies belong to the core group of online direct-selling brand manufacturers.

In a further research step, the sample should therefore be increased in order to obtain a better overview of the behavior of the large mass of brand manufacturers. In addition, it would be interesting to analyze and evaluate the identified online direct sales strategies in terms of their success. Furthermore, other forms of cooperation could be developed in which brand manufacturers and retailers support each other so that both can benefit.

## References

1. Benjamin, R., Wigand, R.: Electronic markets and virtual value chains on the information superhighway. Sloan Manage. Rev. **36**, 62–72 (1995)
2. Li, G., Huang, F., Cheng, T.C.E., Ji, P.: Competition between manufacturer's online customization channel and conventional retailer. IEEE Trans. Eng. Manag. **62**, 150–157 (2015). https://doi.org/10.1109/TEM.2015.2406913
3. Sarkar, M.B., Butler, B., Steinfield, C.: Intermediaries and cybermediaries. J. Comput.-Mediat. Commun. 1 (1995). https://doi.org/10.1111/j.1083-6101.1995.tb00167.x
4. Wölfle, R., Leimstoll, U.: E-Commerce-Report Schweiz 2016—Digitalisierung im Vertrieb an Konsumenten. Eine qualitative Studie aus Sicht der Anbieter. Institute for Information Systems, School of Business, University of Applied Sciences and Arts Northwestern Switzerland, Basel (2016)
5. Chung, C.: An exploratory model of the determinants of disintermediation: the facilitating role of the manufacturer's virtual presence, San Francisco (2005)
6. Li, X., Troutt, M.D., Brandyberry, A., Wang, T.: Decision factors for the adoption and continued use of online direct sales channels among SMEs. J. Assoc. Inf. Syst. **12**, 1–31 (2011)
7. Peterson, R.A., Wotruba, T.R.: What is direct selling?—definition, perspectives and research agenda. J. Pers. Sell. Sales Manag. **16**, 1–16 (1996)
8. Sa Vinhas, A., Anderson, E.: How potential conflict drives channel structure: concurrent (direct and indirect) channels. J. Mark. Res. **42**, 507–515 (2005)
9. Käuferle, M., Reinartz, W.: Distributing through multiple channels in industrial wholesaling: how many and how much? J. Acad. Mark. Sci. **43**, 746–767 (2015). https://doi.org/10.1007/s11747-014-0406-5
10. Bower, P.: Consumer goods, direct-to-consumer, and S&OP—a doomed marriage? J. Bus. Forecast. 16–29 (2016)
11. Saunders, M., Lewis, P., Thornhill, A.: Research methods for business students. Financial Times, Prentice Hall, Harlow (England) (2012)
12. Gläser, J., Laudel, G.: Experteninterviews und qualitative Inhaltsanalyse als Instrumente rekonstruierender Untersuchungen. VS Verlag für Sozialwissenschaften/Springer, Wiesbaden (2010)
13. Wölfle, R., Leimstoll, U.: E-Commerce-Report Schweiz 2017—Digitalisierung im Vertrieb an Konsumenten. Eine qualitative Studie aus Sicht der Anbieter. Institute for Information Systems, School of Business, University of Applied Sciences and Arts Northwestern Switzerland, Basel (2017)
14. Wölfle, R., Leimstoll, U.: E-Commerce-Report Schweiz 2018—Digitalisierung im Vertrieb an Konsumenten. Eine qualitative Studie aus Sicht der Anbieter. Institute for Information Systems, School of Business, University of Applied Sciences and Arts Northwestern Switzerland, Basel (2018)
15. Wölfle, R., Leimstoll, U.: E-Commerce-Report Schweiz 2019—Digitalisierung im Vertrieb an Konsumenten. Eine qualitative Studie aus Sicht der Anbieter. Institute for Information Systems, School of Business, University of Applied Sciences and Arts Northwestern Switzerland, Basel (2019)
16. Höld, R.: Zur Transkription von Audiodaten. In: Buber, R. und Holzmüller, H.H. (hrsg.) Qualitative Marktforschung: Konzepte, Methoden, Analysen. S. 655–668. Gabler, Wiesbaden (2009)
17. Mayring, P.: Qualitative Inhaltsanalyse: Grundlagen und Techniken. Beltz, Weinheim und Basel (2010)
18. Albers, S., Peters, K.: Die Wertschöpfungskette des Handels im Zeitalter des Electronic Commerce. Mark. ZFP. **19**, 69–80 (1997)
19. Peters, K., Albers, S., Schäfers, B.: Die Wertschöpfungskette des Handels im Zeitalter des Electronic Commerce – Was eingetreten ist und was dem Handel noch bevorsteht. ZBW - Deutsche Zentralbibliothek für Wirtschaftswissenschaften, Leibniz-Informationszentrum Wirtschaft, Kiel und Hamburg, Kiel (2008)

20. Leimstoll, U., Wölfle, R.: Auswirkungen des E-Commerce auf die Wertschöpfungsstrukturen im B2C-Einzelhandel: Erkenntnisse aus der Schweiz. In: Tagungsband Multikonferenz Wirtschaftsinformatik 2014 (MKWI 2014). S. 520–532. Paderborn University, Paderborn (2014)
21. Haushalte und Bevölkerung: E-Commerce und E-Banking, E-Commerce-Ausgaben der Privathaushalte, Entwicklung. Federal Statistical Office, Bern (2019)

# The Digital Marketing Toolkit: A Literature Review for the Identification of Digital Marketing Channels and Platforms

**Marc K. Peter and Martina Dalla Vecchia**

**Abstract** Digital transformation, driven by technological advances and changing customer requirements, is stimulating the use of digital marketing. 11% of Swiss organizations regard digital marketing as a key investment area as part of their overall digital transformation strategy, with over one third of Swiss organizations currently investing in new sales and marketing tools. Unfortunately, there are implementation gaps between Swiss small and medium-sized enterprises (SME) and large enterprises (LE). In short, SME are lagging behind LE and generally do not use digital marketing tools, channels, and platforms. Barriers that prevent SME from adopting higher digital marketing tools are cultural change, limited resources/high costs, technology, and expertise. The objective of this study is to close the knowledge gap and provide SME with an overview of the most important digital marketing tools based on a literature review in order to leverage the opportunity of digital technology in the marketing discipline and reduce the distance to LE. The literature review identified nineteen relevant articles. These articles include 162 citations of tools, channels, platforms, and methods, which can be used by SME to close the knowledge gap and thus take advantage of a new, digital marketing portfolio. The twenty-four unique digital marketing tools are presented based on a comparative analysis, with the eleven most often cited tools being defined and described. Potential for further research was identified.

**Keywords** Digital marketing · Digital marketing strategy · Digital marketing toolkit · Digital communication · Digital marketing tools · Internet marketing · Digital transformation · Marketing communication

M. K. Peter (✉) · M. Dalla Vecchia
School of Business, Institute for Information Systems, FHNW University of Applied Sciences and Arts Northwestern Switzerland, Peter Merian-Strasse 86, 4052 Basel, Switzerland
e-mail: marc.peter@fhnw.ch

M. Dalla Vecchia
e-mail: martina.dallavecchia@fhnw.ch

 
R. Dornberger (ed.), *New Trends in Business Information Systems and Technology*, Studies in Systems, Decision and Control 294,
https://doi.org/10.1007/978-3-030-48332-6_17

## 1 Introduction

The Internet has fostered new strategy and business management frameworks that have also changed marketing practices. Here, digital marketing has transformed the traditional marketing mix and the existing integrated marketing communication model [8, 43]. In the literature, this is sometimes referred to as the fourth marketing revolution [3].

Digital marketing drives customer engagement and involvement in both the setup of the marketing mix and the application of marketing communications. In addition, in order to remain competitive, businesses of all sizes—including SME—are forced to take advantage of digital marketing [8, 34]. Hence, the adoption of new digital marketing technology is essential for the survival of SME [1]. Unfortunately, marketing strategies and the adoption of digital marketing in SME are often not formalized due to limited resources, lack of expertise, and other factors specific to SME [9, 12, 30, 42, 44].

In this chapter, the goal is to understand the broader context of digital marketing in business strategy (and especially with regard to digital transformation) with a focus on Switzerland, and to identify the most important digital marketing technology (i.e. channels and platforms) in order to support SME in the development and implementation of digital marketing.

An introduction to SME and digital marketing as well as data from a recent Swiss study will provide contextual information in Sects. 2 and 3, followed by a literature review to identify digital marketing channels and platforms in Sect. 4. To support business practice, the synthesis of the literature review is presented in the form of the digital marketing toolkit in Sect. 5. Finally, Sect. 6 provides the conclusion with suggestions for future research.

## 2 SME and Digital Marketing

SME cannot continue to sell their products and services in the traditional way and need to adapt their digital opportunities and tools [45]. Here, the Internet provides a new, very different environment for international marketing and new paradigms, which need to be considered. The Internet has become an important marketplace: New frameworks for business and marketing strategies have emerged and the number of Internet users continues to grow globally [3, 8, 28].

The potential for online or digital marketing has long been recognized: Poon and Jevons [38] stated that "many small businesses [...] are compelled to compete beyond their comfort zone (local markets) due to the globalization and internationalization of the marketplace. At the same time, the commercialization of the Internet has created unprecedented opportunities for small businesses [...]". The goal of this opportunity is in fact digital marketing: It tries to achieve marketing objectives through the applications and tools of the Internet [11]. Digital marketing has especially grown

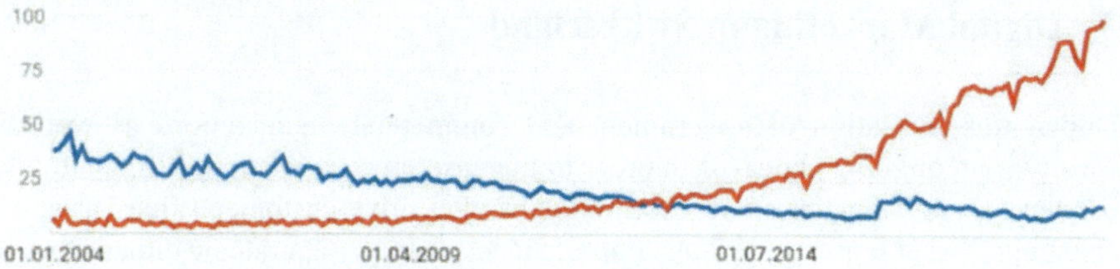


**Fig. 1** Global utilization of the terms internet marketing (blue, high starting point) and digital marketing (red, low starting point) since 2004 [20]

with a large number of websites in the past two decades [28, 46]. While in the past, internet marketing (e.g. [21, 31]) was the predominant term, digital marketing (e.g. [5, 24, 44]) has been the preferred term since around 2013 (Fig. 1; [20]).

As mentioned earlier, marketing strategies in SME are often less formal than in LE. This is due to the manner in which the owner/manager runs their business, limited resources, lack of technology or the use of outdated technology, and also by the lack of expertise [9, 12, 30, 42, 44].

Since SME have identified digital marketing as an important business function and are investing more in new systems and channels, the impact of digital marketing is seen as essential and is now an important element of a modern organization [45, 46]. However, SME still seem to ignore the many benefits of digital marketing and the utilization of digital marketing still has potential for improvement [44, 46]. Digital marketing can help SME to market and promote their products and services. It also provides cost efficiency/savings as well as faster communication and broader market coverage, and creates customer awareness. With the advancement of technology, digital channels are standardized, interactive, ubiquitous, and cheap. These benefits motivate SME in the adoption of new digital channels and most importantly, help reduce the distance to LE in the introduction of digital technology [22, 30, 42, 44].

On the other hand, organizations need to cope with multiple challenges, including the organization's ability to generate customer insights, data security and privacy, to manage brand health and reputation in the age of social media, and to assess the effectiveness of their digital marketing activities [22, 28]. Another challenge for organizations is to align their marketing channels, especially since they need to be tailored to the needs of customers. Here, the selection of channels is an important strategic factor, as is channel orchestration, i.e. the integrated marketing or omni-channel approach [29]. As the Swiss study has shown, SME lack the knowledge and expertise how to use these tools. Therefore, a marketing toolkit would support organizations, including SME, not only to determine their strategic selection of channel, but also to consider an alignment among these tools, channels and platforms.

## 3 Digital Marketing in Switzerland

Digital transformation of government and commercial organizations as well as society is an ongoing endeavor in order to increase process efficiency, create transparency, take advantage of new technologies, and drive customer value based on changing market requirements. An important action field (or strategic dimension) of digital transformation, as identified in a large Swiss study of 2,590 participants, is digital marketing [35, 36].

According to the Swiss study, digital marketing is a focus topic of digital transformation in about 11% of Swiss organizations [35, 36]: There is a clear trend away from traditional marketing activities and tools towards digital marketing. Marketing strategies are directed towards new technological opportunities offered by digitalization, such as responsive marketing communication (with a direct feedback-loop from customers back to the organization) or the multi-channel approach, where digital and analog channels are served in a coordinated manner (including distribution, sales, and customer service). This allows several target groups to be addressed in parallel, which means that the market can be served more effectively. In the field of digital marketing, most organizations focus on the new requirements of customers (e.g. e-mail support), online platforms, website visibility, and social media.

However, barriers affect digital transformation initiatives in Swiss firms. The lack of knowledge and expertise is the main barrier in 48% of the participating firms, which also affects the success of digital marketing strategies and activities.

Digital marketing is the logical consequence of the general digitization of business processes, of which the marketing and sales division is an important part. The aim is to be able to react flexibly to market changes and customer needs. There is an awareness that mobile devices such as tablets and smartphones have changed the consumption and communication behavior of customers, and that technical developments will continue to influence the purchasing behavior in the future. In addition to e-commerce, some organizations rely on search engine marketing (SEM) instead of advertisements, and digital newsletters (e-mail marketing) instead of direct mail campaigns. Many of the organizations are aware that a large part of commercial transactions is conducted online and that the supply and sale of goods and services is shifting to the digital marketplace. Organizations know that the traditional sales behavior (where customers go to the store, obtain advice and buy the product locally) has changed; towards a purchasing behavior in which customers like to inform themselves online (and perhaps by visiting a local store to see and touch the product) and then buy the products (and services) online. Therefore, there is a shift of many components of the physically accessible world to the virtual one on the Internet. For this reason, many organizations also see the need to open up new digital sales and procurement channels and to operate digital platforms, which are often more comprehensive than traditional online shops. The advantage of this lies in an increase of sales efficiency, in the reduction of the time it takes a product to reach customers, and in the additional customer data that can be obtained and further evaluated in this process.

### *3.1 Important Concepts of Digital Marketing*

The study [35–37] identified core concepts of digital marketing in Switzerland. Addressing and connecting customers via digital channels seems to be important for many organizations in the age of digital marketing. Digital marketing should not only open up new sales channels in order to address new markets; it should also increasingly address customer needs. As a result, some organizations develop mobile apps for their customers, offer online payment options, provide as much information as possible about their products and services online, and tailor their websites to their customers' needs. These customer needs and preferences can be further optimized by analyzing search and purchase behavior.

Social media is seen as an integral part of digital marketing and serves to position an organization's products and services in the online community market and through social media platforms. Social media is used by organizations for customer loyalty, customer communication, social selling, and market expansion. In any case, the advantage is seen in the relatively simple networking paradigm between different stakeholders. In many cases, social media is already part of the communication strategy and serves to increase the online presence. Organizations state that they benefit from the fusion of private and business life through social media and use it as a platform for their market offers. By developing valuable and creative content, which is then shared with a social media community according to the specific use cases, organizations try to turn customers into fans and employees into proud hosts/ambassadors. The goal is to serve customers on the digital channels that are available to them. In some organizations, social media, such as Yammer, is also used internally to inform employees about new products and services, thus facilitating internal communication and interaction. The surveyed organizations in the Swiss study also use social media, including Facebook, Instagram, LinkedIn, Twitter, and XING, as a means of acquiring new customers and maintaining contact with existing customer groups.

Finally, the participating organizations in the Swiss study are observing an increasing need for strong online visibility. A good web presence, i.e. website, is therefore essential to remain competitive in the market. A successful website helps organizations to achieve stronger visibility in a market that has shifted significantly to the online sector. In this sense, the digital presence secures their existence in the market. At the same time, organizations are aware that not the whole population is digitally savvy, and that a multi-channel approach is therefore needed to reach and communicate with all groups. Analogue sales are combined with digital marketing tools such as online advertising and newsletters. The awareness of the shift of the marketing function to the digital sector is present among the organizations surveyed. The focus is on expanding their presence in the web through extensive and up-to-date websites as well as using social media and marketing technology to communicate with customers and expand their own market. There is also a strong awareness that the online offers must be strongly geared to the changing needs of customers, which include mobile applications, personalized information and services, and the aforementioned multi-channel approach.

### 3.2 Digital Marketing in Swiss Organizations

In terms of digital marketing tools, channels and platforms (Table 1), a number of gaps exist between small and medium-sized enterprises (SME) with 1–249 employees (including micro enterprises with 1–9 employees, small enterprises with 10–49 employees, and medium-sized enterprises with 50–249 employees) and large enterprises (LE) with over 250 employees [35, 36].

The marketing approach in the digital environment with the highest relevance is search engine marketing (SEM): This tool/approach is highly relevant in 57.5% of Swiss organizations, followed by e-mail and social media marketing (both with a high relevance in 50% of organizations). The major difference between SME and LE lies in display advertising. Many SME do not seem to use this form of marketing communication (only 17.7% of SME), while LE (43.9%) actively use it. This is probably due to the work involved, as this form of digital marketing requires a graphic designer and, in many cases, an advertising agency. With regard to SEM, especially search engine advertising (SEA), the work can often be carried out by the SME themselves or by an agency with manageable effort. It also seems that SME have not yet discovered the advantages of content marketing, as this is relevant for only 40.2% of SME (and 61.2% of LE). Especially in connection with social media marketing, which is relatively high among SME at 47.6%, content marketing could contribute to higher chances of success with this particularly interesting digital marketing tool.

**Table 1** Digital marketing tools, channels, and platforms usage in Swiss organizations (n = 278; Scale from 1 (low relevance to 5 (high relevance), with high relevance defined as ≥3) from Peter et al. [35, 36]

| **Digital tools, channels and platforms with high relevance** | Company size | | | | | Total |
|---|---|---|---|---|---|---|
| | **SME (%)** | Micro enterprises (%) | Small enterprises (%) | Medium sized enterprises (%) | **Large enterprises (%)** | |
| Search engine marketing (SEM) | **58.20** | 45.60 | 65.30 | 66.70 | **55.40** | 57.50 |
| E-Mail marketing | **51.60** | 45.10 | 49.30 | 65.90 | **45.60** | 50.00 |
| Social media marketing | **47.60** | 32.90 | 57.90 | 53.50 | **56.70** | 50.00 |
| Content marketing | **40.20** | 30.40 | 37.10 | 62.50 | **61.20** | 45.90 |
| Display advertising | **17.70** | 10.60 | 18.60 | 28.20 | **43.90** | 24.90 |

# 4 Literature Review

The objective of this chapter is to provide SME with an overview of the most important digital marketing tools (for later reference, tools also include go-to-market channels and digital platforms) in order to close the knowledge gap, leverage the opportunities of digital technology in this field, and reduce the capability gap to LE. An initial literature review revealed that only a small number of researchers had the objective to identify and describe such digital marketing tools. Overall, there are only a small number of publications on SME digital marketing that cover more than just one specialized topic, e.g. social media, content marketing, or marketing automation [9, 12].

## *4.1 Research Design*

A detailed literature review was conducted with the goal to identify and analyze the current body of knowledge. Based on pre-defined keywords (digital marketing platforms, online marketing platforms, SME digital marketing, SME digital marketing channels, SME online marketing, social media platforms, SME online marketing, online marketing tools, internet marketing, digital marketing, usage digital marketing, usage digital platforms), four databases were used: EBSCO [15], ProQuest [39], Google Scholar [19], and Google [18]. This resulted in a collection of almost 140 publications from these four sources.

All academic publications from the first three sources were analyzed to evaluate their academic standing. One of the criteria was that the literature was based on survey results. Other publications, such as whitepapers (from the fourth source) were included if the findings were based on survey results and/or if the source was a known source (e.g. larger consulting firms). While many articles are based on a literature review or general business practice, others have an empirical basis (e.g. [7, 13, 24, 26, 27, 30]). In order to focus on articles that provided insights into tool sets or frameworks, only articles that took a multi-dimensional approach with at least two digital marketing tools or methods were included in the analysis, as the focus was on publications that included comparisons and/or linked tools. Articles with a focus on one single topic (e.g. social media) were excluded from the literature review.

Following a review of the remaining literature of nineteen sources, a comparative analysis was conducted that identified all digital marketing tools described or analyzed in the literature. Following the completion of the comparison, all tools were grouped and included in the digital marketing toolkit. Those which appeared in more than 25% of all articles (i.e. have five or more citations), are briefly defined and described in Sect. 5 to provide contextual information and definitions. The threshold was set at the 25% value because a significant differentiation could be observed at this point.

## *4.2 Digital Marketing in Literature*

The literature review identified nineteen articles, which described and/or analyzed two or more digital marketing tools (Table 2). These articles include a total of 162 citations of tools and methods that can potentially be used and leveraged by organizations to benefit from the new, digital marketing portfolio.

On average, 8.5 tools were mentioned in each article. In terms of number of tools described, Demishkevich [14] and Jensen [26] list thirteen tools each, Chaffey [7] lists twelve tools; & Bhayani and Vachhani [4], Borges Tiago and Verissimo [5] & Trivedi [45] eleven tools each. Only for the past five to six years (since 2014), the industry seems to have considered the utilization of multiple digital marketing tools in organizations and thus has tried to articulate its benefits, challenges, and opportunities.

It is worth pointing out that some of the fundamentals of marketing, e.g. analytics and monitoring, brand management, customer relationship management (CRM) and marketing strategy were not explicitly mentioned as tools or functions of digital marketing. In some publications, they were discussed in an introduction, but not individually surveyed or defined as a component or success factor of digital marketing (except in [2, 4, 14, 25, 45]).

The literature with a large number of digital marketing tools are a consulting white paper with empirical data [7], a dissertation [14], a literature review [4], an empirical study [5], and two literature reviews in combination with an empirical study [26, 45].

Section 5 provides an illustration with the twenty-four referenced digital marketing tools and describes the most frequently cited tools. The following tools with less than five citations are not discussed in detail, i.e.

- Analytics and monitoring: The tracking, monitoring, and analysis of campaign metrics, customers, and website data;
- Brand management: Branding in the digital age;
- CRM and big data: The utilization of customer insights and smart data to drive digital strategy;
- Directories and listings: To further promote a business on specialized websites;
- E-commerce and web shops: Most likely because they are seen as a driver of digital businesses, rather than digital marketing;
- Promotions and e-coupons: Price driven promotions/digital campaigns;
- Marketing automation: A more recent topic; the technology which combines process engineering and digital marketing best practice;
- Marketing strategy: The analysis of the market and development of a strategic plan;
- New technology: This includes, for instance, wearables as well as augmented and virtual reality;
- Video marketing: This might be regarded as a subset or format of content marketing; and
- Wikis (knowledge databases) and e-learning programs to further educate the market.

**Table 2** Digital marketing tools, channels, and platforms in the literature (own illustration)

| Digital Marketing Tools, Channels and Platforms | Total | [2] | [4] | [5] | [6] | [7] | [10] | [13] | [14] | [16] | [24] | [25] | [26] | [27] | [28] | [30] | [33] | [44] | [45] | [46] |
|---|---|---|---|---|---|---|---|---|---|---|---|---|---|---|---|---|---|---|---|---|
| Affiliate Marketing | 7 | | x | | x | x | x | | x | | | | x | x | | | | | | |
| Analytics & Monitoring | 4 | x | | | | | | | x | | | x | | | | | | | x | |
| Brand Management | 2 | | x | | | | | | | | | | | | | | | | x | |
| Content Marketing | 10 | x | x | x | | x | x | x | x | | | x | | | | | x | | x | |
| CRM / Big Data | 4 | x | | | | x | | | x | | | | | | | | | | x | |
| Customer Centricity/Experience | 4 | | | x | | x | | | | | | x | | | | | | | x | |
| Directories / Listings | 3 | | | | | | | x | | | | | | | | | | x | | x |
| Display / Online Advertising | 15 | | x | x | x | x | x | x | x | x | x | | x | | x | x | | x | x | x |
| E-Commerce / Webshops | 2 | | | | | | | | | | | | | | x | | | | x | |
| E-Coupons / Promotions | 3 | | | | | | | x | x | | | | x | | | | | | | |
| E-Mail Marketing / Newsletters | 16 | x | x | x | x | | x | x | x | x | x | x | x | x | x | | | x | x | x |
| Games | 1 | | | | | | | | | | | | x | | | | | | | |
| Marketing Automation | 2 | x | | | | x | | | | | | | | | | | | | | |
| Marketing Strategy | 1 | | | | | | | | | | | | | | | | | | x | |
| Mobile Marketing / Apps | 10 | | | x | | x | | x | x | x | x | x | x | | x | | | | x | |
| New Technology | 4 | x | | x | | x | | | | | | | | | x | | | | | |
| Online Public Relations | 7 | | x | | | x | x | | x | | | | x | x | | | | | | x |
| Search Engine Advertising | 12 | | x | | | x | x | x | x | | x | | x | x | x | | x | x | | x |
| Search Engine Optimization | 14 | x | x | x | x | x | x | x | x | | x | | x | x | x | | x | x | | |
| Social Media / Communities | 16 | x | x | x | x | x | x | x | | x | x | x | x | x | x | | x | x | | x |
| Video Marketing | 3 | | | x | | | | | x | | | x | | | | | | | | |
| Viral Marketing | 5 | | x | x | x | | | | | | | | x | x | | | | | | |
| Website | 15 | x | x | x | | | | x | x | x | x | | x | x | x | x | x | x | x | x |
| Wikis / E-Learning | 2 | | | | | | | | | | | x | x | | | | | | | |
| **Total** | **162** | 9 | 11 | 11 | 6 | 12 | 8 | 10 | 13 | 5 | 7 | 8 | 13 | 8 | 9 | 2 | 5 | 7 | 11 | 7 |

## 5 The Digital Marketing Toolkit

The complete digital marketing toolkit available to SME (Fig. 2) to reduce the gap to LE and to develop and implement a digital marketing strategy includes twenty-four tools, of which eleven were mentioned in more than a quarter of all articles. These are affiliate marketing, content marketing, display/online advertising, e-mail marketing, mobile marketing, online public relations, search engine advertising (SEA), search engine optimization (SEO), social media and its communities, viral marketing, and the company website.

The digital marketing toolkit is divided into seven categories, based on the context used in the original literature, namely (1) marketing strategy, (2) overall customer experience, and (3) brand management. These three categories have a strategic focus and have a direct impact on the following categories four to seven.

Channels, platforms, and formats (4) include a large collection of tools available to marketers. Here, content marketing (5) plays a crucial, all-encompassing role because it provides valuable and entertaining content for the organization's digital channels and platforms. Analytics and monitoring (6) provide the feedback-loop and learning capabilities to the organization. Finally, all digital marketing capabilities are supported by marketing technology (7), including CRM and marketing automation.

The most frequently cited tools include the following propositions:

**Affiliate Marketing**

The goal of affiliate marketing is to increase the number of channels and the organization's market coverage through partnerships with other organizations/websites. The company will provide a commission for the promotion and sale of products and services through these third-party partner websites, or for leads (potential new customers) generated by third-party partners [6, 27]. In this category, Bhayani and

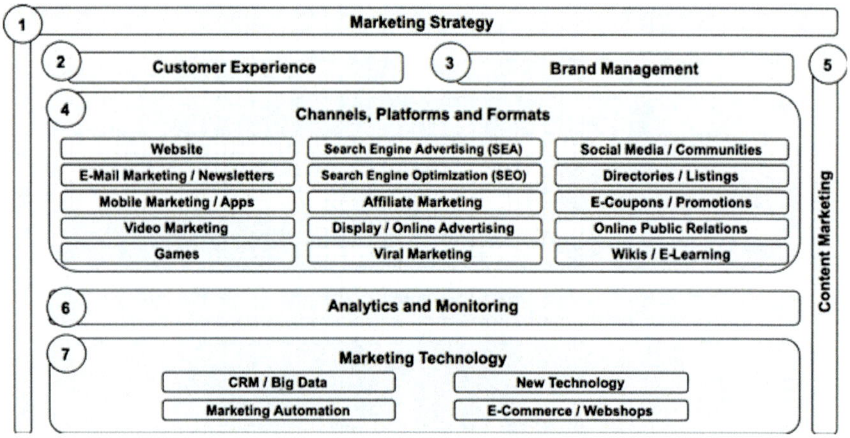


**Fig. 2** The digital marketing toolkit. The eleven most often cited tools are highlighted in checks (own illustration)

Vachhani [4] also include link building. In addition, Demishkevich [14] refers to lead aggregators, Jensen [26] to sponsorships, and Leeflang et al. [28] to third-party online stores.

**Content Marketing**

Content marketing, including blogs, whitepapers etc., is seen as an effective strategy for branding, developing customer trust and loyalty, and to generate leads through targeted, valuable content [5, 14, 23, 33]. Here, Bhayani and Vachhani [4] mention blogs as an important component of content marketing, Demishkevich [14] highlights the need for video marketing; and Järvinen et al. [25] and Jensen [26] refer to online events and webinars as a potential content format.

**Display/Online Advertising**

Display and online advertising are two of the oldest and most common channels/formats of digital marketing. They utilize virtual space to promote marketing messages and to drive user activity; they create a relatively low-cost, agile, and interactive channel between advertisers and consumers [6, 14, 24]. New forms of online advertising include video marketing and games [26, 30, 45, 46].

**E-Mail Marketing**

E-mail marketing is also one of the oldest, yet most effective marketing channels and methods, especially due to the low execution cost and the relatively high response rate. However, in recent years, the response rate has dropped because consumers tend to ignore e-mail marketing campaigns [6, 24, 27]. In addition, aggressive e-mail tactics (i.e. spam) could affect the organization's reputation [17].

**Mobile Marketing**

Mobile marketing includes the whole spectrum of digital marketing tools available to organizations in order to target and engage mobile consumers (with a focus on smartphone users). This category covers mobile apps (applications): They can be used as a service tool, but also to interact with the market. Some researchers also highlighted the ongoing importance of SMS (short message service) and MMS (multimedia messaging service). The strength of mobile marketing is reflected in the direct, personalized relationship it can build with the market and its customers [5, 14, 24–26, 45].

**Online Public Relations**

The aim of online PR is to engage organizational stakeholders through digital tools. Online PR takes advantage of content marketing and social media (referred to as social media relations); some dedicated marketing automation solutions are also available for online PR processes. Content formats include blogs, media statements, dedicated landing pages, and newsletters [4, 7, 10, 14, 26, 27, 46].

**Search Engine Advertising**

In most publications, SEA is part of SEM [4, 27], which is used as an umbrella term that also includes search engine optimization (SEO). Others (e.g. [13]) apply the term SEM as a synonym for SEA. SEA is considered a necessary tool to build brand awareness and drive traffic to the website. It includes mostly text-based online ads, which are presented to a user based on a match between the keywords associated with an ad (and other criteria, in most cases) and the search word/s from a user entered on a search engine. Because of the many targeting options, SEA is an attractive digital tool [4, 24]. Instead of SEM, some authors use the terms paid search [28] or pay per click [14, 33].

**Search Engine Optimization**

The goal of SEO is to increase the visibility (in most cases this is measured by the page rank and the appearance of the organization's website on the first couple of pages of a search engine) of a website and drive web traffic to that specific website. It includes strategies and techniques (some technical, some content related) which require expertise and ongoing optimization (hence the term). It is a cost effective option for SME because it requires mostly human power, as opposed to a monetary budget as is the case with SEA [6, 24, 27]. Some authors use the terms natural search [28] or organic search [32, 40] as a synonym for SEO.

**Social Media/Communities**

Social media is a cost-effective digital marketing tool. Its benefits are reflected in brand exposure, increased targeted traffic, and search engine optimization (SEO), word-of-mouth and lead generation. In addition, social media analytics provide information about customers and competition. Social media includes strategies and tactics (for instance, content marketing and online PR) to create and share information about an organization, its products and services. Various stakeholders can create and share content, including individual users, the community, and the organization. Marketing technology is available to facilitate the process of sharing, listening, and responding to social media posts [6, 12, 24, 27, 41, 46]. Some authors, e.g. Jensen [26] and Kilmartin [27], also mention community building in the context of social media.

**Viral Marketing**

Viral marketing (also referred to as viral advertising) is a technique that takes advantage of word-of-mouth marketing and uses the user's personal social media network (or e-mail list) to re-distribute an organization's brand, product or service messages, with the ultimate goal of creating a hype or buzz. Content formats include text, images and videos [4–6, 27]. Because of its narrative capabilities, video is a recommended format by Jensen [26].

**Website**

The website itself is an important digital tool to promote the organization and attract customers. It enables organizations to provide information, interact with customers, and generate sales. It also provides options for brand building, customer service and communication efficiency. Hyperlinking allows that the collection of web pages and multimedia content can be retrieved and consumed by users. This includes dedicated micro-sites and landing pages, for instance for specific campaigns [4, 26]. Here, onsite marketing refers to the discipline of optimizing an organization's website to drive activity and achieve digital marketing goals [13, 24, 27, 30, 45]. According to Taiminen and Karjaluoto [44], it is the most actively used digital marketing tool by SME.

## 6 Conclusion

Digital marketing is an important strategic dimension of digital transformation. Unfortunately, the potential of digital marketing has not yet been leveraged by SME to the extent that LE have implemented it. Naturally, this provides meaningful opportunities for SME. Digital marketing is distinctively different to traditional marketing as it is based on new technology, responsive and measurable marketing communication, and a multi-channel approach, which allows the data-driven orchestration of digital tools. As is evident from the literature, traditional marketing topics such as analysis and monitoring, brand management, and marketing strategy appear as well, but to a lesser extent than the purely digital tools.

The study identified 162 citations of digital marketing tools from nineteen sources. The trend for a digital marketing toolset is supported by an average of 8.5 cited tools per source. The developed digital marketing toolkit includes twenty-four tools, with eleven dominating tools: Affiliate marketing, content marketing, display/online advertising, e-mail marketing, mobile marketing, online public relations, search engine advertising (SEA), search engine optimization (SEO), social media, viral marketing, and the company website are the most cited, analyzed, and discussed digital marketing tools.

Unfortunately, neither the existing literature nor this literature review succeeded in providing a sophisticated decision-making tool, which would allow SME to plan, select, and implement the most appropriate digital marketing tools for their individual market situation. This is considered a major research gap that could arouse interest in further enquiries. In addition, it could be beneficial to analyze the actual application of these tools in the Swiss (and other) markets to validate the toolkit. Finally, the low number of citations around traditional marketing activities (e.g. analysis, brand management, and strategy) could highlight the need for a combination of traditional and digital disciplines, which is another potential future research field.

As a first practical step for SME, the digital marketing toolkit will close the knowledge/expertise gap and provide a basis and guidance to take advantage of proven digital marketing tools, channels, and platforms in the age of market and organizational transformations.

## References


1. Alford, P., Page, S.J.: Marketing technology for adoption by small business. Service Ind. J. **35**(11–12), 655–669 (2015)
2. Ballou, E.: How small businesses invest in digital marketing in 2018. https://clutch.co/agencies/digital/resources/small-business-digital-marketing-survey-2018
3. Balmer, J.M.T., Yen, D.A.: The internet of total corporate communications, quaternary corporate communications and the corporate marketing Internet revolution. J. Market. Manage. **33**(1/2), 131–144 (2017)
4. Bhayani, S., Vachhani, N.V.: Internet marketing vs traditional marketing: a comparative analysis. FIIB Bus. Rev. **3**(3), 53–63 (2014)
5. Borges Tiago, M.T., Verissimo, J.M.: Digital marketing and social media: why bother? Bus. Horiz. **57**, 703–708 (2014)
6. Bostanshirin, S.: Online marketing: challenges and opportunities. In: Proceedings of SOCIOINT14—International Conference on Social Sciences and Humanities, 8–10 September, pp. 783–792. Istanbul, Turkey (2014)
7. Chaffey, D.: 8 Business-critical digital marketing trends for 2019. Smart insights. https://www.smartinsights.com/managing-digital-marketing/marketing-innovation/business-critical-digital-marketing-trends/ (2019)
8. Chauhan, P., Agrawal, M., Chauhan, G.: Understanding the critical aspect of digital marketing for meaningful strategic marketing perspective. J. Market. Commun. **11**(3), 21–29 (2015)
9. Clarke, C.G.: Online marketing in a small-medium enterprise (SME). MBA research project. Saint Mary's University, Halifax, Nova Scotia, Canada (2013)
10. Colwyn, S.: UK SME usage and attitudes to online marketing. http://www.smartinsights.com/managing-digital-marketing/resourcing-digital-marketing/ali-cort/ (2014)
11. Chaffey, D., Ellis-Chadwick, F., Johnston, K., Mayer, R.: Internet marketing: strategy, implementation and practice, 4th edn. Financial Times (Prentice Hall), Upper Saddle River, NJ, USA (2008)
12. Dahnil, M.I., Marzuki, K.M., Langgat, J., Fabeil, N.F.: Factors influencing SMEs adoption of social media marketing. Procedia—Soc. Behav. Sci. **148**, 119–126 (2014)
13. Darga, A.L.: Use of digital and social media marketing among SMEs in Mauritius. Mauritius Research Council, April (2018)
14. Demishkevich, M.: Small business use of internet marketing: findings from case studies. Walden Dissertations and Doctoral Studies Collection. Walden University, Minneapolis, MN, USA (2015)
15. EBSCO. https://www.ebsco.com (2019)
16. Experian: The 2016 digital marketer. Experian information solutions (2016)
17. Gilmore, A., Gallagher, D., Scott, H.: E-marketing and SMEs: operational lessons for the tuture. Europ. Bus. Rev. **19**(3), 234–247 (2007)
18. Google. https://google.com (2019)
19. Google Scholar. https://scholar.google.com (2019)
20. Google Trends. https://trends.google.de/trends/ (2019)
21. Hamill, J., Gregory, K.: Internet marketing in the internationalisation of UK SMEs. J. Market. Manage. **13**(1–3), 9–28 (1997)
22. Kimiloglu, H.: The "E-Literature": a framework for understanding the accumulated knowledge about internet marketing. Acad. Market. Sci. Rev. **1** (2004)
23. Harn Chua, A.P., Deans, K.R., Parker, C.M.: Exploring the types of SMEs which could use blogs as a marketing tool: proposed future research agenda. Australas. J. Infor. Syst. **16**(1) (2009)
24. Isohella, L., Oikarinen, E., Saarela, M., Muhos, M., Nikunen, T.: Perceptions of digital marketing tools in new microenterprises. Management challenges in a network economy. In: Management, Knowledge and Learning International Conference, 17–19 May, Lublin, Poland (2017)

25. Järvinen, J., Tollinen, A., Karjaluoto, H., Jayawardhena, C.: Digital and Social media marketing usage in B2B industrial section. Market. Manage. J. **22**(2), 102–117 (2012)
26. Jensen, M.B.: Characteristics of B2B adoption and planning of online marketing communications. J. Target. Measure. Analy. Market. **14**(4), 357–368 (2006)
27. Kilmartin, J.: Internet marketing: a study within SMEs in the Northwest of Ireland. Master Thesis. Letterkenny Institute of Technology, School of Business (2013)
28. Leeflang, P., Verhoef, P.C., Dahlström, P., Freundt, T.: Challenges and solutions for marketing in a digital era. Eur. Manag. J. **32**, 1–12 (2014)
29. Lichtenthal, J.D., Eliaz, S.: Internet integration in business marketing tactics. Ind. Mark. Manage. **32**, 3–13 (2003)
30. Mokhtar, N.F.: Internet marketing adoption by small business enterprises in Malaysia. Int. J. Bus. Soc. Sci. **6**(1), 59–65 (2015)
31. Morgan, R.F.: An internet marketing framework for the World Wide Web (WWW). J. Market. Manage. **12**(8), 757–775 (1996)
32. Murphy, H.C., Kielgast, C.D.: Do small and medium-sized hotels exploit search engine marketing? Int. J. Contemp. Hospita. Manage. **20**(1), 90–97 (2008)
33. Negoita, O.D., Purcarea, A.A., Popescu, M., Niculescu, A., Coman, C.N.: Digital marketing tools used by companies. FAIMA Bus. Manage. J. **6**(4), 66–76 (2018)
34. Patrutiu-Baltes, L.: Inbound marketing—the most important digital marketing strategy. Bull. Transilvania Uni. Brasov. Econo. Sci. Series V. **9**(2), 61–68 (2016)
35. Peter, M.K., Kraft, C., Streitenberger, M.: Forschungsresultate zum Stand der Digitalen Transformation. In: Peter, M.K. (ed.) KMU-Transformation: Als KMU die Digitale Transformation erfolgreich umsetzen. Forschungsresultate und Praxisleitfaden, pp. 29–55. FHNW Hochschule für Wirtschaft, Olten (2017)
36. Peter, M.K., Kraft, C., Schwaferts, D.: Praxisleitfaden: Handlungsfelder und Vorgehensmodell der Digitalen Transformation. In: Peter, M.K. (ed.) KMU-Transformation: Als KMU die Digitale Transformation erfolgreich umsetzen. Forschungsresultate und Praxisleitfaden, pp. 57–131. FHNW Hochschule für Wirtschaft, Olten (2017)
37. Peter, M.K., Kraft, C., Lindeque, J.: Strategic action fields of digital transformation: an exploration of the strategic action fields of swiss SMEs and large enterprises. J. Strategy Manag **13**(1), 160–180 (2020)
38. Poon, S., Jevons, C.: Internet-enabled international marketing: a small business network perspective. J. Market. Manage. **13**(1–3), 29–41 (1997)
39. ProQuest. https://www.proquest.com (2019)
40. Quinton, S., Khan, M.A.: Generating web site traffic: a new model for SMEs. Direct Market. Inter. J. **3**(2), 109–123 (2009)
41. Rugova, B., Prenaj, B.: Social media as marketing tool for SMEs: opportunities and challenges. Acad. J. Bus. Admin. Law Soc. Sci. **2**(3), 85–97 (2016)
42. Stankovska, I., Josimovski, S., Edwards, C.: Digital channels diminish SME barriers: the case of the UK. Econo. Res.—Ekonomska Istraživanja **29**(1), 217–232 (2016)
43. Steinhoff, L., Arli, D., Weaven, S., Kozlenkova, I.V.: Online relationship marketing. J. Acad. Mark. Sci. **47**(3), 369–393 (2019)
44. Taiminen, H., Karjaluoto, H.: The usage of digital marketing channels in SMEs. J. Small Bus. Enter. Develop. **22**(4), 633–651 (2015)
45. Trivedi, J.Y.: Study on marketing strategies of small and medium sized enterprises. Res. J. Manage. Sci. **2**(8), 20–22 (2013)
46. Van Scheers, L., Botha, J., Van Scheers, J.: Small and medium enterprise (SME) internet marketing challenges in the Tshwane Area, South Africa. J. Gover. Regul. **4**(4), 590–595 (2015)

# How the Internet of Things Drives Innovation for the Logistics of the Future

**Herbert Ruile**

**Abstract** The Internet of Things (IoT) is considered one of the most important trends, drivers, and enablers for business transformation. The field of logistics is affected manifold by this novel trend promising competitive advantages and growth. This chapter addresses an innovation value chain framework that relates advanced technology developments to business innovation in the field of logistics. The proposed framework contributes to clarify the complexity within the technology-driven innovation chain and helps organizations define, discuss, and develop their own IoT-driven business model. Based on three case studies, the framework is explained and discussed in terms of its relevance and value for an organization.

**Keywords** Business transformation · Internet of things · Innovation · Logistics · Case study

## 1 Introduction

Logistics is one of the major and still growing markets in the world. IMARC, a leading market research company providing detailed industry analysis, estimates that the global logistics market reached a value of US$ 1,171 Billion in 2017 [1]. They have identified three major factors that drive the growth of the global logistics market (a) the rapidly expanding e-commerce industry, (b) the growing focus on sustainability and compliance, especially environmental issues and corporate social responsibility (CSR), and (c) the rise of international trade agreements coupled with an increasing demand for reverse logistics services. The increasing use of technologies such as RFID (Radio Frequency Identification), Bluetooth, and newly used technologies such

Formerly Institute for Information Systems, School of Business, FHNW University of Applied Sciences and Arts Northwestern Switzerland.

H. Ruile (✉)
Logistikum Schweiz GmbH, c/o Business Help Point, Dätwylerstrasse 27, 6460 Altdorf, Switzerland
e-mail: herbert.ruile@fhnw.ch



R. Dornberger (ed.), *New Trends in Business Information Systems and Technology*, Studies in Systems, Decision and Control 294,
https://doi.org/10.1007/978-3-030-48332-6_18

as drone delivery and driverless vehicles today increases the chances of efficiently enhancing logistics and its services.

Due to the constant competitive pressure and the limitation of infrastructure in industrialized regions logistics is becoming a major bottleneck in the value chains and requires innovative approaches [2].

Although innovation has become one of the key economic indicators, a guarantee of social prosperity and health, the logistics industry still lacks innovation [3]. We distrust this, because we see Schumpeter's five types of innovations are also applicable to logistics: product/service and process innovation, demand and supply innovation, and new industry structures [4]. We see huge efforts in science and technological developments that are perceived as prerequisites and key elements for radical innovation. Since the increased use of technologies in the late 18th century, various phases of industrialization have followed. At the very least, the introduction and use of the internet as a commercial platform has radically changed the way of doing business: digital networking and autonomous interaction of distant objects, information and people became possible [5] and has steadily changed the way logistics business is done.

The latest digital developments offer new opportunities for growing prosperity and have therefore become popular for major investments by governments and economies [3]. These developments are described through a common and least differentiated use of terms such as "Industry 4.0", "Internet of Things", "Cyber Physical Internet", "Digital Transformation" or "Digitalization". The World Economic Forum investigated the impact of digitalization on logistics and identified the overall economic impact: "*$ 1.5 trillion of value on stake for logistics players and a further $2.4 trillion worth of social benefits as a result of digital transformation of the industry up until 2025*" [6]: p. 4.

Based on literature reviews and expert panels, Kersten et al. [7] studied the most important trends for logistics, which are essentially caused by digitization: digitization of business processes, increased transparency in the supply chain, networking, business analytics, automation and decentralization that give regional and urban transport platforms a more important role. These trends are still effective [6, 8].

Growing markets, competitive pressure, limited infrastructure, and technological advances require further innovations in logistics. How does it work? What are the right approaches and processes that logistics can rely on? What does it mean for logistics organizations to take advantage of the opportunities offered by advanced technology?

To provide a contribution to technology-driven innovation in logistics, this chapter introduces an integrated conceptual framework for IoT-driven innovation value chains in logistics. It is organized in three sections: (a) description of the fundamental terms of logistics and IoT, (b) the introduction and explanation of an IoT-driven innovation framework and (c) the application and discussion of the framework.

# 2 Fundamental Terms

## *2.1 Logistics as the Object of Innovation*

More than five decades have passed since business logistics was first scientifically addressed in the mid-fifties, without finding a satisfactory answer to the question of the identity of logistics (What is logistics?) [9]. From an academic point of view, the impetus for the progress of knowledge has its origins in the discussion of the terms that describe the object of analysis [10]. From the point of view of management science, with increasing functions, tasks, and responsibilities in the field of logistics, the conceptual clarifications and definitions of logistics become increasingly important. The understanding of logistics in science and practice has developed in recent years. In the following, empirical-inductive explanatory approaches will be pursued that take up and summarize concrete problems of logistics practice (see Table 1).

- Logistics is defined as the management of transport, warehousing and trans-shipments, which covers the functional management, technical, and organizational design and optimization of transportation, fleet management, warehousing, handling of materials, order fulfillment, logistics network design, inventory management, and management of third-party logistics services providers [11]. Logistics can be experienced physically and technically. This view of the industry is characterized by technical advances in the areas of automation, autonomy, energy consumption, and pollution.
- Logistics is understood as a contemporary leadership concept for the development, design, management, and realization of effective and efficient flows of objects (goods, services, information, and financing) in the company-wide value-added system. Logistics links important business functions and business processes between the point of entry and delivery [9, 11]. Process integration, lean management, and IT integration approaches play an important role. The administration and modern use of master and transaction data of processes, products customers and/or suppliers in a more or less closed system (e.g. business data warehouse) will be managed via data mining, big data analytics, and deep learning approaches.

**Table 1** Perspective of logistics and terms used based on [9]

| | Terms in use | Describes logistics as management of |
|---|---|---|
| 1 | Transport, warehouse, trans-shipment | isolated business function |
| 2 | Logistics management | firm's integrated flow of material and information |
| 3 | Supply chain management | the overarching value chains from supplier to customer |
| 4 | Value creating networks | interconnected partners of a value creation network |

- Logistics as Supply Chain Management (SCM) represents a further and innovative development stage. The intra-organizational perspective is replaced by a multi-tiered value chain with a consistent focus on the ultimate customer requirements and demand. In essence, supply chain management integrates supply and demand management within and across companies. SCM is a system approach to viewing the value-adding channel as a whole [8, 9, 11, 12]. The future of SCM will lie in its unique agility based on real-time capabilities. Suppliers, manufacturers, and customers can rely on information and decide on the availability of materials, capacities, and flexibility.
- Logistic as management of a value-creating network (VCN). This perspective will replace supply chains with strategic business networks. Sydow [13] defines: A strategic network is a polycentric form of organization of economic activities between market and hierarchy, which aims at the realization of competitive advantages and yet is strategically managed by one or more enterprises. It is characterized by complex reciprocal, rather cooperative than competitive and relatively stable relationships between legally independent, but mostly economically dependent enterprises.

Younger schools of thought expand the linear SCM approach into a network [14]. The holistic view of value creation networks allows development, design, realization, and optimization of economic systems [15]. Networks are characterized by more or less free and dynamic interaction between those involved: suppliers, service providers, manufacturers, retailers, customers, consumers, competitors, regulators, and institutions. With an increasing number of participants and dynamics of reverse interaction, digital networking creates a highly complex economic system. For further discussion, we use the general term of logistics but consider all four perspectives.

## 2.2 *Internet of Things*

The enabling technology in our investigations is termed the "Internet of Things", which goes back to Mark Weiser's work at the computer science lab at Xerox PARC where he formulated the ubiquitous computing vision and described a world, where algorithms are closely embedded in our daily life [16]. Later, in 2009, the term "Internet of Things" was coined by Kevin Ashton, RFID pioneer and co-founder of the Auto-ID center at the Massachusetts Institute of Technology [17]. The subsequent discussion will follow the IoT definition of the International Telecommunication Union as "the global infrastructure for the information society, enabling advanced services by interconnecting (physical and virtual) things based on existing and evolving interoperable information and communication technologies" [18]. "Sensors and algorithms offer a system of interconnected smart devices, which enable real-time and intelligent communication from man to machine, machine to machine, and enterprise to enterprise. The term "thing" with regard to IoT is defined as an object of the physical world (physical things) or the information world (virtual things), which is capable of being identified and integrated into communication networks" [18]. "With regard to the 'Internet of Things', this [device] is a piece of equipment with the mandatory capabilities of communication and the optional capabilities of sensing, actuation, data capture, data storage and data processing" [18].

A related term for IoT is "Industry 4.0" which was introduced to the public at the Hanover Trade Fair in 2011, and presented as part of Germany's high-tech strategy [19]. Meanwhile, the 4.0 extension is being used in almost all economic areas and for all functions to express the significant digitally induced changes that are expected (e.g. logistics 4.0, government 4.0, health 4.0, etc.). Although the term was coined in Germany, industry 4.0 shares some commonalities with developments in other regions where it has been labeled as "Internet of Things" or "Cyber physical Internet". The latter terms reflect primarily the technical content and not the resulting business and social transformation.

Meanwhile, the German term industry 4.0 has become a transitional term for the broader understanding of digital transformation, which is defined as the digitalization of analog machine and service operations, organizational tasks, and managerial processes [20] to create added value for customers and employees [21–23]. Or in other words: "Digital transformation is the evolving pursuit of innovative and agile business and operational models—fueled by evolving technologies, processes, analytics, and talent—to create new value and experiences for customers, employees, and stakeholders" [21].

Therefore, the terms IoT, digital transformation, and logistics will be amalgamated to "logistics 4.0" considering the aspects of IoT driven transformation processes. Thus, the definition of Logistics 4.0 by Hofmann and Rüsch [24]: p. 25 seems to be helpful, because it includes the aspect of business model innovation [25, 26]:

- *Products, services, infrastructure, and environment are flexibly connected* via *the internet or other network applications*;
- *The digital connectivity enables an automated and self-optimized production of individual goods and services including the delivery without human interventions*;
- *The value-creating networks are controlled decentralized while system elements make autonomous decisions*;
- *Data-driven, networked business models are transforming industry structures, organization, the way of collaboration and required roles, skills and capabilities of network members.*

## 3 IoT Driven Innovation Framework

### *3.1 IoT Solution Architecture*

Because IoT and related business transformation have been defined so far, we propose an IoT architecture that integrates the definitions and consists of the five building blocks described below. The IoT architecture becomes an overarching and integrative

**Table 2** IoT solution architecture based on [17, 21]

| | Layer | Descriotion | Examples |
|---|---|---|---|
| 1 | Business application | Business process | Resource planning; source, make, and delivery planning, execution and control (system) |
| 2 | Analytics | Analysis of data in order to create usable knowledge | Big data analysis, artificial intelligence, machine learning |
| 3 | Connectivity | Data enrichment by integrating data fro different applications and sources | Cloud computing, data crawler, sementic web, block chain |
| 4 | Communication network | Interoperable information and communication technologies | LAN/WAN, 5G, TCP/IP |
| 5 | Device | Equipment with the mandatory capabilities of communication | Drone, self-driving vehicles |

framework that logically connects technologies and organization. The architecture presented in Table 2 combines a technical IoT architecture [27] with a business model approach [28].

The IoT architecture consists of five complementary layers that transform technology building blocks into new business processes and business models. In layer 1–3 we rely on the definition of ITU [17]. However, in the ITU definition the business layers are not considered. Therefore, we added the definition of digital transformation to create an integrated framework.

- Layer 1 represents devices: front-end technology such as sensors, actuators and RFID chips, smart computers embedded in mobile devices, or intelligent autonomous objects (e.g. drones, robots, vehicles, smart stores, smart infrastructure). Information technology on level 1 creates, stores, processes and transfers data anytime and anywhere autonomously.
- Layer 2 consists of internal and external communication infrastructures. Networks that allow collecting, process, and disseminate valuable information, gathered from distributed devices. The required hardware consists of a secure data network infrastructure of scalable nodes (access points, storage) and linkages (wired and wireless).
- Layer 3 represents cloud-enabled services that provide SaaS (Software-as-a-Service), PaaS (Platform-as-a-Service), IaaS (Infrastructure as a Service), DaaS (Data-as-a-Service), and more. Layer 3 aggregates the "big data" and makes them available to create business applications.
- Layer 4 encompasses the business applications. Such an application uses the available data to simplify and improve existing processes; for instance, block chain and data analytics can help generate and capture value along the full product life by ensuring better coordination between all partners.
- Layer 5 defines the application in existing business functions such as purchasing, logistics, transport, warehousing, and production. The business functions embody

methods and tools that are used to efficiently fulfill their tasks and responsibilities (e.g. planning and execution systems, risk management, etc.).

## 3.2 *Integrated Innovation Value Chain Model*

Business development, logistics, purchasing, and marketing functions are continuously integrated into new products, services, and business models through the use of new digital technologies. The transformation of technology into logistic applications becomes difficult due to the cross-functional character. The objective of the innovation model is to describe the overarching value chain and impact of IoT technology into business applications in order to enhance existing or create new business models within firms. The development and availability of IoT technologies is not sufficient to trigger innovation in products, processes, services, or business models. Technology has no value in itself. The value of technology comes with the width and depth of the application. However, how many value-added steps are necessary in between?

Groher and Ruile [29] proposed an integrated innovation system for logistics. The innovation value chain model describes a stepwise transformation of technology into business values within a business-to-business environment. They identified the following four parties across the value chain: research and technology development, tool development and tool integration, application at logistics service providers, and value creation on the side of shippers. The innovation value chain model does not consider the internal transformation: how people in organizations get to know the tools and how they use them in a specific context to achieve organizational advantage. A prerequisite for the successful use of a new technology and the tool is a learning-oriented organization [30]. "*The adoption of new systems and processes is likely to improve effectiveness in the delivery of the logistics service. Organizational learning will also lead to reductions in the costs of transaction that will contribute to greater effectiveness in the delivery of the logistics service*" [31]: p. 71. We integrate organizational learning into the structured problem-solving cycle [32] to obtain the IoT-driven value chain.

The integrated innovation value chain model describes a stepwise transformation of technology into business value (see Fig. 1):

- Step 1: from technology to digital tools. Digital tools are software programs assisting people in their functional task and responsibility. Software tools for planning and controlling are well known: transport management system (TMS), warehouse management systems (WMS), resource/material planning (ERP), customer and supplier relationship management systems (CRM, SRM), and so on.
- Step 2: from digital tools to learning organization. Management tasks are based on a structured problem-solving cycle: identify, analyze, find and select a solution, plan and implement, and finally evaluate. Logistics management is supported by these advanced tools and should increase efficiency of the management processes.
- Step 3: from a learning organization to specific supply chain processes. According the SC reference model [33], we define the application areas within supply chain

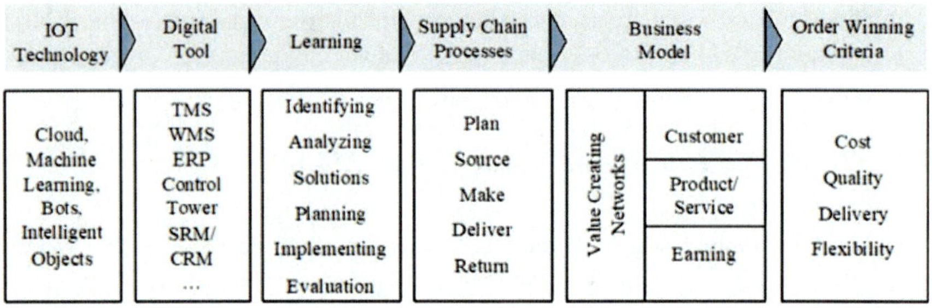


**Fig. 1** Integrated value chain framework for IoT driven innovation in logistics based on [31, 32]

processes: plan, source, make, deliver, and return. Each of the process functions have developed their own expertise and tools. It is expected that cross-functional integration and competitive networking enable organizations to adapt and leverage IoT-Solutions to achieve logistics objectives [34].

- Step 4: from SC processes to the business model. The business model integrates supply chain management into the overall strategic alignment with the product, customer, and business improvement model [35]. According to Gassmann [35], business model innovation is achieved when two or more elements of the business model have been changed in a coordinated and integrated way. The alignment of supply chain design and market requirement is well investigated and documented [36]. Organizations with high SC alignments achieve higher capitalization in the market.
- Step 5: from the business model to order-winning criteria. Criteria for winning customer orders are cost, delivery, quality, and flexibility. A tailor-made business model (product, customer and value chain) is essential to meet the criteria for winning customer orders [37].

## 4 Application

The following sections represent three applications using the IoT system description and the innovation value chain framework. Both elements are necessary to understand IoT technology and business value to enable innovation. Innovation is understood as the successful implementation of novelty in the market.

### *4.1 Case 1: Drive Net*

DRIVE net is an IoT-based prototype proposed for a region wide web-based real-time transportation decision system that adopts digital roadway maps and integrates

**Table 3** Drive net: IoT system

| | |
|---|---|
| Level 5 | Planning transportation for optimized delivery and mobility |
| Level 4 | Travel time estimation and prediction, dynamic routing, incident-induced delay calculation, statistical analysis |
| Level 3 | Incident tracking system, highway safety information system |
| Level 2 | Satellite server with hardware, software, and data processing tools, Bluetooth-based travel time detectors |
| Level 1 | Sensors on vehicles, roads, freeway loop sensor, real-time traffic data, tracking data (GPS) |

**Table 4** Drive net: innovation value chain

| | |
|---|---|
| Tool | Intelligent transportation system with real-time data collection from various data sources, data qualification analysis |
| Learning | Support of traffic analysis and decision making for alternative vehicle routings: sharing, visualizing, modeling, and analyzing transportation data |
| Application | Optimized real-time delivery planning |
| Business model | Customer: road users (private, business)<br>Product: traffic information platform<br>Value chain: data processing from multiple source<br>Earning: **not yet defined** |
| Order winning criteria | Road users (private, business, people, and cargo transport): flexibility, quality (real-time), faster delivery |

multiple data sources (e.g., traffic sensors, incidents, accidents, and travel time) [38]. The objective is to improve traffic flow and transportation time.

The application of the case uses a functional understanding of logistics: transport management (Tables 3 and 4).

## *4.2 Case 2: Used Car Trade*

An exemplary blockchain implementation took place in used car trading in Zurich [39]. The blockchain is used to create trusted data throughout the life of the car, even if car owners change over time. A digital Curriculum Vitae (CV) records all vehicle data in a file. The car importer supplies the new car data for the vehicles, the insurance company includes proof of insurance, and the Road Traffic Licensing Office takes care of traffic licensing. Anonymous usage and event data for the vehicle come from a car-sharing provider. The repair and maintenance services of the workshops are also recorded on the blockchain in unalterable form. Dealers and buyers receive much more data than before, and they can still trust the information.

**Table 5** Used car trade: IoT solution

| | |
|---|---|
| Level 5 | Re-selling platform |
| Level 4 | Data collection and history analysis, check of compliancy<br>**Further analysis not defined** |
| Level 3 | Blockchain for data collection from insurances, licensing offices, car service workshops, etc. |
| Level 2 | Bluetooth, public telecom network (4G) |
| Level 1 | Used car identifier, GPS location |

**Table 6** Used car trade: innovation value chain

| | |
|---|---|
| Tool | Digital platform, cloud-based control system |
| Learning | Identification of invisible, undeclared incidents |
| Application | Support of the sourcing process of the car buyer |
| Business model | Customer: used car buyers and sellers (private, business)<br>Product: transparency<br>Value chain: data processing from multiple sources and partners<br>Earning: pay-per-use, license, member fee |
| Order winning criteria | Car buyer: quality, transparency and avoidance of fraud |

The composition of this ecosystem is not accidental: it is the smallest survivable value system. However, the choice of the right partners is key to the survival of this ecosystem.

The application of the case uses a network understanding of logistics: value-creating network through efficient information flows between partners (Tables 5 and 6).

## 4.3 Case 3: Smart Farming

The example of smart farming [40] shows the digital transformation in the primary economic sector through the creation of networked farms for smart farming [40]. A network of small farmers, urban farmers and gardeners is integrated with their food or flower production. Production is controlled by multiple sensors (humidity, temperature, etc.) and actors (water pumping, fertilizer, shadow system, etc.). Monitoring and sharing of these data will provide insight into the farming practice and

**Table 7** Smart farming: IoT solution

| | |
|---|---|
| Level 5 | Monitor and control on mobile devices |
| Level 4 | **Not yet developed** |
| Level 3 | IoT service platform (called Mobius). Each device is registered and allows machine-to-machine communication |
| Level 2 | Wirelessly connected with the Raspberry-Pi installed with the &Cube through a ZigBee-based network |
| Level 1 | (a) compound sensor (i.e., temperature, humidity, and CO2),<br>(b) photosynthetic photon flux density (PPFD) sensor, and<br>(c) soil moisture sensor<br>Devices are intake and exhaust fans, an air conditioner, sprinklers, LED lights, a cover controller and an irrigation and nutrient management system |

**Table 8** Smart farming: innovation value chain

| | |
|---|---|
| Tool | Expert data base, knowledge management system |
| Learning | Problem identification, scenarios, evaluation of experienced farmers |
| Application | Optimized source and manufacturing process (e.g. which seed and plant protection to buy, how to care for plants, how to harvest them, etc.) |
| Business model | Customer: (experienced and unexperienced) farmers, industry<br>Product: expert system<br>Value chain: data processing from multiple devices and sensors from various partners<br>Earning: no clear description (we estimate: platform) |
| Order winning criteria | Quality (in terms of reliability, actuality and scope) |

should improve its productivity. The parties estimate that agriculture could be further developed by using the Internet of Things.

The application of the case uses a network perspective of logistics: a value-creating network by providing a more effective agriculture for the cultivation of their crops (Tables 7 and 8).

## 5 Conclusion

The objective of the chapter was to describe how IoT drives innovation in logistics. We developed an IoT-driven integrated framework that considers a) the technical and business architecture and b) an integrated innovation value chain for logistics application. We applied the frameworks to three IoT cases from the literature.

The application of the framework to the three cases shows commonalities und differences within the setup of the solution and the value chain. The usage of the IoT architecture allows a systematic analysis of the technological structure, and the application of the innovation value chain provides insight into the value added stream and the applied business model. We see some commonalities and differences in the IoT architecture:

- Advanced IoT solutions provide an interconnection of sensors, devices, and value chain partners by using internet technologies.
- IoT solutions create much more transparency by collecting and analyzing data from different sources.
- The data collection opens opportunities for expert systems and machine learning systems.
- Not all of the cases presented have fully completed the IoT architecture. The missing layers can provide opportunities to add functionalities that can add value to the overall system.

The application of the innovation value chain framework provides insight into the following

- Due to the technical structure, the preferred business model is a platform. Platforms are new players across the value chain and provide additional shared information to optimize the source, create or deliver processes.
- The partners of the platform do not change their business model (transport company, car dealer, farmer), but they have the opportunity to improve their operations.

The IoT architecture and the innovation value chain framework provide advantages for a better understanding of the proposed IoT cases. The mutual view of technology and economy expands our understanding of Logistics 4.0, which is an amalgam of IoT, industry 4.0 and digital transformation. The framework makes it possible to identify technical and business opportunities. In order to successfully implement an IoT-driven radical innovation, we need to experience value adding networked designs. The framework will help to analyze, design, and evaluate such technical and economic networks that are closely related.

The framework presented in this chapter is descriptive. The model is able to analyze, describe, and evaluate (to a certain extent) existing IoT solutions in logistics. However, the framework is not able to predict future solutions, nor to give advice on a process for developing innovative solutions.

Nevertheless, the framework was tested in three cases, and requires further validation and proof in practice. Our future research work will follow the raised questions: (a) what are the success factors, driving forces, and barriers for building successful IoT-driven innovations in logistics? (b) how do companies and networks manage the transitory steps in the cross functional innovation value chain? and (c) how do companies deal with the growing IoT-driven platform economy that will substitute their traditional business model?

## References


1. IMARC: logistics market: global industry trends, share, size, growth, opportunity and forecast 2019–2024, (2018). https://www.imarcgroup.com/logistics-market Accessed 17 Sept 2019
2. Maibach, M., Ickert, L., Sutter, D.: Volkswirtschaftliche Aspekte und Auswirkungen des Projekts Cargo Sous Terrain (CST). Bundesamt für Verkehr (2016). https://www.newsd.admin.ch/newsd/message/attachments/46311.pdf
3. Schwab, K.: The global competitiveness report 2018, WEF (2018). www.weforum.org/gcr
4. Schumpeter, J.A.: The theory of economic development: an inquiry into profits, capital, credit, interest, and the business cycle. Trans. Publishers **1**, 244 (1982)
5. Kagermann, H., Wahlster, W., Helbig, J.: Recommendations for implementing the strategic initiative industriE 4.0. Final Report of the Industrie 4.0 Working Group (2013). http://www.acatech.de/. Accessed 10 July 2015
6. World economic forum (wef): digital transformation of industry —logistics industry. White papper. WEF (2017). www.weforum.org/gcr. Accessed 17 Sept 2019
7. Kersten, W., Seiter, M., von See, B., Hackius, N., Maurer, T.: Trends und Strategien in Logistik und Supply Chain Management-Chancen der digitalen Transformation, 2017. BVL, DVV, Hamburg (2017)
8. Stölzle, W., Hofmann, E., Mathauer, M.: Logistikmarktstudie Schweiz—Band 2018-Autonom und vernetzt: Supply Chain-Entwicklungen im digitalen Zeitalter. Bern (2018). https://shop.gs1.ch/
9. Göpfert, I.: Anwendung der Zukunftsforschung für die Logistik. in Göpfert, I. (ed.): Logistik der Zukunft-Logistics for the future. Springer Gabler (2019)
10. Luhmann, N.: Der Funktionsbegriff in der Verwaltungswissenschaft. Die Bedeutung der Organisationssoziologie für Betrieb und Unternehmung. In: Schriften zur Organisation 1, pp. 3–15. Springer VS, Wiesbaden (2018)
11. Council of supply chain management professionals (CSCMP): CSCMP supply chain management definitions and glossary. https://cscmp.org. Accessed 17 Sept 2019
12. Min, H., Mentzer, J.: Defining Supply Chain Management within Marketing Strategy, University of Tennessee (1998)
13. Sydow, J.: Strategische Netzwerke: Evolution und Organisation, Wiesbaden (1993)
14. Voegele, A.R., Zeuch, MP. (eds.): Supply Network Management: Mit Best Practice Der Konkurrenz Voraus. Springer-Verlag (2013)
15. Peppard, J., Rylander,, A.: From value chain to value network: insights for mobile operators. Eur. Manage. J. **24**, 2–3 (2006)
16. Weiser, M.: The computer for the 21st century. Sci. Am. **265**(9), 66–75 (1991)
17. Ashton K.: That 'internet of things' thing. RFID J. 22 June 2009. http://www.rfidjournal.com/articles/view?4986
18. International Telecommunications Union (ITU): Overview of the internet of things (ITU-T Y.2060). In: Series Y: Global Information Infrastructure, Internet Protocol Aspects and Next-Generation Networks. ITU-T Y. 2060 (06/2012)
19. Liao, Y., Deschamps, F., Loures E.D.F.R., Ramos L.F.P.: Past, present and future of Industry 4.0—a systematic literature review and research agenda proposal. Int. J. Prod. Res. **55**(12), 3609–3629 (2017)
20. Iansiti, M., Lakhani, K.R.: Digital ubiquity: how connections, sensors, and data are revolutionizing business. HBR **92**(11), 91–99 (2014)
21. Solis, D., Digital transformation—the six stages of digital transformation (2017). http://www.briansolis.com/2017/01/definition-of-digital-transformation. Accessed 17 Sept 2019
22. de Man, J.C., Strandhagen, J.O.: An industry 4.0 research agenda for sustainable business models. Procedia Cirp **63**, 721–726 (2017)
23. Maresova, P., Soukal, I., Svobodova, L., Hedvicakova, M., Javan-Mardi, E., Selamat, A., Krejcar, O.: Consequences of industry 4.0 in business and economics. Economies **6**(3), 46 (2018)

24. Hofmann, E., Rüsch, M.: Industry 4.0 and the current status as well as future prospects on logistics. Comput. Ind. **89**, 23–34 (2017)
25. Teece, D.J.: Business models, business strategy and innovation. Long Range Plan. **43**(2–3), 172–194 (2010)
26. Turber, S., Vom Brocke, J., Gassmann, O., Fleisch, E.: Designing business models in the era of internet of things. In: International Conference on Design Science Research in Information Systems, pp. 17–31. Springer, Cham (2014)
27. Gubbi, J., Buyya, R., Marusic, S., Palaniswami, M.: Internet of things (IoT): a vision, architectural elements, and future direc-tions. Future Gener. Comput. Syst. **29**(7), 1645–1660 (2013)
28. Dijkman, R.M., Sprenkels, B., Peeters, T., Janssen, A.: Business models for the internet of things. Int. J. Inf. Manage. **35**(6), 672–678 (2015)
29. Groher, W., Ruile, H.: Vernetztes innovations system für die Logistik 4.0. Logistics Innovation 2017/1. VNL Zürich (2017)
30. Calantone, R.J., Cavusgil, S.T., Zhao, Y.: Learning orientation, firm innovation capability, and firm performance. Ind. Mark. Manage. **31**, 515–524 (2002)
31. Panayides, P.M.: The impact of organizational learning on relationship orientation, logistics service effectiveness and performance. Ind. Mark. Manage. **36**(1), 68–80 (2007)
32. Liang, K., Zhang, Q.: Study on the organizational structured problem solving on total quality management. Int. J. bus. Manage. **5**(10), 178–183 (2010)
33. Council, S.C.: Supply chain operations reference model. Overview of SCOR Version 5.0 (2008)
34. Durugbo, C., Tiwari, A., Alcock, J.R.: Modelling information flow for organisations: A review of approaches and future challenges. Int. J. Inf. Manage. **33**(3), 597–610 (2013)
35. Gassmann, O., Frankenberger, K., Csik, M.: Geschäftsmodelle entwickeln: 55 innovative Konzepte mit dem St. Carl Hanser Verlag GmbH Co KG, Galler Business Model Navigator (2017)
36. Wagner, S.M., Grosse-Ruyken, P.T., Erhun, F.: The link between supply chain fit and financial performance of the firm. J. Oper. Manag. **30**(4), 340–353 (2012)
37. Pongpak, P., Sadler, I., Shee, H.: Supply chain collaboration aligns order-winning strategy with business outcomes. IIMB Manag. Rev. **29**(2), 109–121 (2017)
38. Cui, Z., Zhang, S., Henrickson, K.C., Wang, Y.: New progress of drive net: an e-science transportation platform for data sharing, visualization, modeling, and analysis. In: 2016 IEEE International Smart Cities Conference (ISC2), pp. 1–2. IEEE (2016)
39. Stamm, E.: Die blockchain schafft neue netzwerke. https://www.nzz.ch/wirtschaft/die-blockchain-schafft-neue-netzwerke-ld.1404007. Accessed 16 July 2018
40. Ryu, M., Yun, J., Miao, T.T., Ahn, I., Choi, S., Kim, J.: Design and implementation of a connected farm for smart farming system. IEEE Sensors, pp. 1–4, (2015)

# Recommendations for Conducting Service-Dominant Logic Research

**Joachim C. F. Ehrenthal, Thomas W. Gruen, and Joerg S. Hofstetter**

**Abstract** In the digital age, companies compete on how well their service meets customer needs and solves customer problems, interacting with ever more actors to fulfill their promise. A well-suited mid-range theory to conduct research in this "new" world is service-dominant (S-D) logic. However, there are no guidelines or a commonly agreed framework for systematically conducting S-D logic research. To fill this gap, this chapter reviews and analyzes the use of S-D logic in research. It provides recommendations for conducting S-D logic in research and proposes an organizing framework for applied research settings and beyond.

**Keywords** Service-dominant logic · Literature review · Value co-creation · Service

## 1 Introduction

Recently, a major shift in business towards service-driven business models and value co-creation in networks was proposed, and termed service-dominant (S-D) logic [1]. It transcends the linear goods-dominant (G-D) logic of value production with value co-creation through exchange of service [1], thus reflecting the ongoing change in the business world towards digitally enabled service provisioning, and multi-actor service ecosystems [2].

J. C. F. Ehrenthal (✉)
School of Business, Institute for Information Systems, FHNW University of Applied Sciences and Arts Northwestern Switzerland, Peter Merian-Strasse 86, 4002 Basel, Switzerland
e-mail: joachim.ehrenthal@fhnw.ch

T. W. Gruen
Department of Marketing, The Peter T Paul College of Business and Economics, University of New Hampshire (UNH), 370A University of New Hampshire, Durham, NH 03824, USA
e-mail: tom.gruen@unh.edu

J. S. Hofstetter
KEDGE Business School, Department of Operations Management & Information Systems, 680 Cours Libération, 33405 Talence, France
e-mail: joerg.hofstetter@kedgebs.com



R. Dornberger (ed.), *New Trends in Business Information Systems and Technology*, Studies in Systems, Decision and Control 294,
https://doi.org/10.1007/978-3-030-48332-6_19

Real-life examples of this shift towards S-D logic include: iRobot cleaning robots that provide the service of a "clean household"; Bayer Crop Science that guarantee successful harvests to farmers through remote management of the entire farming process (rather than selling seeds, fertilizers, pesticides, and analytics); Risch Shoes that use digital twins to offer the service of "perfect fit shoes" (in cooperation with retailers); or Klara, a free-to-use integration service that connects customers to accounting, payroll, and other business services, as well as Amazon cloud services (infrastructure and software-as-a-service).

S-D logic was quickly adopted throughout the world of marketing and services research, and also many related research domains, such as supply chain and information systems (e.g. through the IBM Almaden Research Center), leading to a body of research of several thousand articles (see Sect. 4.1).

This rapid spreading has led to a cluttered and unstructured research landscape, with no clear guidance on how to use S-D logic in research.

To alleviate this issue, this research comprehensively analyzes the use of S-D logic in research and derives recommendations on how to conduct research using S-D logic. It also finds evidence that the previous lack of methodological foundation has fostered the development of an increasingly unstructured research landscape.

This chapter extends previous reviews on S-D logic and its impact on thought [2–5]. Examining over 1,700 publications, this work provides evidence for the dissemination and institutionalization of S-D logic and from this distinction reveals good practices for using S-D logic in different research settings.

The remainder of this chapter is organized in four sections. First, a brief primer on S-D logic is provided. Second, the details of the review methodology are presented. Third, the results of the review are presented and discussed. Fourth, recommendations for conducting research using S-D logic are derived.

## 2 A Primer on Service-Dominant Logic

In their seminal article, and with later extensions, Vargo and Lusch [1, 6] propose a service-based (S-D) logic for economic exchange and business based on value co-creation in service networks. Their work is based on observations of the changes in the predominant thought in economics, research, and business over the past century.

The S-D logic views value as being co-created by the reciprocal application of operant resources (knowledge and skills) by resource integrators ("actors") and their networks (e.g. a firm and its suppliers) to the benefit of a receiving entity (e.g. a customer). What they exchange is not "tangible goods and intangible services", but "service", i.e. solutions to problems, sometimes using products as distribution mechanism, sometimes as indirect exchange with currency. Value is defined by the recipient (e.g. customer) in the context and the application (value-in-use). S-D logic defines value as a co-creational, continuous flow of relational interactions in networks of actors integrating resources into their networks to co-create a desired service, eventually forming ecosystems of service offerings and exchange [7, 8].

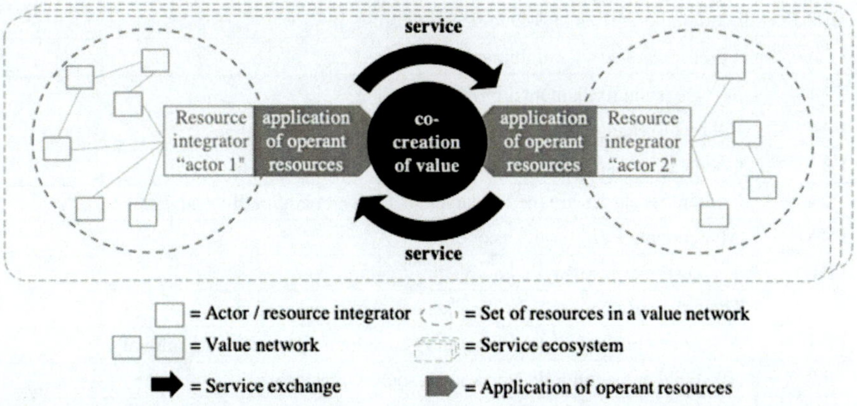


**Fig. 1** An overview of "service-to-service-exchange view of value co-creation" S-D logic (own illustration, extended from [9, p. 10])

Figure 1 summarizes the S-D logic "*service-for-service-exchange view of value co-creation*", in which actors integrate their value-network resources to co-create value through exchange of service in their respective ecosystems.

S-D logic transcends the (neoclassical) goods-dominant (G-D) logic, which is based on the ideas of the first industrial revolution. In the G-D logic, value is built up unidirectionally in a product by the production process, stored in the product, and the predetermined utility level (i.e. value as shown in the price) is released to the product's consumption, where value creation ceases [2].

S-D logic argues that instead of a consumptive goal, goods are bought to solve problems by providing various types of service (e.g. warmth from clothing, cloud servers for ubiquitous computing), that goods and services carry a wide range of intangible goods (e.g. image, meaning); and that they are inputs for experiences and parts of applications in a broader context that leads to a transformative, higher-level representation of a co-creational, network-based approach to business based on service-for-service-exchange. Within this, "service" represents the reciprocal application of operant resources (knowledge and skills) for the benefit of some actor [10].[1]

S-D logic is operationalized in foundational premises (FP), which in turn are simplified into axioms, with axiom1 consisting of FP1-5, axiom2 of FP6-8, axiom3 and axiom4 of FP9 and FP10 respectively. Later, an axiom5/FP11 was added ("value cocreation is coordinated through actor-generated institutions and institutional arrangements"), which is not part of this review. Table 1 lists the FPs of S-D logic as stated in Vargo and Lusch [6].

[1]Please note that "service" thus differs from G-D "services", i.e. intangible goods.

**Table 1** Foundational premises of S-D logic based on [6, p. 7]

| Foundational premises of service-dominant logic | |
|---|---|
| FP 1 | Service is the fundamental basis of exchange |
| FP 2 | Indirect exchange masks the fundamental basis of exchange |
| FP 3 | Goods are a distribution mechanism for service provision |
| FP 4 | Operant resources are the fundamental source of competitive advantage |
| FP 5 | All economies are service economies |
| FP 6 | The customer is always a co-creator of value |
| FP 7 | The enterprise cannot deliver value, only value propositions |
| FP 8 | A service-centered view is inherently customer-oriented and relational |
| FP 9 | All social and economic actors are resource integrators |
| FP 10 | Value is always uniquely and phenomenologically determined by the beneficiary |

# 3 Review Methodology

To uncover and organize the use of S-D logic in research, this study builds on the validity schema of Brinberg and McGrath [11]. Thus, it investigates the use of S-D logic research with respect to the conceptual, methodological and substantive domains. The *conceptual domain* concerns theory, i.e. conceptual causes and their relationships providing the building blocks for a theory or model (the focus of this chapter). The *substantive domain* describes the real-world phenomena under investigation. The *methodological domain* concerns the rigor of an investigation, including research design, data gathering/generation, and analysis and inference. Together, these three domains provide an organizing structure for different paths to scholarly progress, and thus form the basis of this investigation.

## *3.1 Data Gathering and Selection*

This review examines publications that cite either the seminal article by Vargo and Lusch [1] or their [6] clarification paper[2] at least one of these two articles, and a period of four years after the latter article's publication, which offers an observation period without apparent structural breaks or turning points, leading to stable, analyzable results (see below).

The process of publication retrieval was structured as follows: (1) Google Scholar listing of research citing at least one of the two seminal articles on S-D logic[3]; (2) full

[2]Preliminarily trials of full-text searches in academic databases on S-D logic and its concepts, such as "value-in-use" yielded the insight that the identified research consistently cited the seminal papers, hence the approach of citation analysis.

[3]Please note that Google Scholar was used as the leading system and academic databases were accessed consecutively for article retrieval. We found no other database that covered all and "early"

text retrieval for each search result; (3) duplicate entry removal; (4) elimination of search results with "phantom citations", i.e. articles, which Google Scholar retrieved but did not contain a cite to either seminal piece; (5) removing all non-English and non-available publications to ensure consistent coding; (6) removing conference and working papers that were subsequently published in journals; (7) coding (which is outlined in the following section). To ensure data quality, the analysis process followed a consistent four-eye policy (the lead author and one research associate).

## 3.2 Coding of Publication Contents

For coding, each publication was analyzed and catalogued according to the following. The *conceptual domain* is coded based on foundational premises (FPx). Each publication was classified with respect to the foundational premise(s) it addressed. If there was no specific reference to a FP, but only a reference to either seminal paper, this was classified as not referencing to any FP. For the development of the *conceptual domain*, we distinguish between publications that use S-D logic in the following ways: dissemination (mentioning, acknowledging, supporting S-D logic), and institutionalization (applying, extending or testing S-D logic). This distinction became necessary because a large number of publications under review cite S-D logic only to get some kind of inspiration from it (dissemination), while a significantly smaller number of publications actually use and extend S-D logic. Table 2 summarizes the classification of the conceptual domain.

Classifying the *substantive domain*, this review uses the set of the well-established business research areas outlined by Grewal and Levy [12] (e.g. marketing, services, supply chain, information systems, etc.). For the classification of the *methodological domain*, this review builds on the methods analyzed in Brown and Dant's [13] review of scientific methods in research (qualitative, structural equation modeling, analytical

**Table 2** Overview of the classification of the conceptual domain in this research

| Indicator of | Classification | Explanation |
|---|---|---|
| Dissemination | Mentioning | A tangential reference to S-D logic |
| | Acknowledging | Referencing to an element of S-D logic, e.g. "co-creation" |
| | Supporting | A statement of usefulness on S-D logic |
| Institutionalization | Applying | Applying S-D logic to a problem, industry or research field |
| | Extending | Specific proposal to extend S-D logic |
| | Testing | Testing of S-D logic or related idea, e.g. quantitatively |

modes of scholarly communication (such as books and conferences), which is necessary in nascent field reviews as this.

modeling, etc.). In addition, we classified each publication according to research type, i.e. whether it is conceptual, qualitative, or quantitative in nature. Furthermore, publication demographics (authors, journal, lead author institution, etc.) were coded to capture a complete picture of the structure of the publication set.

# 4 Review Results

The following sections describe the body of reviewed publications and give a synthesis of the conceptual, substantive, and methodological domain of current S-D logic research. The remainder of this review refers to the body of reviewed publications as the "set", the findings related to the recommendations are labeled "fx" (e.g. finding f1) to be able to link findings to recommendations.

## *4.1 Description of the Set*

For the analysis, we identified 3,830 publications. Out of these, 268 could not be retrieved (dead links), 498 were double entries, and 34 were "phantom citations". This left a total of 3,030 remaining citations. A total of 346 non-English publications were eliminated, as well as 955 non-available publications (lack of access rights and no response from the authors to provide the publication). In addition, the conference papers that had been subsequently published in journals were removed from the set as to prevent duplicates. This yielded a set of 1,724 publications for analysis, 56.90% of the 3,030 remaining citations, making it likely to capture the major developments in S-D logic research.

The majority of publications are journal articles (65.26%). Other publication types include conference papers (18.56%), as well as working papers and books/edited volumes, and doctoral dissertations (combined 16.18%). Thus, the data stems from various sources of scholarly communications, with a nucleus on journal publications, and reflects the broad scholarly attention to S-D logic.

Overall, the publications include articles from 393 journals. Of these journals, 136 journals are specific to marketing and services (35%), whereas the other 257 journals display a wide variety of foci (65%), but mostly management, operations, as well as information systems, and technology. Hence, the review captures developments on S-D logic published in different research fields.

In respect to the geographical distribution of publications, the set shows the following structure: 49.21% of the publications have a lead author from an institution based in Europe, followed by 30.42% of lead authors from an institution in North America, and 18.15% from Asia and Oceania. The most prolific five countries, as measured by the location of the lead author's institution, are the USA (27.96%), the UK (13.34%), and Australia (7.31%), followed by Sweden (7.13%), and Finland (6.55%). Note that Scandinavian countries account for 17.05% of publications in the set, which meets expectations with their well-known focus on services marketing. Despite an expected predominance of countries with English as their native language (54.64%) in addition to a predominance of European countries (49.19%) the set reflects international development in S-D logic research.

The review identified more than 1,200 unique authors, i.e. the publication base of researchers covered by the set is very broad. In line with our expectations, the originators of S-D logic—Stephen L. Vargo and Robert F. Lusch—are among the top contributors in the set. However, with a combined 4.42%, they account for a low percentage of publications in the set, which signals a broad interest in this field. The most prolific authors in the set are senior researchers, who are well respected in their fields.

Overall, the set of 1,724 publications is diverse. It covers multiple types of scholarly communication, and covers various research areas, geographical regions, countries, and research institutions (including private business research such as IBM), and exhibits a broad author base.

## *4.2 The Conceptual Domain of the Set*

This section investigates the prevalent conceptual domain of S-D logic research as operationalized in the foundational premises (FP).

About two-thirds (65.31%) of the publications examined specifically refer to a Foundational Premise (FP). Out of these, nearly half of them reference to the co-creation of value (FP6, 34.72%), which is by far the most referenced FP. After FP6, the second most referenced is FP4 with 12.87%, which addresses operant resources and competitive advantage. Third in line is FP1, which states that service is the fundamental basis of exchange, and is used by 11.50%. FPs 6, 4 and 1 account for about 60% of S-D logic research. The use of all other FPs ranges between 4.16% (FP2) and 8.21% (FP3). Note that a single publication may address multiple FPs.

However, only a minority of publications using S-D logic is concerned with its institutionalization. The classification of dissemination (mentioning, acknowledging, supporting) that gives specific reference to FPs accounts for 70% of all S-D logic research. The institutionalization classification (applying, extending, testing) accounts for 30% of S-D logic research (finding f1). This 70:30 ratio of dissemination to institutionalization ratio is stable across all substantive domains (±3%). Based on our research, it can be argued that this wide dissemination versus institutionalization ratio appears be one of the root causes of the unstructured landscape of S-D logic research ("many citations", apparent in the use indicators of dissemination, but "limited progress", apparent from institutionalization).

While most dissemination-related publications are based only around the most frequently mentioned FPs (6, 4, and 1), the institutionalization-related publications of S-D logic give all FPs a more balanced attention (finding f2). Table 3 shows the frequencies of the FPs used with respect to the dissemination and institutionalization of S-D logic research for those publications giving special reference to FPs.

**Table 3** An overview of the dissemination and institutionalization of S-D logic

| | Use of S-D logic | FP1 | FP2 | FP3 | FP4 | FP5 | FP6 | FP7 | FP8 | FP9 | FP10 |
|---|---|---|---|---|---|---|---|---|---|---|---|
| Dissemination | Mentioning | 19 | 3 | 10 | 9 | 4 | 32 | 5 | 6 | 5 | 4 |
| | Acknowledging | 56 | 13 | 28 | 47 | 11 | 234 | 21 | 12 | 11 | 21 |
| | Supporting | 68 | 27 | 56 | 108 | 26 | 251 | 56 | 37 | 41 | 41 |
| | Dissemination (%) | 48 | 39 | 44 | 47 | 36 | 57 | 39 | 38 | 40 | 43 |
| Institutionalization | Applying | 109 | 37 | 78 | 130 | 43 | 304 | 77 | 56 | 50 | 53 |
| | Extending | 40 | 26 | 38 | 42 | 26 | 68 | 42 | 28 | 31 | 30 |
| | Testing | 9 | 3 | 5 | 11 | 4 | 20 | 7 | 5 | 5 | 6 |
| | Institutionalization (%) | 52 | 61 | 56 | 53 | 64 | 43 | 61 | 62 | 60 | 57 |

Publications mentioning and acknowledging S-D logic heavily refer to FPs 1 and 6. In the dissemination category "supporting" and the institutionalization categories, other FPs become more common. In addition, we see a linear relationship in institutionalization—nearly the same percentage in the categories of testing from extending from applying across all FPs. The role of FP6 provides great interest: in terms of the percentage of its use, it is the lowest of institutionalization versus dissemination. However, in absolute terms, it provides the largest number of studies that show institutionalization of S-D logic. It is also noteworthy that FP6 stands out in testing, indicating that FP6 may be key to why and how S-D logic is becoming institutionalized through empirical research. A possible explanation would be that FP6, co-creation, has been used and discussed in research streams that existed prior to or in parallel with the formal introduction of S-D logic, but that S-D logic provides context and language for co-creation to become universally discussed, and thus takes a pivotal role in both dissemination and institutionalization of S-D logic (finding f3).

Over time, dissemination publications show no clear path of development with the exception of the rise of FP6. The ratios between the FPs used over time changes, i.e. FPs seem to be cited on an as-needed basis. In sharp contrast, institutionalization publications show a stable use of FPs over time (addendum to finding f2). The ratios in the use of FPs among each other show no significant change, with the exception of the change due to the introduction of FP9 and FP10 in 2008. In addition, the clear distinction between dissemination and institutionalization in the main areas and regions is stable, i.e. research into the application of S-D logic to operations research in Asia is very similar in relation to the use of FP as is research in marketing science from North America. The same applies to research exhibiting dissemination indicators. Similarly, while FP6 increases over time in both dissemination and institutionalization publications, the rise of FP6 accelerates faster in dissemination publications than in institutionalization publications, hinting at different motivations to reference this specific FP in dissemination and institutionalization.

When analyzing the entire set, about 35% do not mention a reference to a specific FP, but generically cite S-D logic. In addition, about 35% of publications reference only a single FP and about 15% use two FPs. Publications that exhibit the dissemination indicators tend to use fewer FPs than publications with institutionalization indicators. Generally, research concerned with institutionalization of S-D logic uses S-D logic as a whole and not just one or more of its parts (finding f4). This finding applies to all areas and geographies investigated, and is time consistent. The use of three or more FPs is much less frequent with only about 15% of the total. Table 4 illustrates the concentration of S-D logic publications on few FPs.

With respect to the absolute number of combinations of FPs, the analysis revealed the following: First, S-D logic research connects FP6 and FP4 (262 instances), i.e. the customer as co-creator of value with operant sources as the fundamental source of competitive advantage. Second, FP6 is frequently used in combination with FP1 (209), FP3 (180), FP9 (132), and FP8 (131). S-D logic research therefore combines these FPs to an argument string: Value creation is an interaction where resources are used to exchange service for service (which may be stored in physical goods) in a multi-actor setting in which value is determined by the beneficiary. Similarly, FP1

**Table 4** Frequency of the use of foundational premises in S-D logic research

| Number of FPs used | Instances absolute | Percentage of total (%) | Thereof dissemination (%) | Thereof institutionalization (%) |
|---|---|---|---|---|
| 1 | 600 | 34.80 | 70.17 | 29.83 |
| 0 | 598 | 34.69 | 94.15 | 5.85 |
| 2 | 263 | 15.26 | 61.22 | 38.78 |
| 3 | 97 | 5.63 | 45.36 | 54.64 |
| 10 | 66 | 3.83 | 27.27 | 72.73 |
| 4 | 43 | 2.49 | 25.58 | 74.42 |
| 8 | 25 | 1.45 | 56.00 | 44.00 |
| 5 | 22 | 1.28 | 13.64 | 86.36 |
| 6 | 5 | 0.29 | 40.00 | 60.00 |
| 7 | 3 | 0.17 | 33.33 | 66.76 |
| 9 | 2 | 0.12 | 0.00 | 100.00 |

is often connected to FP4 (156) and FP3 (142), which leads to the conclusion that S-D logic research has evolved primarily around value, its creation and distribution (addendum to finding f3). FPs 7, 8, 9, and 10 exhibit fewer instances of combinations, which may indicate that they may have been less relevant in S-D logic research. It is important to keep in mind that FP9 and FP10 were introduced later than FP1-8. Table 5 shows the combinations of FPs used in S-D logic research. Note that the main diagonal shows the sum of a FP being referred to in isolation.

Looking at the entire set of publications, some trends emerge that entail potential paths for the development of S-D logic. Relatively few publications have aimed to extend S-D logic or provide some sort of test using S-D logic (finding f5). Based on the classification of this review, the bulk of the articles either supports S-D logic (provide a statement regarding the usefulness of S-D logic to their intended contribution), or applies S-D logic to a problem or industry, or deals with the development of S-D logic itself (finding f6). The latter is the situation in which the potential for the further development and institutionalization of S-D logic may lie. In addition, most studies do not refer to how the application of S-D logic led to different predictions/outcomes compared to the G-D logic, making it hard to see directly whether and how S-D logic was used to innovate and overcome G-D limitations, which is what a new logic should do (finding f7).

## 4.3 *The Substantive Domain of the Set*

This section investigates the substantive domain of the set, i.e. in which fields S-D logic has been used. Most prominently, S-D logic has received attention in the area of

**Table 5** Combinations of foundational premises used in S-D logic research

| Combinations | FP1 | FP2 | FP3 | FP4 | FP5 | FP6 | FP7 | FP8 | FP9 | FP10 |
|---|---|---|---|---|---|---|---|---|---|---|
| FP1 | 301 | 103 | 142 | 156 | 105 | 209 | 124 | 104 | 86 | 82 |
| FP2 | | 109 | 100 | 96 | 93 | 105 | 93 | 91 | 73 | 72 |
| FP3 | | | 215 | 132 | 100 | 180 | 121 | 100 | 82 | 78 |
| FP4 | | | | 347 | 98 | 262 | 127 | 107 | 94 | 83 |
| FP5 | | | | | 114 | 103 | 99 | 91 | 69 | 69 |
| FP6 | | | | | | 909 | 185 | 131 | 132 | 125 |
| FP7 | | | | | | | 208 | 103 | 83 | 85 |
| FP8 | | | | | | | | 144 | 76 | 77 |
| FP9 | | | | | | | | | 143 | 86 |
| FP10 | | | | | | | | | | 138 |

marketing (76.80%), which includes services marketing. In non-marketing research areas, researchers have recently adopted S-D logic. S-D logic has gained attention in the fields of management, technology and operations, lagged and consecutive.

Analyzing the substantive domain of the set with respect to the use of FPs (i.e. with the conceptual domain) yields further insights, as presented in Table 6. Marketing and services research show the more substantial use of all FPs, with FP6 (co-creation) being the most frequently addressed, followed by FP1 and FP4. In management, information technology and operations research, research is most often related to FP6. While management research exhibits a broad use of the FPs, operations research shows the most balanced interest of higher to lower FPs. The developments in the substantive subdomains show that S-D logic is used particularly in research settings that are related to (1) service, (2) innovation/creation, (3) interaction/relation, (4) and networks of actors. Overall, the dissemination of S-D logic research throughout various content areas increases significantly over time, while its institutionalization lags behind, which is consistent with expectations of introducing novel ideas to research.

To further address the goal of understanding the usage of S-D logic and the degree of dissemination versus institutionalization, we compared the relative use of FP1 and FP6 in marketing versus other fields. As the number of publications outside of marketing is comparatively small, we combined the four groups shown in Table 6 into a single group for a meaningful analysis. Table 7 compares the instances of FP1 and FP6 in "marketing & services" and "combined other fields". Note that the similarity of percentage of use of FP1 and FP6 suggests that the degree of dissemination and institutionalization is similar in marketing and all other fields (addendum to finding f3). Given the amount of overall attention of S-D logic in "marketing & services", one might hypothesize that it would be more institutionalized in "marketing & services" versus the combined other fields, but these findings would not provide evidence to support this assumption: Marketing is not necessarily "ahead" of other fields in S-D logic usage.

In addition, we find that all substantive domains in the set display two similar problems: One is conducting S-D logic research while using established, G-D logic, terminology and lexica (e.g. using terms like "end-consumers"), which means risking a "mental relapse" and G-D logic outcomes to S-D logic inspired research and hence falling short of its full potential (finding f8). The other is that only few studies directly address what phenomena or new aspects S-D logic helped to uncover, making it hard to gauge the value of S-D logic to existing and established fields and lines of thought (finding f9).

## 4.4 *The Methodological Domain of S-D Logic Research*

This section explores the nature of the research conducted in our set, the data sources, and the inferential tools used in analysis.

**Table 6** An overview of the use of foundational premises in the substantive domain of service-dominant logic research

| Combinations | FP1 | FP2 | FP3 | FP4 | FP5 | FP6 | FP7 | FP8 | FP9 | FP10 |
|---|---|---|---|---|---|---|---|---|---|---|
| Mkt. and Serv. | 242 | 89 | 175 | 274 | 92 | 710 | 167 | 117 | 116 | 107 |
| Management | 27 | 8 | 13 | 31 | 9 | 86 | 17 | 13 | 14 | 16 |
| Technology | 12 | 2 | 4 | 10 | 2 | 42 | 7 | 4 | 5 | 2 |
| Operations | 10 | 4 | 11 | 18 | 3 | 29 | 5 | 3 | 3 | 5 |
| Other | 10 | 6 | 12 | 14 | 8 | 42 | 12 | 7 | 5 | 8 |

**Table 7** Comparison of the use of FP1 and FP6 in the substantive domain of S-D logic research

| Domain | FP1 | Percentage (%) | FP6 | Percentage (%) |
|---|---|---|---|---|
| Marketing and services | 242 | 78.3 | 710 | 78.1 |
| Combined other | 59 | 21.7 | 199 | 21.9 |
| Total | 301 | 100.0 | 909 | 100.0 |

**Table 8** An overview of the types of research that incorporate S-D logic

| Type of research | Absolute | Percentage (%) | Thereof FP1 (%) | Thereof FP6 (%) |
|---|---|---|---|---|
| Conceptual | 1051 | 60.96 | 20.84 | 51.95 |
| Qualitative | 385 | 22.33 | 13.77 | 56.36 |
| Quantitative | 288 | 16.71 | 10.07 | 50.69 |

S-D logic research is dominated by conceptual publications that indicate that S-D logic is used for innovating or reconsidering research domains rather than testing hypotheses (finding f10). In detail, more than 60% of S-D logic research is conceptual in nature, while 22% of the publications that include S-D logic also contain qualitative analyses, and the remaining 18% focus on quantitative research (mostly using various types of regression models or analytical modeling). 53% of the research relies on secondary data, 28% does not use any substantial data, the remainder uses industry, customer, and employee surveys as data. A common shortcoming in current methods used in S-D logic research lies in reporting: Methodological challenges overcome by S-D logic or innovative approaches to hypotheses testing inspired by S-D logic, are rarely reported (finding f11).

Table 8 shows the distribution of research approaches using S-D logic, followed by the usage of FP1 and FP6 by each of the three categories. The data highlights that all three categories (conceptual, qualitative, and quantitative) reference FP6 more than half of the time, showing a clear focus on value co-creation. However, FP1 is referenced in more than 20% of the conceptual studies, while only 14% in qualitative and 10% in quantitative studies. This suggests that the institutionalization of S-D logic is being carried in conceptual research through FP1, and in empirical research through testing FP6 (finding f12).

## 5 Recommendations

The purpose of this chapter has been to understand the use of S-D logic in research in order to maximize its potential for scholarly discovery and application in business and teaching.

The literature review offers four key contributions: (1) It finds that current S-D logic research falls into two broad categories with a 70:30 split: dissemination (merely talking about S-D logic) and institutionalization (actually using S-D logic). (2) It provides evidence that the dissemination-related type of research has obscured the contributions of institutionalization-related type of research with its sheer mass. (3) It revealed that while the most frequently referenced principle "co-creation of value" (FP6) is the driver of S-D logic, dissemination and empirical institutionalization, and the principle "service exchange" (FP1) is the driver of conceptual institutionalization. (4) It showed that S-D logic is primarily used as an accelerator of conceptual thought. Furthermore, it provides evidence that these four key findings are consistent over time, research domains, and geographies.

Based on the review results, and to guide future research and advancements in S-D logic, we propose a contextual and goal-oriented structuring framework for further research using S-D logic. We derive this framework from the patterns identified in institutionalization-type research (finding f1) as well as from the shortcomings identified in the review (f2-12), and combine these insights with the logical baseline research approach of contrasting the S-D and G-D logic ("changing views"). The relations of the framework to our findings are noted with "fx"). The framework with recommendations is presented in Table 9.

This proposed structure results in numerous targeted applications of S-D logic research, for example: (1) compare G-D logic and S-D logic and see whether outcomes are different. This applies to both conceptual and explorative research as well as hypothesis-driven research (G-D logic prediction versus S-D logic prediction) and analytical modeling (actors behaving according to G-D or S-D logic); (2) investigate what kind of new methods are adopted in S-D logic research and/or (3) what kind of barriers in S-D logic research needs to be overcome, and (4) whether it facilitated theory development elsewhere and how. Future research needs to argue specifically for the use of S-D logic, and to report its specific contribution(s) to the conceptual, substantive and methodological domains, while editors and reviewers need to prevent pure dissemination-related research.

By proposing future research orientations, we hope to clarify the best possible use of S-D logic research given specific research settings, context and goals. We also hope that this framework may be used in applied research projects where business models change from the provision of goods and services to digitally enabled services ("solution propositions") or co-creation ("customization"). Similarly, the presented framework should be tested, refined and extended, maybe also for use in classroom settings, i.e. to investigate case studies rooted in G-D logic to be transferred to S-D logic as a task for students ("what happens to processes and business model if product X becomes service Y").

Future research may build on this chapter to (a) analyze in more detail how, why and to what effect researchers use S-D logic, (b) how S-D logic research should be reviewed in order to ensure rigor, and (c) which S-D logic tools facilitate and accelerate the application of S-D logic in business practice.

**Table 9** Structuring framework for conducting research using S-D logic

| Research goal | Applying | Extending | Testing |
|---|---|---|---|
| Specific goal | Applying S-D logic to some phenomenon | Extending some theory, concept, method | Testing some theory, concept, method |
| Research type | Conceptual | Conceptual/empirical | Empirical |
| Grounding in S-D logic | Transcending of phenomenon into S-D logic, using all FPs, pronouncing FPs as needed, potentially starting with FP6 (f2, f3) | Identify and overcome G-D boundaries by applying S-D logic, using all FPs, pronouncing FPs as needed, with special relation to FP1 (f2, f12) | Holistic framing of the issue using all FPs, pronouncing FPs as needed (f2) |
| Basic approach | Contrast G-D to S-D logic in order to envision, explicate, relate, and debate the phenomenon as per the delta identified (f10) | Contrast G-D logic to S-D logic in order to identify and overcome existing shortcomings (f10) | Contrast G-D logic to S-D logic to derive competing, testable hypotheses to judge which leads to more accurate predictions (f5) |
| Terminology | Define the minimal S-D logic terminology and vocabulary necessary (f8)<br>Explicitly state the core contribution/value of using S-D logic (f6) | | |
| Substantive impact | What phenomena or new aspects did S-D logic uncover/explain and how? (f6)<br>What are implications of S-D logic and to whom do they make a difference? (f6) | | |
| Conceptual impact | How is S-D logic related to/different from the existing logic? (f7, f9)<br>Does S-D logic lead to different predictions/outcomes and how? (f7, f9)<br>How are constructs changed when using S-D logic? (f7, f9) | | |
| Methodological impact | What methodological challenges does S-D logic help overcome and how? (f11)<br>How does S-D logic change the way that research is conducted? (f11)<br>How does S-D logic enable generalizable methodological advances? (f11, f5) | | |
| S-D logic impact | How does the research advance S-D logic itself (f1, f6)?<br>How does thought and behavior change as a result of using S-D logic? (f9) | | |

# References


1. Vargo, S.L., Lusch, R.F.: Evolving to a new dominant logic for marketing. J. Mark. **68**(1), 1–17 (2004)
2. Brust, L., Antons, D., Breidbach, C., Salge, T.O.: Service-dominant logic and information systems research: a review and analysis using topic modeling. In: 38th International Conference

on Information Systems: Transforming Society with Digital Innovation, ICIS 2017. Association for Information Systems (2018)
3. Kryvinska, N., Olexova, R., Dohmen, P., Strauss, C.: The s-d logic phenomenon—conceptualization and systematization by reviewing the literature of a decade (2004–2013). J. Serv. Sci. Res. **5**(1), 35–94 (2013)
4. Olexova, R., Kubickova, V.: The evolution of the s-d logic approach and its impact on service science. J. Serv. Sci. Res. **6**(1), 99–124 (2014)
5. Wilden, R., Akaka, M., Karpen, I., Hohberger, J.: The evolution and prospects of service-dominant logic: an investigation of past, present, and future research. J. Serv. Res. **20**(4), 345–361 (2017)
6. Vargo, S.L., Lusch, R.F.: Service-dominant logic: continuing the evolution. J. Acad. Mark. Sci. **36**(1), 1–10 (2008)
7. Chandler, J.D., Vargo, S.L.: Contextualization and value-in-context: how context frames exchange. Mark. Theory **11**(1), 35–49 (2011)
8. Grönroos, C.: Conceptualising value co-creation: a journey to the 1970s and back to the future. Journal of Mark. Manag. **28**(13–14), 1520–1534 (2012)
9. Akaka, M.A., Vargo, S.L., Lusch, R.F.: The complexity of context: a service ecosystems approach for international marketing. J. Int. Mark. **21**(4), 1–20 (2013)
10. Vargo, S.L., Lusch, R.F.: Why service? J. Acad. Mark. Sci. **36**(1), 25–38 (2008)
11. Brinberg, D., McGrath, J.E.: Validity and the Research Process. Sage, Beverly Hills, CA (1985)
12. Grewal, D., Levy, M.: Retailing research: past, present and future. J. Retail. **83**(4), 447–464 (2007)
13. Brown, J.R., Dant, R.P.: Scientific method and retailing research: a retrospective. J. Retail. **84**(1), 1–13 (2008)

# Blockchain Technologies Towards Data Privacy—Hyperledger Sawtooth as Unit of Analysis

**Pascal Moriggl, Petra Maria Asprion, and Bettina Schneider**

**Abstract** For digital business models data is the most crucial asset—this calls for increased awareness of appropriate privacy protection measures. The European Union *General Data Protection Regulation* is a consequence that followed the discussions and now forces organizations to ensure that their information ecosystems comply with the law. There is currently an emerging trend to apply blockchain technologies to business models that rely on data exchange, because the technology promises to make a centralized data authority redundant. We have taken this as the purpose for our efforts to provide insights that will help decision-makers select a suitable blockchain configuration that complies with data privacy regulatory requirements. By applying design science, we created a morphological box along with a grid, serving as a 'data privacy assessment tool' for the blockchain configuration *Hyperledger Sawtooth*. The research results can potentially be generalized to assess any other blockchain configuration.

**Keywords** Blockchain · Data privacy · Information security · Distributed ledger · Hyperledger Sawtooth · Information ecosystem (IES)

## 1 Introduction

There is a growing interest in the fields of *blockchain technologies* and *data privacy*, which is reflected in an increasing number of publications. We analyzed relevant databases (Scopus, Web of Science, and IEEE Xplore) with the result that

P. Moriggl (✉) · P. M. Asprion · B. Schneider
School of Business, Institute for Information Systems, FHNW University of Applied Sciences and Arts Northwestern Switzerland, Peter Merian-Strasse 86, 4002 Basel, Switzerland
e-mail: pascal.moriggl@fhnw.ch

P. M. Asprion
e-mail: petra.asprion@fhnw.ch

B. Schneider
e-mail: bettina.schneider@fhnw.ch



R. Dornberger (ed.), *New Trends in Business Information Systems and Technology*, Studies in Systems, Decision and Control 294,
https://doi.org/10.1007/978-3-030-48332-6_20

in September 2019 Scopus showed 512 hits, Web of Science resulted in 398 hits and in IEEE Xplore 54 hits were returned; in comparison, in 2015 the identical keyword search in Scopus, Web of Science and IEEE Xplore led to only one hit.

Applicable since 2018, the European Union (EU) *General Data Protection Regulation* (GDPR) is a strong driver for researchers and practitioners to evaluate the impact of privacy regulations on an information ecosystem (IES) or on business processes. Over the past decade, blockchain, as an emerging technology with various configurations (e.g., Bitcoin, Ethereum, and Hyperledger) and with high disruptive potential, has enabled new opportunities to design and execute business processes; this is especially relevant for processes that involve a variety of IESs and associated stakeholders. Although there are some research activities that use blockchain configurations to ensure data privacy in IESs [1–3], there is no standardized assessment available that helps evaluate and compare different possible blockchain configurations for their suitability regarding data privacy capabilities.

One explanation for this lack of research interest might be the circumstance that under certain conditions, data privacy issues can be circumvented for blockchain applications. First, the problem of data privacy can be "outsourced" and therefore classified as not relevant because personal data could theoretically be encrypted and pseudonymized outside the blockchain (off-chain); with this approach, personal data processing risks and data privacy requirements can be by-passed for any blockchain-based solution. Another way to avoid data privacy issues is to store only *hashes* or *checksums* of the personal data on-chain. In addition, any personal data could be encrypted in order to hinder other network participants without a valid private key to see the content. However, with regard to encryption in general, it needs to be discussed whether it will withstand future advancements in technologies that potentially could decrypt its content [4].

Despite these avoidance strategies, the storing and processing of private sensitive data in blockchain configurations occurs in practice and we consider data privacy as highly relevant for our society, driving us to advance this unresolved research issue. Therefore, the goal of this study is to design a conceptual assessment—a morphological box along with an assessment grid—that allows the comparison of blockchain configurations based on their data privacy capabilities. The morphological box and particularly the grid provide a starting point for further discussions and refinements that could lead to a standardized framework for assessing data privacy in a blockchain-based IES. Target audiences are both researchers and decision-makers in organizations that analyze or plan to establish blockchain-based solutions.

This study follows a design-based research approach [5]. First, according to [5] the existing knowledge around data privacy and blockchain solutions is analyzed, condensed, and used to scope this research. Second, the design process is outlined, starting with elaborating blockchain- and privacy-related characteristics to include in our morphological box. This leads to the foundational artifact, an assessment grid, evaluated by applying it to a dedicated blockchain configuration. The artifact makes use of the morphological analysis, a general method for non-quantified modeling [6]. The development of the morphological grid follows privacy audit criteria for IESs [7, 8] adapted to be used for blockchain-based solutions dealing with private data.

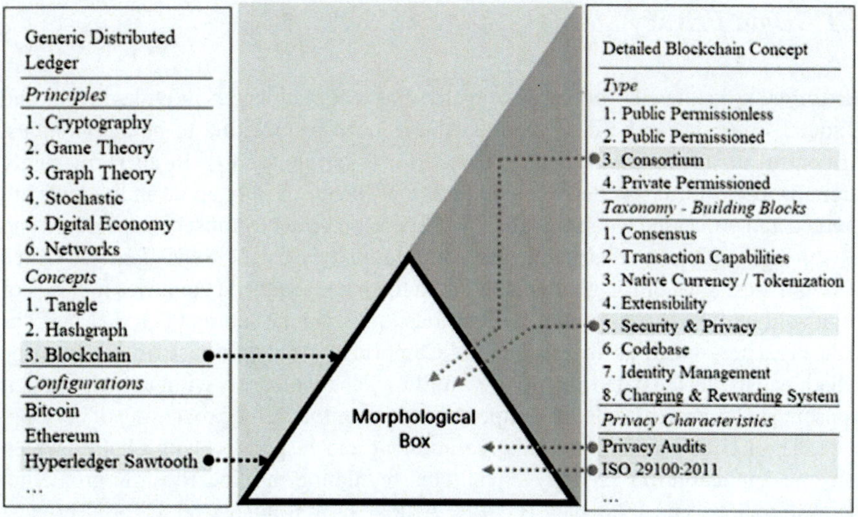


**Fig. 1** Research domain resulting in a morphological box

Figure 1 shows (1) our research domain synthesis, a "Generic Distributed Ledger" with its three abstraction layers, related characteristics/attributes (left box) and (2) the elements from one "Detailed Blockchain Concept" (right box). We selected an exemplary *Hyperledger Sawtooth*, and (3) in the center of Fig. 1, the specific elements that are relevant to design our desired artifact—the morphological box (the triad and related arrows). The characters marked in grey in the left box set the artifact boundaries, whereas the ones marked in grey in the right box set its content.

The remainder of this contribution is structured as follows. The foundations of our research—data privacy and blockchain—are elaborated in Sect. 2. In Sect. 3, we discuss the characteristics forming the morphological box; as baseline, we use blockchain and data privacy characteristics to discuss relevant attributes. These include the application of the morphological box to highlight how *Hyperledger Sawtooth* fulfills data privacy requirements at the technical level. The contribution concludes in Sect. 4 with a short summary of the practicability and further research opportunities in the field of technical data privacy characteristics.

## 2 Foundation

The foundation section covers data privacy and blockchain as central units of analysis. It elaborates firstly the concept of data privacy and highlights its increasing relevance in digitalized IESs. Secondly, relevant essentials of blockchain will be introduced that allow demarcating the boundaries of this research.

## 2.1 Data Privacy

Historically, in the EU, privacy is regarded as a fundamental individual right and a social value; it can be described as "the right to private life, to be autonomous, in control of information about yourself, to be let alone" [9]. In an increasingly digitalized world, it has become common that plenty of data about individuals are created and processed electronically [10]. Therefore, privacy must be considered and established with particular care. According to a study by the Swiss Federal Office for Statistics, 87% of EU residents living in the most advanced countries in terms of digitization provide personal data of various kinds via the Internet [11, 12]. Serious data breaches—such as the scandal of Cambridge Analytica, a British company, which had harvested data from up to 87 million Facebook users without their explicit consent [13]—have revealed the importance of monitoring the processing of personal data. The EU had anticipated this requirement and responded with GDPR, which became applicable on 25 May 2018. This regulation, aiming towards protecting natural persons and their data (GDPR, Article 1), unfolds extraterritorial effects; it refers to any data processing related to individuals who are in an EU member state—irrespective whether or not the data processing itself takes place in the EU [14]. Consequently, this means every organization, across all industries and locations, except non-EU countries without EU trade relationships, must verify whether it is affected and is required to comply with GDPR [9]. As a very special case, non-EU organizations, such as higher education institutions must also consider GDPR [15]. As the new regulation not only replaces but also updates the preceding EU Data Protection Act (DPA) from 1998, several new elements have been introduced. An example is the mandatory data protection impact assessment, "a process to manage risks related to the processing of personal data by assessing and mitigating such information" [16].

In addition, the GDPR leads to a close link between data protection and information security. GDPR Article 32 explicitly obliges organizations to process personal data securely using "appropriate technical and organizational measures", such as encryption and pseudonymization. The Information Commissioner's Office (ICO), a regulatory body used to protect information rights in the public interest, confirms that concepts originally stemm from the field of information security [17], e.g. from the CIA triad [18]—a widely used model for managing information security policies. The abbreviation CIA comes from the attributes (1) *Confidentiality* as a set of rules that limits access to information, (2) *Integrity* as the assurance that the information is trustworthy and accurate, and (3) *Availability* as a guarantee of reliable access to the information by authorized people [17].

The three attributes of the CIA triad are essential information security characteristics and delineate the boundaries in which our research operates. The attributes of the CIA triad will be referred to when building and testing the research results.

## 2.2 Blockchain

The term blockchain first appeared in 2008 after *Satoshi Nakamoto*, a pseudonymous person or group, published an essay about a peer-to-peer electronic cash system using the terms *blocks* and *chain* [19]. Nowadays, blockchain is understood as a concept with the umbrella term *distributed ledger*. According to Burkhardt et al. [20], a distributed ledger can be defined by complementary principles, which focus on mathematical and information technology-related aspects, such as *cryptography, stochastics, graph theory,* and *network* structures. The principles themselves focus on economic aspects: 'crypto-economics' exploiting the *game theory* [21, 22] and strategies to build consensus in untrusted environments and issues of a 'digital economy', e.g. double-spending digital money or dealing with cryptocurrencies [23].

The distributed ledger principles can be assembled in particular ways forming three distinctive distributed ledger services (DLS) [24]: (1) tangle, (2) hashgraph, and (3) blockchain. The three concepts differ in their architecture, where tangle and hashgraph are based on a 'directed acyclic graph', and blockchain is based on its own 'blockchain'-technology [20]. The latter became widely known among the public and in the scientific community when in 2009 the first cryptocurrency whitepaper titled as *Bitcoin* was released [19, 25]. Today, many implementations of the blockchain concept are actively developed by researchers or leading commercial players (e.g., IBM, Intel, Ethereum Foundation).

For this study, Hyperledger is chosen as an analytical unit to test the applicability of the developed assessment instrument, the morphological box. Hyperledger is considered the most appropriate because it is an open-source collaborative effort designed to advance cross-industry blockchain technologies. The underlying design philosophy of Hyperledger fits perfectly with this research as it aims to keep ledgers distributed and make smart contracts safe, in particular for enterprise applications. Its distinctive architecture separates the core system from the application domain, which aims on the one hand to simplify blockchain application development and on the other hand allows developers to choose their preferred programming language [26]. Hyperledger is based on a global collaboration including leaders of different industries. The Linux Foundation, known as a worldwide acting non-profit technology consortium, is hosting it [26]. The selected configuration for this study, *Hyperledger Sawtooth*, in the following abbreviated as *Sawtooth* is a project under the Hyperledger umbrella. It is a modular platform for creating, deploying, and running distributed ledgers [26]. For our desired artifact, an assessment tool in the form of a morphological box, the fundamentals of blockchain set the boundaries. Table 1 consolidates the distributed ledger characteristics, namely the principles, the DLS and the configuration along with respective values (alphabetically ordered). We highlighted the values delineating the scope (italics formatting)—general distributed ledger characteristics, focusing on blockchain such as DLS and Sawtooth as the exemplary unit of analysis.

**Table 1** Generic distributed ledger characteristics setting the scope of this study

| Characteristics | Values | | | | | |
|---|---|---|---|---|---|---|
| Principle | *Cryptography* | *Digital Economy* | *Game theory* | *Graph theory* | *Networks* | *Stochastic* |
| Service | Tangle | | Hashgraph | | *Blockchain* | |
| Configuration | Bitcoin | Ethereum | IOTA | *Hyperledger Sawtooth* | Ripple | … |

# 3 The Morphological Box

In the previous section, we introduced the basics of data privacy, information security, and blockchain to demonstrate research boundaries. This section elaborates the building blocks for the morphological box drawing on existing research work. It starts with the blockchain-related characteristics followed by data privacy characteristics. Currently, there is no conclusive research on blockchain configurations that best correspond to data privacy requirements because of the difficulty of matching technical features with data privacy requirements that do not necessarily dictate configuration choices.

## *3.1 Blockchain Characteristics*

Two relevant characteristics, the "Blockchain Design Elements" and the "Blockchain Type" will be incorporated into the assessment tool. Table 2 shows both characteristics and related values; values that apply to our unit of analysis, Sawtooth, are highlighted (italics formatting).

### 3.1.1 Blockchain Types

When selecting the blockchain-related characteristics to build the morphological box, we first draw on one of the blockchain principles described in Sect. 2.2,

**Table 2** Blockchain-related characteristics of the morphological box

| Characteristics | Values | | | |
|---|---|---|---|---|
| Blockchain design elements | Charging and rewarding | *Consensus* | *Identity management* | Tokenization |
| | *Codebase* | *Extensibility* | *Security/Privacy* | *Transaction capabilities* |
| *Blockchain type* | *Consortium* | Private permissioned | Public permissioned | Public permissionless |

namely the network structure principle that focuses on different participation modes; this principle is used to classify a distributed ledger based on its centralization degree and participation mode. The network principle differentiates between three degrees: *centralized*, *decentralized*, or *distributed*, and two participation modes: *permissioned*, or *permissionless* [27].

Blockchain as a concept aims to achieve a decentralized consensus on valid ledger entries among untrusted participants. The blockchain concepts are compelling and can be differentiated in two classes of blockchain types that are either "*open* or *closed*", as visualized in Table 3 where example configurations are outlined that are currently used in several applications of the concept of blockchain. When it comes to our unit of analysis, Sawtooth, it is considered an enterprise blockchain platform and allows inherently to deploy different blockchain types, but for this project the base is *consortium.*

According to Burkhardt et al. [20], a blockchain is composed of six building blocks (1) transactions, (2) roles, (3) blocks, (4) verification and validation processes, (5) algorithms, and (6) cryptography. In general, transactions (1) are transparent and visible to each participant, depending on the two complementary blockchain types (Table 3). Participants in a blockchain-based network have an identical copy of the current ledger [20], depending on the permission settings and the blockchain design.

Blockchain participants have roles (2) with associated tasks to interact with each other and find consensus. These roles work together and build a network that provides an unalterable history of data exchange in form of transactions between the participants, whereas all transactions are stored in blocks. Roles, for example, are "smart contracts", "endorsers", "committers", "validators", and "orderers" [29]. Blocks (3) are cryptographically concatenated and always chained to the previous block. All

**Table 3** Competing blockchain types [28]

| | | | Read | Write | Commit | Examples |
|---|---|---|---|---|---|---|
| Blockchain type | Open | Public permissionless | Open to everyone | Everyone | Everyone | Ethereum, Ethereum classic |
| | | Public permissioned | Open to everyone | Authorised participants | All or subset of authorized participants | Sovrin |
| | closed | Consortium | Restricted to an authorised set of participants | Authorised participants | All or subset of authorized participants | Multiple banks operating a shared ledger |
| | | Private permissioned | Fully private or restricted to a limited authorised set of participants | Network operator only | Network operator only | Internal bank ledger shared between parent companies |

blocks together form the blockchain (ledger) and depict a single point of truth to all nodes that have access to it. In this context, a node can be either a full node or a light node that participants use to access the network. A full node downloads and stores the whole blockchain and is designed to verify and validate (4) transactions back to the very first block ("genesis block"), whereas light nodes only download the block header of the previous block. Light nodes rely on full nodes for operations requiring the complete blockchain, thus, the data storage is less consuming/expensive for the participant of light nodes [29]. Blockchain participants have an identical copy of the current ledger [20], depending on permission settings and technical design,. The participants agree on the current ledger state by relying on a (consensus) algorithm (5), and today many different algorithms exist [30]. The encryption (6) as the last building block additionally increases the difficulty for attempts that want to alter the ledger state history and thus assures a high degree of immutability [31].

### 3.1.2 Core Design Elements

Tasca and Tessone [32] developed a taxonomy for blockchain which can be used as a blockchain reference standard. The reference standard covers, among other things, the smallest required technical elements that are part of a blockchain configuration; they allow to differentiate the configurations at a technical level, which reflects the different purposes they target. The taxonomy provides a clear structure to categorize the relationships of the inherent blocks and is a sufficient tool, when applied rigorously, to compare different, heterogeneous configurations [21].

The taxonomy encompasses the identification, description, nomenclature, and hierarchical classification of blockchain components. In addition, it groups them in a hierarchical structure that highlights functional relations and design patterns. The taxonomy focuses on generic blockchain design choices and consists of eight core design elements:

(1) *Consensus*: The consensus describes the validation and verification process that leads to mutual trust among participants in a blockchain network. The consensus directly affects the reliability, authenticity, and accuracy of the stored data within the blockchain [30]. The consensus algorithms are different regarding immutability and failure tolerance, latency, and finality and relate "to the set of rules and mechanics that allows for the maintenance and updating of the ledger and guarantees the trustworthiness of the records in it." [32].

(2) *Transaction Capabilities*: Transaction capabilities determine transaction scalability and transaction usability. In this context, a transaction initiates a ledger state change. Transaction scalability relates to quantitative measures such as transactions per second (TPS) that allow a performance comparison between a blockchain and other solutions (e.g., MasterCard Payment Gateway) that serve similar functionality, while transaction usability relates to the degree, to which the transactions are suitable to be used in IESs [32].

(3) Native Currency/Tokenization: The 'digital economy' (value in Table 1) includes the so-called *cryptonomics* empowered through tokens. The Swiss Financial Market Supervisory Authority (FINMA) classifies tokens as blockchain-based units that can be categorized as either (1) payment tokens, (2) utility tokens, or (3) asset tokens. The first, payment tokens, are native currencies such as Bitcoin—they have no other functions aside from the currency. The second, utility tokens, provide digital access to services or applications; the third, asset tokens, represent ownership of an underlying asset, which can be both physical or digital [33].

(4) Extensibility: Extensibility is the degree to which a blockchain network can be extended to interact with elements outside the network. Extensibility stands for interoperability, intraoperability, governance, and scripting language components as part of a blockchain. As an example, the governance can be performed through an open-source community (e.g. Hyperledger), through a technical solution provider (e.g. Microsoft), or through an alliance (e.g. R3) [32].

(5) Security and Privacy: Security and privacy design differentiate between data encryption and data privacy settings. Data encryption stands for the encryption algorithms that are used in a blockchain to ensure (1) integrity, (2) authenticity and (3) the correct order of events. (1) and (2) can be aligned with two of the attributes from the CIA triad—*integrity* and *confidentiality* [32]. Data privacy can either be an integral part of the blockchain network ('by design') or an add-on relying on external solutions [32].

(6) Codebase: The coding language, code licensing, and software architecture define the codebase. The latter can be described as a collection of source code used to build a system, application, or one of its components. A blockchain supports either single or multiple languages, open-source or closed-source license and either a monolithic or a polythic software architecture design [32].

(7) Identity Management: The blockchain identity management builds on three layers: (1) identity, (2) access, and (3) control. (1) and (2) can be aligned with the attributes *confidentiality* and *integrity* from the CIA triad [32]. The access and control layer refers to the blockchain type (Table 3) [32].

(8) Charging and Rewarding System: A blockchain depends on computing power using hardware and electricity. A cost model that organizes resources and rewards contributors is essential for a blockchain to ensure the ongoing service and the *availability*—the third CIA triad attribute [32]. The incentives that root in the charging and reward system do not apply to all blockchain architectures, because such modalities could also be defined outside the blockchain system. A simple solution would be a consortium with participating members that contribute to the same amount of resources (e.g. a full node).

The eight core design characteristics and their outlined specifics are an excellent theory-based foundation to map and compare complementary blockchain configurations. However, the taxonomy does not provide guidance for comparison and, finally, the selection of a specific blockchain configuration based on dedicated requirements, such as data privacy to comply with a specific law such as GDPR. Consequently,

**Table 4** Data privacy-related characteristics of the morphological box

| Characteristic | Values | | |
|---|---|---|---|
| Data operation | Store | Transfer | Use |
| Permissions, policies and roles | Least privilege | Logical access | Segregation of duties |
| Encryption | Data encryption | | |
| Security mechanism | Monitoring and logging | Third-party controls | Transparent changes |

we claim that it is necessary to add guiding factors that explicitly target the privacy design attributes. This will lead to a more detailed taxonomy and a prioritization scheme.

## 3.2 *Data Privacy Characteristics*

Four technically relevant audit characteristics, "Data Operation", "Permissions, Policies, Roles", "Encryption" and "Security Mechanism" will be incorporated to enrich our morphological box. Similar to the previous section, these characteristics draw on existing contributions and will be explained in more detail in the following paragraphs. Table 4 summarizes the data privacy-related characteristics.

As a specific contribution to closing the derived research gap, we united the data privacy-related characteristics of the morphological box into a novel assessment matrix—we call it 'morphological grid' (Fig. 2). Instead of highlighting values in Table 4, this time the newly developed morphological grid will be applied to our Sawtooth analyses unit. The parameters are based on audit guidelines.

| **Safeguarding Controls** | | **Data Operations** | | |
|---|---|---|---|---|
| | | **Use** | **Transfer** | **Store** |
| **Permissioning, Policies, Roles** | | **1** | **2** | **3** |
| Logical Access Controls | **a** | + | + | ++ |
| Least Privilege | **b** | + | ++ | ++ |
| Segregation of Duties | **c** | + | ++ | ++ |
| **Encryption** | | | | |
| Data Encryption | **d** | + | ++ | ++ |
| **Overall Security Mechanism** | | | | |
| Transparent Changes | **e** | + | ++ | + |
| Third-party Controls | **f** | + | - | - |
| Monitoring & Logging | **g** | ++ | ++ | ++ |

++ default
+ on demand
- missing

**Fig. 2** Morphological grid for assessing blockchain-privacy applied to Sawtooth

An audit is a sufficient procedure to review relevant activities in an organization that affects the IES and thus data privacy [34]. The IES and the related information life cycle ranges from data creation/collection to deletion and can be assessed regarding its data privacy maturity [35]. To conduct a data privacy audit, an audit framework/standard should be used that enables organizations to assess their IES in a structured and recommended way. The results should reflect an organization´s ability to comply with given privacy requirements. Some leading audit-related institutions work on common privacy assessment standards that include specific guidelines to audit IESs [34]. For example, the International Standard Organization (ISO) provides the "ISO 29100:2011 Information technology—Security techniques—Privacy framework" [36], which includes essential elements of a data privacy audit [37]. To apply ISO 29100:2011 for the assessment of blockchain-based IESs, the standard needs to be adopted since it covers business logics, e.g., legal requirements related to the business case and not to the enabling IES, and because it handles data operations that are usually not part of blockchain configurations but of 'classical' (relational) databases. The ISO 29100:2011 provides privacy safeguarding controls, which are techniques or practices that can be aligned with data operations within IESs. The related controls are broadly categorized into (1) permissioning, (2) policies and roles, (3) encryption, and (4) overall security mechanism, and are adopted from an exemplary data privacy controls audit [38].

In order to judge the privacy capabilities of a blockchain configuration, the relevant data privacy characteristics must first be established. When technical data privacy attributes are defined, a blockchain configuration's privacy capabilities can be assessed by analyzing which defined privacy attributes it fulfills. The configuration has these attributes that are built in either by default or as needed, or they do not existent and cannot be implemented with this configuration.

For our study, we build on two existing works: Firstly, the data operations [39] from ISO 29100:2011 and secondly, detailed privacy safeguarding controls for a privacy control audit based on [38]. Other ISO 29100:2011 privacy framework elements were not considered because they are directly relevant for technical features and therefore ignored.

The developed morphological grid to assess blockchain configurations is shown in Fig. 2, with its three data operation-related columns "Use", "Transfer", and "Store"; these three operations match the blockchain data operations, where a state is stored ("store"), its status is transferred ("transfer") or accessed ("use"). In the columns, the attributes "Permissioning, Policies, Roles" refer to the more detailed safeguarding controls that are not described in the ISO framework. The controls are (a) "Logical Access Controls" for each data operation step (b) "Least Privilege" for the principles in place, meaning system components (user, processor, program), which must be able to access only the information and resources that are necessary for its legitimate purpose of either using, transferring, or storing data. Next, (c) "Segregation of Duties" forces the concept of having more than one person or system component to process the task of either using, transferring or storing the data. The second attribute part assesses the (d) "Data Encryption". The last attribute group looks at the "Overall Security Mechanism" with the perspective on (e) "Transparent Changes" that affect data operations. The attribute (f) "Third-party Controls" refers to controls that allow software from outside a blockchain network to control internal changes. In this context, a blockchain network is a closed system with only limited interaction with the outside world. Therefore, there are no possible specific third party controls over key roles (validators, processors, peers). The only interface that theoretically allows third-party controls is the client itself. The last security mechanism that is part of the morphological grid relates to (g) "Monitoring & Logging" procedures, also applied to the three data operations.

Figure 2, the morphological grid, is—in addition to the characteristics of the morphological box (Fig. 1) our main research contribution, the artifact with its relevant data operations on the x-axis, and seven outlined privacy safeguarding controls on the y-axis. The safeguarding controls correlate with *confidentiality*, one of the three attributes from the CIA triad [18]. Although a blockchain configuration often provides other attributes from the CIA triad (e.g. consensus) or availability (e.g. transaction capabilities), confidentiality is most relevant from a privacy perspective and therefore should be emphasized with a specific focus. The term *confidentiality* stands for a set of rules that limits access to information or according to the ISACA glossary is vital for "preserving authorized restrictions on access and disclosure, including means of protecting privacy and proprietary information" [40]. In the blockchain context, confidentiality is dedicated to authorized restrictions which relate to participants, their identities, permissions, and roles; access and disclosure of information relate to transactions that contain payloads (information), their flow through the system and their encryption.

The following three mutations are the base for setting the horizontal and vertical morphological grid axes. They describe the changes made to the original ISO 29100:2011 framework:

(1) Environmental factors including privacy principles, legal requirements, and the business case defining involved personal identifiable information (PII) are not part of the morphological grid and were removed. They belong to the business logic and define the steps that must be undertaken before deciding to use a blockchain configuration. The "collect" data operation depends on the business logic that specifies what data needs to be collected and how this task can therefore be performed by applications other than a blockchain. The data operation "destroy" is, by default, not foreseen to be implemented on a blockchain. In contrast to a blockchain configuration, "destroy" or "delete" is a standard operation for a regular database.
(2) "Privacy Safeguarding Controls" were specified by listing control elements that are part of privacy audits for IT system controls [38].
(3) Each grid indicates whether the configuration has the privacy capability integrated by default, on-demand, or whether it is missing and cannot be implemented on the configuration in question.

## 4 Conclusion and Outlook

The goal of this study was to develop a tool that helps organizations make decisions for a blockchain configuration that matches their use case. The developed morphological box and the grid allow to identify Sawtooth's technical capabilities that secure data and support data privacy at different levels. In order to apply the morphological box and the assessment grid, thorough knowledge about the assessed blockchain configuration is required, in terms of the architecture and the different component roles and not least how they work together. For this reason, our theoretical approach is perfect for open-source configurations. It does, however—and this is one limitation of our research—potentially not work for closed-source configurations because of their lack of transparency. Another limitation is that the morphological box and the related grid do not distinguish between diverse encryption methods, and therefore inherently assume that any encryption method relies on a proper encryption algorithm. Nevertheless, the research output can be used (on a high/medium level) to evaluate each blockchain configuration, in particular with regard to its data privacy capabilities.

Future research opportunities could be firstly a more detailed breakdown of the outlined privacy safeguarding controls based on different or complementary security frameworks. Secondly, a thorough guide could be developed for different target groups (more or less experienced with blockchain and data privacy) that explains the application of the developed artifact and related assumptions. Finally, the developed artifact could be enhanced with information on other blockchain configurations (e.g., Stellar, Quorum). Sawtooth is a modular platform that comes—by default—with robust security functionalities and offers various customizing options. The validation on Sawtooth worked well because Sawtooth is designed to be secure and appropriate for enterprise applications that naturally have various data privacy requirements. However, it remains unclear to what extent the artifact is able to reflect the actual data privacy capability of a blockchain configuration because many factors from regular privacy audits are not considered in this study (e.g., hardware controls). Further research on data privacy and blockchain configurations will hopefully path the way for organizations to adopt the technology at a faster rate, and eventually develop feasible applications that protect personal data in a better and auditable way.

## References

1. Gur, A.O., Oksuzer, S., Karaarslan, E.: Blockchain based metering and billing system proposal with privacy protection for the electric network, pp. 204–208 (2019)
2. Jiang, Y., Wang, C., Wang, Y., Gao, L.: A privacy-preserving e-commerce system based on the blockchain technology. In: IWBOSE 2019—2019 IEEE 2nd International Workshop Blockchain Oriented Software Engineering, pp. 50–55 (2019). https://doi.org/10.1109/IWBOSE.2019.8666470
3. Marsalek, A., Kollmann, C., Zefferer, T., Teufl, P.: Unleashing the full potential of blockchain technology for security-sensitive business applications. In: 2019 IEEE International Conference

Blockchain Cryptocurrency, pp. 394–402 (2019). https://doi.org/10.1109/bloc.2019.8751444
4. Chang, H.: Blockchain: disrupting data protection? Priv. Laws Bus. Int. Rep. (2017)
5. Hevner, A.R., Chatterjee, S.: Design science research in information systems (2010)
6. Ritchey, T.: Adapted from "Fritz Zwicky, morphologie and policy analysis. In: General Morphological Analysis A general method for Non-quantified Modelling, pp. 2002–2013 (2013)
7. Bakis, Bruce J., J.S.M.: How to conduct a privacy audit. http://www.mitre.org/sites/default/files/pdf/HowToConductPrivacyAudit.pdf
8. Snedaker, S., Russ, R.: IT Security Project Management Handbook. In: Syngress IT Security Project Management, pp. 196–197, Canada (2006)
9. EUFRA: The EU's independent data protection authority. In: Handbook on European Data Protection Law, pp. 1–402. Publications Office of the European Union, Luxembourg (2018)
10. Tankard, C.: What the GDPR means for businesses. Netw. Secur. **2016**, 5–8 (2016). https://doi.org/10.1016/S1353-4858(16)30056-3
11. Federal Statistical Office: Erhebung zur Internetnutzung 2017. Digitale Kompetenzen, Schutz der Privatsphäre und Online-Bildung: die Schweiz im internationalen Vergleich. https://tinyurl.com/bfs-study-2017
12. eurostat: Digital economy and society statistics—households and individuals. https://tinyurl.com/eurostat-survey
13. BBC: Facebook fined GPB 500,000 for Cambridge Analytica scandal (2018). https://www.bbc.com/news/technology-45976300
14. European Union: Regulation (EU) 2016/679 (General Data Protection Regulation—GDPR). Off. J. Eur. Union. EN 1–88 (2016)
15. Habbabeh, A., Schneider, B., Asprion, P.M.: GDPR assessment instrument an exemplary case for higher education institutions. Int. J. Manag. Knowledge, Learn. 311 (2019)
16. Data Protection Working Party: Guidelines on data protection impact assessment (DPIA) and determining whether processing is "likely to result in a high risk" for the purposes of Regulation 2016/679 (WP29). Artic. 29 Data Prot. Work. Party. WP 248 rev 22 (2017). https://doi.org/10.2139/ssrn.2972855
17. Information Commissioner's Office: Guide to the general data protection regulation (GDPR). https://tinyurl.com/GDPR-UK
18. Perrin, C.: The CIA triad (2008). https://www.techrepublic.com/blog/it-security/the-cia-triad/
19. Nakamoto, S.: Bitcoin (2008). https://bitcoin.org/bitcoin.pdf
20. Burkhardt, D., Werling, M., Lasi, H.: Distributed ledger. In: 2018 IEEE International Conference on Engineering, Technology and Innovation. ICE/ITMC 2018—Proceedings, pp. 1–9 (2018). https://doi.org/10.1109/ICE.2018.8436299
21. Shapley, L.S.: A value for n-person games. In: Contributions to the Theory of Games II, Annals of Mathematics Studies. Princeton University Press (1953)
22. Kuhn, H.W.: Lecturers on the Theory of Games—Annals of Mathematics Studies. Princeton University Press, Princeton and Oxford (2003)
23. Bonneau, J., Miller, A., Clark, J., Narayanan, A., Kroll, J.A., Felten, E.W.: SoK: research perspectives and challenges for bitcoin and cryptocurrencies. In: Proceedings of IEEE Symposium on Security and Privacy, pp. 104–121 (2015). https://doi.org/10.1109/SP.2015.14
24. Margolis, E., Laurence, S.: The ontology of concepts—abstract objects or mental representations? Nous. **41**, 561–593 (2007)
25. Hobson, D.: What is bitcoin? XRDS crossroads. ACM Mag. Stud. **20**, 40 (2013). https://doi.org/10.1145/2510124
26. Hyperledger: Hyperledger Sawtooth project page (2019). https://www.hyperledger.org/projects/sawtooth
27. Sawnson, T.: Consensus-as-a-service: a brief report on the emergence of permissioned, distributed ledger systems (2015). http://www.ofnumbers.com/wp-content/uploads/2015/04/Permissioned-distributed-ledgers.pdf
28. Tasca, P.: Swiss Blockchain Research Symposium (2019)

29. Palai, A., Vora, M., Shah, A.: Empowering light nodes in blockchains with block summarization. In: 2018 9th IFIP International Conference on New Technologies, Mobility and Security NTMS 2018—Proceedings, pp. 1–5 (2018). https://doi.org/10.1109/NTMS.2018.8328735
30. Alsunaidi, S.J., Alhaidari, F.A.: A survey of consensus algorithms for blockchain technology. In: 2019 International Conference on Computer and Information Sciences, ICCIS 2019, pp. 1–6 (2019). https://doi.org/10.1109/ICCISci.2019.8716424
31. Hyperledger: Hyperledger Sawtooth documentation. https://sawtooth.hyperledger.org/docs
32. Tasca, P., Tessone, C.J.: A taxonomy of blockchain technologies: principles of identification and classification. Ledger **4**, 1–39 (2019). https://doi.org/10.5195/ledger.2019.140
33. FINMA: ICO Guidelines, pp. 1–11 (2018). https://doi.org/10.1515/9783598440397.13
34. Cooke, I.: IS audit basics: auditing data privacy. ISACA J. **3** (2018)
35. Riffat, M.: Privacy audit—methodology and related considerations. ISACA J. **1** (2014)
36. International Organization for Standardization: ISO/IEC 29100:2011. https://www.iso.org/standard/45123.html
37. Lachapelle, E., Ajvazi, B., Rama, F.: ISO 29100 how can organizations secure its privacy network? https://tinyurl.com/y2l4on83
38. Photopoulos, C.: Managing Catastrophic Loss of Sensitive Data. Syngress (2011)
39. Andress, J.: The Basics of Information Security: Understanding the Fundamentals of InfoSec in Theory and Practice. Syngress (2011)
40. ISACA: Glossary (2019). https://tinyurl.com/ISACAGlossary

# Leadership in the Age of Artificial Intelligence—Exploring Links and Implications in Internationally Operating Insurance Companies

**Sarah-Louise Richter and Dörte Resch**

**Abstract** This study examines the effect of artificial intelligence (AI) on leadership in internationally operating insurance companies. Because insurance is a model example of a data-intensive industry, companies are already applying AI-powered technology and searching for opportunities for further use. However, any important step in automation may trigger a leadership shift and a research gap has been identified in the development of leadership positions in insurance companies confronted with the use of AI. Specifically, the objective is to investigate how leadership could change due to the introduction of AI as an example of digitalization. For this study within an interpretive paradigm, qualitative data were collected in 19 semi-structured interviews, with interviewees representing five insurance companies headquartered in Western Europe. The findings suggest that the use of AI and its implications for leadership are closely linked to the underlying structures of the industry, which has led to the existing leadership discourse and organizational metaphor in the first place. The implications of AI, in turn, depend on the leadership discourse and existing structures. Thus, if AI is used only in accordance with the current discourse, the implications for leadership are minimal. Therefore, it can be concluded that the use of AI-powered software itself is unlikely to trigger change in leadership. Nevertheless, AI holds significant potential for insurers. For example, AI could support the insurer's core competencies, and connect companies with broader ecosystems and customer communities.

**Keywords** Leadership · Artificial intelligence · Insurance industry

S.-L. Richter (✉)
Institute for Information Systems, FHNW University of Applied Sciences and Arts Northwestern Switzerland, Riggenbachstrasse 16, 4600 Olten, Switzerland
e-mail: sarahlouise.richter@fhnw.ch

D. Resch
Institute for Human Resource Management, FHNW University of Applied Sciences and Arts Northwestern Switzerland, Riggenbachstrasse 16, 4600 Olten, Switzerland
e-mail: doerte.resch@fhnw.ch



R. Dornberger (ed.), *New Trends in Business Information Systems and Technology*, Studies in Systems, Decision and Control 294,
https://doi.org/10.1007/978-3-030-48332-6_21

# 1 Introduction

This study explores the impact of artificial intelligence (AI) applications on leadership at the middle management level in internationally operating insurance companies. In insurance, data-driven decision-making plays an important role and AI holds further potential for insurers [1] as the technology becomes affordable [2, 3].

However, every major step in automation triggers a change in leadership. This can be explained by social science approaches such as the Actor-Network-Theory (ANT), suggesting that technology changes workplace structures, collaboration, and relationships [4, 5]. Thus, the foundations of leadership are transformed [6, 7], which is reflected in leadership discourses that emerge as technology progresses [8]. This raises the question how leadership might be affected by the adoption of AI in its various forms and use cases. Hence, the research question that we seek to answer in this study is: *How might leadership at middle management levels in internationally operating insurance companies be influenced by the adoption of AI?*

Surprisingly enough, while the impact of AI on economies is widely debated, there is little research specifically focused on leadership. This research gap is worth addressing to generate insights that inform further academic research, while also providing practitioners with information from a non-commercial perspective. In general, this research aims to explore how leadership could change given the use of AI as a popular example of digitalization. Furthermore, this study contributes to the broader discussion about the impact of AI on business and society.

To approach this topic, we extract findings from existing literature and theory, examining the role of AI in insurance as well as the link to leadership. The methodology is then explained before the analysis of primary data is presented. The results are then discussed and linked to broader literature before conclusions are drawn.

We delimit the topic to the example of international insurance companies in the primary insurance industry. The reinsurance market, obligatory and supplementary health insurance, or pension funds are not examined. Moreover, questions of data security and privacy when applying AI are not addressed.

# 2 Artificial Intelligence in the Insurance Industry

Because insurance is a model example of a data-intensive industry, knowledge-based expert systems marked a first success in commercial applications of AI [3]. One elegant definition suggests that "Artificial Intelligence is the study of how to make computers do things at which, at the moment, people are better" [9], which implies that the definition changes as AI progresses. To assess how AI can be applied in the insurance industry, three categories are considered.

First, AI can be used to automate or support tasks such as managing customer data from online self-service portals, updating records, and extracting information from documents, e.g. using Natural Language Processing (NLP) [10]. This is particularly useful to manage information from multiple backend systems, as is often the case with international insurance companies. The required investments for such technology are comparably low, implementation is not overly complex and the system is transparent and auditable [11]. This option is often useful for reducing a company's cost base and improving the efficiency of routine work [12, 13].

Second, AI can be applied as a means to cognitive insights [11], to identify patterns in data and interpret their meaning. In insurance, this can support risk assessment, fraud prevention and detection [14]. In addition, predictive analytics for natural disasters, life expectancy estimations, and actuarial modelling are use cases for AI. Such applications require reliable and correct data to come to fruition [15, 16] and people need to judge and evaluate the results [13]. However, the technology is frequently considered a black box, which means that it is not transparent how the results were derived [11], which is problematic in a highly regulated industry [10].

Third, AI can be applied for cognitive engagement, e.g. using NLP [10, 11]. This can be chatbots handling customer requests [1]. The advantages include easy access to reliable and consistent information and increased speed of responding to requests. However, contemporary AI can provide data and identify patterns, but not explain them [13, 17] or abstract new ideas [18]. Thus, a human interpreter needs to add expertise, social and moral values [17, 19].

## 3 Leadership in the Age of Artificial Intelligence

A definition of leadership provided by Alvesson et al. [6], referring specifically to modern corporations and purposefully not separating leadership from management, states that "leaders are interpersonally trying to define meaning/reality for others who are inclined to (on a largely voluntary basis) accept such meaning-making and reality-defining influencing acts" [6]. However, scholars criticize the implied passive role of the followers and an exaggerated focus on the person of the leader [6, 20]. Moreover, Western's [8] eco-leadership discourse suggests that leadership does not necessarily reside within one person, but can be shared among people or be inherent in an organizational system. Hence, this study understands leadership as a system of relations and influences which might not be defined by hierarchy [6–8]. Complementing leadership itself, leadership discourse plays an important role. Western [8] defines discourse as "how we act and think in 'normative' or taken-for granted ways", both influencing and being influenced by social norms, behavioral patterns, power dynamics, and other accepted, yet subconscious factors. Western's [8] leadership discourses implicitly draw from the seminal work of Morgan's [21] organizational metaphors. Morgan [21] defines metaphors as "a way of thinking and a way of seeing that pervades how we understand our world", suggesting that the imagery used to depict organizations is indicative of this and shapes how leadership works in a given context.

When assessing the leadership discourse [8] and organizational metaphors [21] in German and Swiss insurance companies, the main characteristics are emphasis on rules, policies, and predictability. In addition, insurance companies show hierarchical orientation and an aversion to change, ambiguity, and risk-taking [22]. Typically, there is a controller leadership discourse and a machine metaphor at play, rooted in industry regulation and a history of little competition in a stable market where insurance products and prices were mostly standardized [23]. Hence, the deregulation of the industry, declining interest rates, and the need to remain profitable have left the industry struggling. At the same time, stricter regulation was introduced to protect customer interests after the 2008 financial crisis [22, 24]. Therefore, methods such as lean management and six sigma, aiming to reduce inefficiencies and achieve operational excellence, became popular in the financial services industry [24–26]. Reengineering through strict measurement and monitoring accounts for the prevalence of the controller leadership discourse and machine metaphor in insurance [21]. However, it causes companies to suffer from organizational inertia [21, 22]. This reveals an interesting contrast to contemporary leadership being understood as a network of relationships, people and organizational systems that develop, learn, and change in tune with its broader environment [27]. Within this setting, the use of AI raises a number of questions.

First, questions of ethics arise, which Western [8] puts at the heart of his so-called eco-leadership discourse, reflecting the organization being "rethought as 'ecosystems within ecosystems'" [8]. Additionally, scholars highlight ethical and moral issues related to AI [28, 29]. For example, fraud and misuse can occur when individuals manipulate outcomes, hide or over-emphasize information, and distort data interpretation. Moreover, individuals are prone to confirmation bias, which means that they typically interpret information by substantiating their personal view while ignoring contradictory data [30, 31], which increases the risk of data being misinterpreted or misused [12]. However, AI might also mitigate bias in human decision-making by providing data where otherwise a decision would have been made by gut feeling [19]. In any case, AI informs decisions that could have far-reaching influence, which will raise questions about ethical standards, moral judgement, and fairness. Hence, social and emotional skills will need to compliment technological skills in the workplace [32].

Another related issue for leadership is trust, which provides the basis of any social interaction [33, 34] and is crucial for leadership as a process of creating meaning. However, trusting a human manager or co-worker differs from trusting a software. Nevertheless, employees need to develop relevant knowledge and trust in AI to use it effectively and reduce the perceived risk and skepticism [35, 36]. In addition, dependency on technology might raise concerns. Insurers typically take a scientific approach to business and people in the insurance industry are likely to rely on data [22]. However, the use of AI could limit exposure to contradictory information, which, in interplay with confirmation bias [30, 31], could reduce triggers for critical thinking and reflection.

A related issue is the decrease in human interaction that could be caused by AI. Today, decision-making in international companies usually requires interpersonal coordination. However, if information is readily available, it can reduce the time spent together, thus limiting the opportunities for informal interpersonal contact.

This has an impact on leadership, as it relies to a great extent on network-based peer influence [6, 7]. Therefore, one challenge of leadership in the age of AI might be to find a balance that integrates human and digital interaction [37].

Hence, the literature review revealed questions that mark the starting point for this study. The current leadership discourse and organizational metaphor in the insurance industry works best for standardized tasks and processes. The use of AI is particularly attractive for the insurance industry, as it adheres strictly to predefined guidelines in compliance with regulation and specific terms and conditions. In that regard, AI excels and outranks people. However, this strength in consistency is also AI's core weakness. If a case deviates from the norm, the AI's result may be flawed and AI fails to consider ethical aspects. This reveals a key issue for leadership. Today, accountability and competencies are usually determined by written guidelines. Future issues will call for more holistic consideration. Especially in middle management, the focus is likely to shift from the assessment of industry expertise to emphasis on emotional, creative, and social skills. Therefore, leadership will need to facilitate a new quality of compass, facilitating ethical decision-making, relationship building in a digitalized workplace and trust between AI and humans. Hence, AI is expected to change and challenge leadership as the implications of AI in the workplace go beyond the technological perspective. In scenarios of self-organization and discourses such as eco-leadership, these issues become even more pressing, as there is not a single leader to be held responsible. This is particularly true for leadership in global organizations that require international coordination.

## 4 Research Methodology

This study features a qualitative research design, since the study aims to explore a contemporary phenomenon dealing with individual perceptions rather than measurable frequencies or incidences [38]. It is recognized that there is not a single truth to be found about the potential implications of AI on leadership, which is why this study is understood from a relativist ontology. In addition, a contextual epistemology is used, suggesting that knowledge can be true in a certain context, but is not universally transferable [38, 39], since leadership can hardly be understood outside of its setting [7]. The context of the study are five international insurance companies with headquarters or major business units in Germany and Switzerland. Primary qualitative data were collected in 19 semi-structured and problem-centered interviews [38–40] in July and August of 2018 with middle and senior managers, industry and AI experts, and employees. Due to the exploratory nature of the research and epistemological orientation, strictly purposive sampling is neither achievable nor desirable. Thus, elements of convenience and snowball sampling are reflected in the sample [38]. The interviewees are persons from 30 to 60 years of age, with a higher academic or vocational education and several years of experience in the insurance industry. They come from different countries, such as Germany, Switzerland, India, Canada, USA, and the United Kingdom, among others. The interviewees work in a variety of business areas and functions, such as customer service, claims management, strategy development, project management, HR, IT, and marketing. This diverse sample supports the acquisition of different views and balance perspectives, while also strengthening transferability. The method of thematic analysis was used to find patterns and capture meaning by encoding the interview transcripts [39, 41]. The final inductive and deductive codes were developed into overarching categories. Then, the key points of the categories were described and links to other categories examined [38, 39, 41].

## 5 Results of Analysis

The categories *artificial intelligence, trust in technology and processes,* and *interpersonal relationships and identification* are described in this section.

**Artificial Intelligence (AI)** The category *artificial intelligence* describes results concerning *application and opportunities of AI* as well as *risks and challenges concerning AI*. AI-powered tools are already applied in the insurance industry, e.g. for knowledge sharing, risk assessments or customer relationship management. In general, AI is understood as an enabler for increasing customer value, delivering better services, and efficient organization. However, the interviewees are aware of potential issues and claim that AI should be carefully programmed and monitored to prevent misconduct. There is also consensus that companies should be mindful about where and how they use AI while taking into account the customer's perspective.

**Trust in Technology and Processes** The category *trust in technology and processes* captures notions about the quality of processes and technology and the trust that is put into these by summarizing the codes *guiding decisions*, *institutionalized collaboration and communication*, and *bad experiences with IT*. The data shows that decisions are guided by mostly explicit guidelines and bypass processes are punished. People in the industry trust in processes with formal routines of collaboration and communication that are defined by hierarchy, silos, and legal requirements. However, technology raises issues. According to the data, insurance employees are frustrated with the IT landscape that consists of sub-standard legacy systems. Hence, there is a general skepticism towards new IT being implemented, as such efforts usually create problems instead of providing solutions. Nevertheless, interviewees report that they would appreciate IT systems that help them organize more efficiently and provide outstanding service to customers.

**Interpersonal Relationships and Identification** The category *interpersonal relationships and identification* aggregates the codes *family/team spirit and people orientation*, *trust in people* and *noble cause and company heritage*. With these aspects, the category refers directly to the definition of leadership established in Sect. 3. According to the data, the teams have established a strong sense of family and team spirit. In addition, interviewees highlight the noble cause of insurance and are proud to work for companies with a rich history. Interviewees also emphasized that the industry attracts a specific type of character that values stability, predictability,

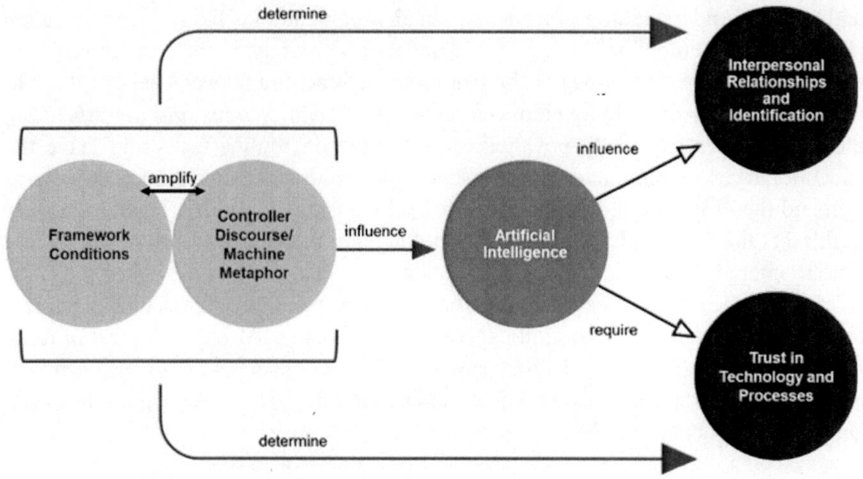


**Fig. 1** Relationships between categories (author's own illustration)

and long-term orientation. Thus, interviewees report that it is easy to connect and build trust within this homogenous workforce. The data indicates also that trust is built through personal interaction, by delivering good work, living up to one's promises, and demonstrating expertise. However, trust in other parts of the organization and senior management is reported to be low. According to the data, turnover in senior and C-level management is high, while middle managers and employees stay with the company for long tenures. Thus, the employees together with their direct managers went through times of painful changes together, which strengthened their mutual trust and gave them a sense of belonging. This implies that middle managers and their teams influence leadership culture much more than senior management.

Figure 1 illustrates the relationships between the categories and the status quo. Major links and tensions, which contribute to answering the research question, are examined hereinafter.

The data indicates that the industry's framework conditions, discourse, and metaphor determine *trust in technology and processes.* To fulfil requirements set up by the framework conditions, internal processes guide decisions and institutionalize collaboration and communication. Hence, there is perceived value and trust in processes which are organized along the formal hierarchy. However, interviewees highlight that, because of the prioritization of shareholder interests, investments in IT were neglected and the development of suitable state-of-the-art technology in insurance companies was impeded. Hence, trust in technology is low. However, a positive example of a company introducing an AI-powered knowledge-sharing platform illustrates that employees accept a technology that works well and adds value.

The data further suggest that all categories connect to the topic of AI. The framework conditions determine how insurers may use AI from a legal perspective. In addition, the understanding of leadership in an organization is likely to influence the purpose of implementing AI, e.g. if it is understood merely to cut costs or rather to move decision-making power to the front line and encourage peer-based collaboration. Concerning the relationship to *interpersonal relationships and identification*, the interviewees expect AI-powered tools to take on routine tasks and leave the exceptional cases to humans. These cases would require interdisciplinary collaboration and therefore strengthen interpersonal relationships and trust. The data further highlights that AI would eliminate the inefficiency of collaboration, leaving more time to interact with customers and live up to the noble cause of the business. Thus, referring back to the eco-leadership discourse, interviewees see AI as a tool to connect to customers and local communities. Nevertheless, using AI requires *trust in technology and processes* and adequate supervision to prevent issues for the company and customers. In turn, robust processes and explicit decision-making guidelines can support implementation and acceptance of AI.

## 6 Discussion of Results

After having elaborated on the results of the analysis, this section connects the findings to theory and extracts the key points for answering the research question. Based on the insights from the analysis it can be stated that AI-powered software is used in the insurance industry mainly in the form of isolated systems, each serving its specific purpose. In accordance with previous research, insurance companies predominantly seek to leverage AI to improve efficiency, gain insights, model risks through advanced analytics and engage with customers and employees [11–13, 42]. In addition, the analysis supports the fact that insurance companies are aware of the risks and challenges such as compliance and auditability of AI, lack of data quality, or potential social and moral issues. The issues for leadership that could be triggered by AI are discussed below.

Concerning relationships and trust, as the basis for leadership [6, 7], interpersonal relationships can be strengthened by the need for collaboration in cases beyond the capabilities of AI. However, the analysis revealed an interesting categorization of personal interaction. Some interviewees framed collaboration as being good or bad based on its efficiency, which could be enhanced by AI. However, this perspective disregards the basic principles of human interaction and trust-building, which is not efficient by design. Rather, the creation of meaning as a core element of leadership requires spending time together without pursuing a specific purpose and allowing serendipity to occur [6, 43]. Nevertheless, experts who are currently trapped in organizational silos could use AI to access data and contextual information by clicking a button, thus broadening their perspective and understanding their role in the customer journey. Looking at relationships with customers, while the network of sales persons is shrinking, AI could support reestablishing connections to communities. For instance, insurance companies could leverage their core competency of risk management to create value for customers [44, 45]. In addition, insights from predictive analytics could help individualize services and anticipate customers' life-changing moments which might trigger the need for new or revised insurance coverage [46]. However, the analysis has highlighted low levels of trust in technology in the sampled companies. Nevertheless, interviewees stated that they are looking forward to workplace technology being brought up to standard. That being said, if employees have experienced technology as reliable and well-working, studies suggest higher levels of trust in technology [47, 48]. This raises the question whether people would then rely too much on a software's output [49].

On the issue of accountability, the analysis showed that interviewees advocate strict monitoring of AI and claim that a human decision-maker should ultimately check the results and take responsibility. This means that AI should influence decisions, but should not interpret results or have the final say, which is also claimed in theory [50]. In addition, it is well known that basic knowledge of technology will be an important skill for every employee [51]. As a result, insurers are investing in the recruitment of tech-savvy talent and training the existing workforce with technical know-how [52], which benefits the industry's expert culture.

Interviewees, however, expect that monitoring of AI activities by monitoring key performance indicators (KPIs) is possible. This notion operates on the positivist assumption that reality can be defined, measured and expressed through data and processes. Yet, this is barely achievable as people construct reality through a variety of filters set up by societal norms, culture, and context [53]. This is not only reflected in decision-making, but provides the basis for coding and training of AI, as well as defining KPIs in the first place. Hence, regulation, organizational procedures and guidelines are in place to define adequate basic structures. On a more granular level, leadership needs to address questions of AI use and monitoring. In this respect, peer-based forms of leadership are a good fit because they can facilitate four-eye due diligence among experts, represent a form of social control and relieve the manager of micro-managing.

Regarding the ethics of data use and interpretation, this issue is not specific to using AI, but to working with data in general. In an industry such as insurance, which is built on data, evaluating and interpreting data is a core competency. However, human decision-making, even when following business guidelines, is not as rational and straight-forward as generally assumed [54]. In contrast, decisions made by AI would not be subject to arbitrary human somatic states and profanities, such as being hungry or having a headache [55]. Thus, the requirements and concerns about AI making decisions might be exaggerated compared to the factitious rationality of human decision-making [56].

This is related to the topic of dependency on technology and critical evaluation. The focus on technical knowledge and expertise provides a promising basis, because employees in an expert-driven culture are likely to have the necessary know-how to assess information. In fact, an insurance actuary is a data scientist who manages, evaluates, and interprets volumes of data, e.g. for risk or natural disaster modeling [1, 57]. As a result, the insurance business itself has a track record of successfully using data to manage activities. However, this talent is currently not fully leveraged. The corporation's headquarters traditionally requires the reporting of pre-defined micro-KPIs, thus leaving little room for assessment and interpretation from a holistic perspective. Moreover, the contextualization of data is impeded by organizational silos. Hence, the industry today incentivize compliant behavior, reporting routines and written guidelines, which leads to a lack of critical thinking. Nevertheless, the existing expert culture and risk-averse mindset would be a good starting point for critically evaluating information and questioning technology output. By revising reporting structures, applying peer-based forms of consultation and collaboration, as well as training insurance experts on the potential technical shortcomings, the industry could leverage the status quo to address these challenges.

## 7 Conclusion

This study has set out to examine the effect of AI on leadership in internationally operating insurance companies. The literature review suggested that the insurance industry exhibits a controller leadership discourse and machine metaphor. However, a research gap was identified concerning how leadership qualities in insurance that are confronted with the use of AI could develop.

Based on the discussed findings, the research question can be answered as follows. Insurance companies are already using AI-powered technology and seeking opportunities for further use. However, the use of AI and its implications for leadership are closely linked to the fundamental machine metaphor of the industry, which led to the existing leadership discourse and organizational metaphor in the first place. In turn, the findings suggest that the implications of AI for leadership largely depend on the lens of leadership discourse and pre-existing structures through which they are understood. Thus, if AI is only employed in accordance with what the current discourse allows, the impact is minimal. Concerning the potential effect on leadership, it can be concluded, therefore, that the use of AI-powered software in itself is unlikely to trigger change in the leadership discourse. Yet, to truly tap into the potential of AI, the underlying structures would need to be reconsidered [58]. Nevertheless, AI holds significant potential for insurers. For example, AI could support an insurer's core competencies, as well as link companies to broader ecosystems and customer communities.

The theoretical implications of these findings are that, despite much attention to the impact of AI, the effect on organizations might not be revolutionary. Moreover, studies often fail to acknowledge factors such as an existing leadership discourse and thus ignore the influence of underlying structures. The implications for organizations could be an increased awareness of the discourse and its impact, which could prevent initiatives, such as investing in AI projects, from being neglected. In addition, insurance companies could use these results to derive learnings about the status quo, and about how they should approach the digital transformation.

Nevertheless, due to the small sample of interviewees, the transferability of the results to other contexts is limited. Specific regulations applying exclusively to the insurance industry, as well as the contextual orientation of this study, make it difficult to formulate legitimate claims for other industries. However, the findings make a valuable contribution by examining leadership discourse and the potential impact of AI in a widely under-researched industry. Furthermore, insights from an academic perspective are provided, which balance grandiose claims from practitioner-oriented literature about leadership and the impact of AI. For further research, we suggest conducting studies in insurance companies that are not primarily rooted in western-European countries. In addition, case studies specifically concerned with the use of AI in insurance companies would be interesting to extract key learnings.

## References

1. Riikkinen, M., Saarijärvi, H., Sarlin, P., Lähteenmäki, I.: Using artificial intelligence to create value in insurance. Int. J. Bank Mark. **36**, 1145–1168 (2018). https://doi.org/10.1108/IJBM-01-2017-0015
2. Davenport, T.H., Bean, R.: Big companies are embracing analytics, but most still don't have a data-driven culture. Harv. Bus. Rev. Digit. Artic. 2–4 (2018)
3. Ertel, W.: Grundkurs Künstliche Intelligenz. Springer Fachmedien, Wiesbaden (2016)
4. Knorr Cetina, K.: Sociality with objects: social relations in postsocial knowledge societies. Theory Cult. Soc. **14**, 1–30 (1997). https://doi.org/10.1177/026327697014004001
5. Latour, B.: Social theory and the study of computerized work sites. In: Information Technology and Changes in Organizational Work, pp. 295–307. Springer, Boston, MA (1996)
6. Alvesson, M., Blom, M., Sveningsson, S.: Reflexive Leadership: Organising in an Imperfect World. Sage Publications Ltd, London (2016)
7. Linstead, S., Fulop, L., Lilley, S.: Management and Organization: A Critical Text. Palgrave Macmillan Education, Basingstoke, UK, New York (2009)
8. Western, S.: Leadership. Sage Publications Ltd, Los Angeles; Thousand Oaks (2013)
9. Rich, E.: Artificial Intelligence. McGraw-Hill, New York (1983)
10. Wilson, H.J., Daugherty, P.R.: Human + Machine: Reimagining Work in the Age of AI. Harvard Business Review Press, Boston, Massachusetts (2018)
11. Davenport, T.H., Ronanki, R.: Artificial intelligence for the real world. Harv. Bus. Rev. **96**, 108–116 (2018)
12. Lu, H., Li, Y., Chen, M., Kim, H., Serikawa, S.: Brain intelligence: go beyond artificial intelligence. Mob. Netw. Appl. N. Y. **23**, 368–375 (2018). https://doi.org/10.1007/s11036-017-0932-8
13. Wilson, H.J., Bataller, C.: How people will use AI to do their jobs better. Harv. Bus. Rev. Digit. Artic. 2–5 (2015)
14. Quest, L., Charrie, A., du Croo de Jongh, L., Roy, S.: The risks and benefits of using AI to detect crime. Harv. Bus. Rev. Digit. Artic. 2–5 (2018)
15. Nagle, T., Redman, T.C., Sammon, D.: Only 3% of companies' data meets basic quality standards. Harv. Bus. Rev. Digit. Artic. 2–5 (2017)
16. Redman, T.C.: If your data is bad, your machine learning tools are useless. Harv. Bus. Rev. Digit. Artic. 2–5 (2018)
17. Khalil, O.E.M.: Artificial decision-making and artificial ethics: a management concern. J. Bus. Ethics Dordr. **12**, 313–321 (1993)
18. Brynjolfsson, E., McAfee, A.: The Second Machine Age: Work, Progress, and Prosperity in a Time of Brilliant Technologies. W. W. Norton & Company, New York London (2016)
19. Osoba, O., Welser, W.: An Intelligence in Our Image: The Risks of Bias and Errors in Artificial Intelligence. RAND Corporation, Santa Monica, California (2017)
20. Western, S., Garcia, É.-J.: Global Leadership Perspectives: Insights and Analysis. Sage Publications Ltd, Thousand Oaks (2018)

21. Morgan, G.: Images of Organization. Sage Publications Ltd, Thousand Oaks (2006)
22. Zimmermann, G., Richter, S.-L.: Gründe für die Veränderungsaversion deutscher Versicherungsunternehmen. In: Zimmermann, G. (ed.) Change Management in Versicherungsunternehmen, pp. 11–35. Springer Gabler, Wiesbaden (2015)
23. Berry-Stölzle, T.R., Born, P.: The effect of regulation on insurance pricing: the case of Germany. J. Risk Insur. Malvern. **79**, 129–164 (2012)
24. Altuntas, M., Uhl, P.: Industrielle Exzellenz in der Versicherungswirtschaft. Springer Fachmedien, Wiesbaden (2016)
25. Leyer, M., Moormann, J.: How lean are financial service companies really? Empirical evidence from a large scale study in Germany. Int. J. Oper. Prod. Manag. Bradf. **34**, 1366–1388 (2014)
26. Sarkar, S.A., Mukhopadhyay, A.R., Ghosh, S.K.: Improvement of claim processing cycle time through Lean Six Sigma methodology. Int. J. Lean Six Sigma Bingley. **4**, 171–183 (2013). https://doi.org/10.1108/20401461311319347
27. Resch, D., Steinkellner, P.: Diskursive und Systemische Ansätze der Führung. In: Heimerl, P., Sichler, R. (eds.) Strategie, Organisation, Personal, Führung, pp. 533–558. Facultas.wuv, Wien (2012)
28. Bendel, O.: 300 Keywords Informationsethik: Grundwissen aus Computer- Netz- und Neue-Medien-Ethik sowie Maschinenethik. Springer Gabler, Wiesbaden (2016)
29. Hurlburt, G.F., Miller, K.W., Voas, J.M.: An ethical analysis of automation, risk, and the financial crises of 2008. IT Prof. Mag. Wash. **11**, 14–19 (2009). https://doi.org/10.1109/MITP.2009.2
30. Nickerson, R.S.: Confirmation bias: a ubiquitous phenomenon in many guises. Rev. Gen. Psychol. **2**, 175–220 (1998)
31. Plous, S.: The Psychology of Judgment and Decision Making. McGraw-Hill, New York (1993)
32. Di Fiore, A.: Why AI will shift decision making from the C-suite to the front line. Harv. Bus. Rev. Digit. Artic. 2–4 (2018)
33. Hasel, M.C., Grover, S.L.: An integrative model of trust and leadership. Leadersh. Organ. Dev. J. **38**, 849–867 (2017). https://doi.org/10.1108/LODJ-12-2015-0293
34. Joseph, E.E., Winston, B.E.: A correlation of servant leadership, leader trust, and organizational trust. Leadersh. Organ. Dev. J. Bradf. **26**, 6–22 (2005)
35. Martinho-Truswell, E.: 3 questions about AI that nontechnical employees should be able to answer. Harv. Bus. Rev. Digit. Artic. 2–4 (2018)
36. Rousseau, D.M., Sitkin, S.B., Burt, R.S., Camerer, C.: Not so different after all: a cross-discipline view of trust. Acad. Manage. Rev. **23**, 393–404 (1998). https://doi.org/10.5465/amr.1998.926617
37. Avolio, B.J., Kahai, S., Dodge, G.E.: E-leadership: implications for theory, research, and practice. Leadersh. Q. **11**, 615–668 (2000). https://doi.org/10.1016/S1048-9843(00)00062-X
38. Braun, V., Clarke, V.: Successful Qualitative Research: A Practical Guide for Beginners. Sage Publications Ltd, Los Angeles (2013)
39. Mayring, P.: Einführung in die qualitative Sozialforschung. Beltz, Weinheim, Basel (2016)
40. Witzel, A.: The Problem-centered interview. Forum Qual. Sozialforschung Forum Qual. Soc. Res. 1 (2000)
41. Guest, G.S., Macqueen, K.M., Namey, E.E.: Applied Thematic Analysis. Sage Publications Ltd, Los Angeles (2011)
42. Eling, M., Lehmann, M.: The impact of digitalization on the insurance value chain and the insurability of risks. Geneva Pap. Risk Insur. Lond. **43**, 359–396 (2018). https://doi.org/10.1057/s41288-017-0073-0
43. Svensson, G., Wood, G.: The serendipity of leadership effectiveness in management and business practices. Manag. Decis. Lond. **43**, 1001–1009 (2005)
44. Grislain-Letrémy, C.: Assurance Et Prevention Des Catastrophes Naturelles Et Technologiques. Vie Sci. Entrep. Rueil-Malmaison. 60–81 (2014)
45. Grislain-Letrémy, C., De Forges, S.L.: The benefits of uniform flood insurance. Geneva Risk Insur. Rev. Lond. **40**, 41–64 (2015). https://doi.org/10.1057/grir.2014.14

46. Hair, J.F.: Knowledge creation in marketing: the role of predictive analytics. Eur. Bus. Rev. Bradf. **19**, 303–315 (2007). https://doi.org/10.1108/09555340710760134
47. Lankton, N.K., McKnight, D.H., Tripp, J.: Technology, humanness, and trust: rethinking trust in technology. J. Assoc. Inf. Syst. Atlanta. **16**, 880–918 (2015)
48. McKnight, D.H., Carter, M., Thatcher, J.B., Clay, P.F.: Trust in a specific technology: an investigation of its components and measures. ACM Trans. Manag. Inf. Syst. TMIS **2**, 12:1–12:25 (2011). https://doi.org/10.1145/1985347.1985353
49. Hardré, P.L.: When, how, and why do we trust technology too much? In: Tettegah, S.Y., Espelage, D.L. (eds.) Emotions, Technology, and Behaviors, pp. 85–106. Academic Press, San Diego (2016)
50. Smith, R.E.: Idealizations of uncertainty, and lessons from artificial intelligence. Econ. Kiel. **10**, 1–40 (2016). https://doi.org/10.5018/economics-ejournal.ja.2016-7
51. Hunt, C.S.: Leading in the digital era. Talent Dev. Alex. **69**, 48–53 (2015)
52. Wilson, H.J., Daugherty, P.R.: Why even AI-powered factories will have jobs for humans. Harv. Bus. Rev. Digit. Artic. 2–5 (2018)
53. Hoffman, D.D., Prakash, C.: Objects of consciousness. Front. Psychol. 5 (2014). https://doi.org/10.3389/fpsyg.2014.00577
54. Martino, B.D., Kumaran, D., Seymour, B., Dolan, R.J.: Frames, biases, and rational decision-making in the human brain. Science **313**, 684–687 (2006). https://doi.org/10.1126/science.1128356
55. Miller, A.P.: Want less-biased decisions? Use algorithms. Harv. Bus. Rev. Digit. Artic. 2–5 (2018)
56. Zerilli, J., Knott, A., Maclaurin, J., Gavaghan, C.: Transparency in algorithmic and human decision-making: is there a double standard? Philos. Technol. 1–23 (2018). https://doi.org/10.1007/s13347-018-0330-6
57. Frees, E.W., Derrig, R.A., Meyers, G.: Predictive Modeling Applications in Actuarial Science. Cambridge University Press, New York (2014)
58. Dey, P., Resch, D.: Discourse analysis as intervention: a case of organizational changing. In: Steyaert, C., Nentwich, J., Hoyer, P. (eds.) A Guide to Discursive Organizational Psychology, pp. 313–332. Edward Elgar Publishing, Cheltenham, UK (2016)